2/79

Advanced Physics:
Fields, Waves and Atoms

Advanced Physics: Fields, Waves and Atoms

T. DUNCAN B Sc M Inst P

Senior Lecturer in Education, University of Liverpool
Formerly Senior Physics Master, King George V School, Southport

JOHN MURRAY · LONDON

Printed in Great Britain by The Garden City Press Limited,
Letchworth, Hertfordshire SG6 1JS

0 7195 3212 4

Preface

This is a companion volume to *Advanced Physics: Materials and Mechanics* and covers the remainder of ' A ' level requirements. Subject matter has been grouped so that there is greater emphasis on such unifying ideas as fields, waves and atoms which cut across the traditional boundaries of physics. Also, whenever possible, topics have been related to experiments and demonstrations that can be performed in a school laboratory.

In the treatment of electricity in Part 1 only B and E are used. B is introduced by the ' force on a current ' approach and E by way of ' point charges and Coulomb's law '. Rationalized SI units are employed throughout. In Part 2, waves in general are considered first and then the ideas applied to sound waves and light waves. Part 3 discusses the kinetic theory of gases, elementary thermodynamics and those topics regarded as ' modern physics '. In electronics, the thermionic diode and triode are treated briefly but the main emphasis is on semiconductor devices. The influence of the Nuffield Advanced Physics course will be evident in the sections on thermodynamics and electronics as well as in other ways.

I am again much indebted to Dr J. W. Warren and Mr J. Dawber for their very thorough reading of the manuscript and for their helpful comments and suggestions. My sincere thanks are also due to Mr B. Baker who once more kindly constructed and tested the objective-type questions, to Drs B. L. N. and H. M. Kennett who checked the answers to the numerical problems and to Mr D. Norris of Tektronix U.K. Ltd for information about oscilloscopes. The onerous task of preparing the typescript was undertaken by my wife.

For permission to use questions from recent examinations, and to convert them into SI when necessary, grateful acknowledgement is made to the various examining boards, indicated by the following abbreviations: *A.E.B.* (Associated Examining Board); *C.* (Cambridge Local Examination Syndicate); *J.M.B.* (Joint Matriculation Board); *L.* (University of London); *O.* (Oxford Local Examinations); *O. and C.* (Oxford and Cambridge Schools Examination Board); *S.* (Southern Universities Joint Board); *W.* (Welsh Joint Education Committee).

T.D.

Acknowledgements

Thanks are due to those who have kindly permitted the reproduction of copyright photographs:

Figs. 1.3, Aerofilms Ltd; 1.8a,b, Education Development Center, Inc.; 2.25, Unilab Ltd; 2.37, United Kingdom Atomic Energy Authority; 3.23a,b, Leybold-Heraeus; 3.30b, National Physical Laboratory; 3.35c, Walden Precision Apparatus Ltd; 4.2b, Royal Institution; 4.14b, Mullard Ltd; 4.16, Science Museum, London; 4.23a, GEC Turbine Generators Ltd; 4.32a, Central Electricity Generating Board; 4.32b, Electricity Council; 4.35, Central Electricity Generating Board; 4.48a,b, Mullard Ltd; 4.51, McGraw-Hill Book Co.; 6.3a, A. M. Lock & Co. Ltd; 6.8b, Griffin and George Ltd; 6.10b, 6.15b and 6.20a,b,c, W. Llowarch, *Ripple Tank Studies of Wave Motion*, (Clarendon Press, Oxford); 6.19, Decca Navigator Company Ltd; 6.21a,b, from *PSSC Physics* (D. C. Heath and Company); 6.34, Decca Radar Ltd; 7.2a,b, United States Information Service; 7.4, 7.17, GLC Architect's Department; 8.1a, Philip Harris Ltd; 8.3b, D. G. A. Dyson; 8.12, Bausch and Lomb Optical Company Ltd; 8.19, C. B. Daish; 8.21, 8.24a, Addison-Wesley Publishing Company; 8.53, 8.54, Open University; 8.57, Barnes Engineering Company; 9.8, BBC Publications; 9.20, Rolls-Royce (1971) Ltd; 10.16a,b, from *Nuffield Advanced Physics, Teachers' Guide, Unit 1* (Longman Group Ltd); 10.23, AEI Scientific Apparatus Ltd; 10.31a, Professor H. Hill; 10.31b, G. Mollenstedt and H. Duker; 11.3b,c, Tektronix UK Ltd; 11.9, 11.49, Mullard Ltd; 12.7a,b, Panax Equipment Ltd; 12.11a,b, C. T. R. Wilson; 12.20a, 12.28a,b, and 12.29, United Kingdom Atomic Energy Authority; 12.20b, UKAEA (Courtesy National Hospital); 12.22 AERE Harwell; 12.34a,b, Photo CERN; 12.35a,b, Lord Blackett's estate; A9.1b, A. G. Gaydon.

The following photographs, taken by the author, show equipment or experiments from these sources:

Figs 3.16 and 7.30a,b, Philip Harris Ltd; 6.3b, A. M. Lock & Co. Ltd; 6.4, 6.28 and 11.29b, Unilab Ltd; 12.11c, University of Liverpool.

Contents

Part 1 | FIELDS

1 Electric fields

Simple electrostatics

(*a*) *Electric charges*. In general, when any two different materials are rubbed together they exert forces on each other and each is said to have acquired an ' electric charge '. Electrostatics is the study of electric charges at rest. Experiments show that there are two kinds of charge and that *like charges repel, unlike charges attract*. The two kinds cancel one another out and in this respect are opposite. One type is taken to be positive and the other negative.

The allocation of signs to charges was made quite arbitrarily many years ago and, with the materials used today, the choice makes *polythene* rubbed with wool *negatively* charged and *cellulose acetate* (and also Perspex) rubbed with wool *positively* charged. Previously ebonite (rubbed with fur) and glass (rubbed with silk) were used to obtain negative and positive charges respectively. The forces between charged strips can be investigated as in Fig. 1.1.

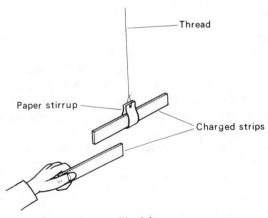

Fig. 1.1

3

The sign convention adopted leads to the electron having a negative charge and the proton a positive one. According to modern theory, an atom normally contains equal numbers of electrons and protons, making it electrically neutral. Electrification by rubbing may be explained by supposing that electrons are transferred from one material to the other. For example, when cellulose acetate is rubbed with wool, electrons go from the surface of the acetate to the wool, thus leaving the acetate deficient of electrons, i.e. positively charged, and making the wool negatively charged. Equal amounts of opposite charges should therefore be produced (see p. 63).

(b) *Insulators and conductors.* On the electron theory, all the electrons in the atoms of electrical insulators (such as polythene, cellulose acetate, Perspex, ebonite and glass) are considered to be firmly bound to their nuclei and the removal or addition of electrons at one place does not cause the flow of electrons elsewhere. That is, the charge is confined to the region where it was produced (e.g. by rubbing) or placed. Electrical conductors (e.g. metals) have electrons that are quite free from individual atoms (although fairly strongly bound to the material as a whole) and if such materials gain electrons, these can move about in them. Loss of electrons by a conductor causes a redistribution of those left. A charge on a conductor therefore spreads over the entire surface. The ' free ' electron theory is adequate for our present purposes but later we will see that it has been extended by the more advanced ' band ' theory which explains the behaviour of insulators, semiconductors and conductors in terms of energy levels.

The human body and the earth are comparatively good conductors and if we try to charge a metal rod by rubbing, it must be well insulated and not held in the hand. Otherwise any charge produced is conducted away through the body to earth. Water also conducts and its presence on the surface of many materials (e.g. glass) that are otherwise insulators accounts for the charge leakage which often occurs. Many modern plastics (e.g. polythene, Perspex, cellulose acetate) are water-repellent.

(c) *Electrostatic induction.* A negatively charged polythene strip held close to an insulated, uncharged conductor such as a small aluminized expanded–polystyrene ball, attracts it. This may be explained by saying that electrons are repelled to the far side of the ball leaving the near side positively charged, Fig. 1.2a. The attraction between the negative charge on the polythene strip and the induced positive charge is greater than the repulsion between the strip and the more distant negative charge. The effect is called *electrostatic induction*. It accounts for the attraction of scraps of paper by a plastic comb, charged by being drawn through the hair.

(d) *Electrophorus.* This is a device for producing charges by electrostatic induction, Fig. 1.2b. It consists of a circular metal plate with an insulating handle, placed on an insulating sheet (e.g. of polythene) previously charged by rubbing.

Fig. 1.2

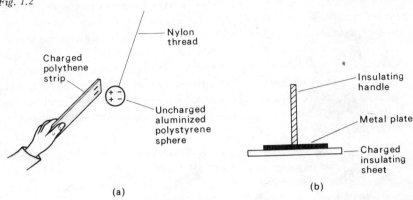

(a)

(b)

When the plate is earthed by touching it with the finger and then removed, it has a large charge of opposite sign to that on the insulating sheet. This charge can be transferred to another conductor and the electrophorus plate recharged as before, again and again.

It may seem strange that the metal plate does not become charged by contact with the insulating sheet and have the same sign of charge as it. However, it appears that contact between even plane surfaces occurs at only a few points so that, in fact, the metal plate is charged by induction. This would account for it becoming oppositely charged.

(*e*) *Static and current electricity.* Static charges produced by rubbing insulators (or insulated conductors) give the same effects when they move as do electric currents due to a battery. However, in electrostatics we are usually dealing with quite a small charge (a few microcoulombs) but a large p.d. (thousands of volts); in current electricity the opposite is usually true.

Electrostatics today

Electrostatics was the first branch of electricity to be investigated and for a long time was regarded as a subject of no practical value. In recent years this has changed and it now has important industrial applications.

The electrostatic precipitation of flue-ash that would otherwise be discharged into the atmosphere from modern coal-fired power stations is a vital factor in the reduction of pollution. An average power station produces about 30 000 kg (30 tonnes) of flue-ash per hour. Power station precipitators are shown in Fig. 1.3 between the chimney and the main building which contains the coal bunkers, boiler house and turbine hall. The precipitators are built to remove 99 per cent of the ash from the flue gases before they reach the power station chimney. A precipitator is made up of a number of wires and plates. The wires are negatively charged and give a similar charge to the particles of ash which are then attracted

5

Fig. 1.3

to the positive plates. These are mechanically shaken to remove the ash which is collected and used as a by-product.

Electrostatic precipitation is also important in the steel, cement and chemical industries where flue gas outputs are high. Electrostatic spraying of paints, plastics and powders is also possible and lends itself to automation.

In nuclear physics research, electrostatic generators of, for example, the van de Graaff type (p. 65) are employed to produce p.d.s of up to 14 million volts for accelerating atomic particles. Their use in this field has done much to renew interest in electrostatics.

A knowledge of electrostatics is important in the design of cathode-ray tubes for radar and television, in electrical prospecting for minerals and in surveying sites for large structures. Electrostatic loudspeakers and microphones are in common use as are electrostatic office copying machines.

Electric charges can build up due to friction on aircraft in flight and on plastic sheeting in industry, creating a potential explosion hazard unless preventive steps are taken. In the case of aircraft the rubber tyres are made slightly conducting so that the charge leaks away harmlessly at touch-down. The crackling which occurs when a nylon garment is removed from the body or when someone steps from a car with plastic seat covers is also due to static charges causing the insulation of the surrounding air to break down. A flash of lightning is nature's most spectacular electrostatic event.

Coulomb's law

(a) *Statement.* A knowledge of the forces that exist between charged particles is necessary for an understanding of the structure of the atom and of matter. The magnitude of the forces between charged spheres was first investigated quantitatively in 1785 by Coulomb, a French scientist. The law he discovered may be stated as follows:

The force between two point charges is directly proportional to the product of the charges divided by the square of their distance apart.

The law applies to point charges. Sub-atomic particles such as electrons and protons may be regarded as approximating to point charges. Later we shall see that a *uniformly* charged conducting sphere behaves—so far as external effects are concerned—as if the charges were concentrated at its centre. It is therefore sometimes considered to be a point charge but there must not be any charges nearby to disturb the uniform distribution of charge on it, i.e. it must be an *isolated* charged spherical conductor. In practice two small spheres will only approximate to point charges when they are far apart. A point charge, like a point mass, is a convenient theoretical simplification.

Coulomb's law may be stated in mathematical terms as

$$F \propto \frac{Q_1 Q_2}{r^2}$$

where F is the electric (or Coulomb) force between two point charges Q_1 and Q_2, distance r apart.

(*b*) *Experimental test.* Two small metallized spheres X and Y are used. X is glued to the bottom of a ' V ' of a metre length of fine nylon thread so that it can only swing at right angles to the plane of suspension, Fig. 1.4*a*. The position of the centre (or edge) of the shadow of X on the scale on the screen is noted. Sphere Y is glued to the end of an insulating rod.

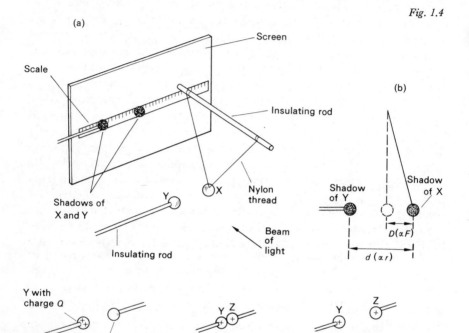

Fig. 1.4

Both spheres are given the same charge by touching each in turn with, say, the plate of an electrophorus. When Y is brought up to X, repulsion occurs, Fig. 1.4*b*. The positions of the centres (or edges) of the shadows of X and Y are noted. Distance d is proportional to the separation (r) of the spheres and it can be shown that the deflection D (the distance between the centres—or edges—of the first and second positions of the shadow of X) is proportional to the force between X and Y. If a few readings of d and D are taken and $D \times d^2$ found to be constant, then $F \propto 1/r^2$.

To test if $F \propto Q_1 Q_2$, Y is touched by another exactly similar but uncharged sphere Z, Fig. 1.4*c*, so that its charge is halved (Z taking the other half). Using one of the previous separations (d) between X and Y, the force should therefore become half what it was before. It will be one quarter of its original value if the charge on X is also halved by sharing its charge with the uncharged Z.

For reasonable success this experiment requires the insulators to be thoroughly dry (if necessary by using a hair drier) to prevent loss of charge by leakage. The readings should therefore be taken in quick succession. The most convincing evidence for Coulomb's law, however, is provided not directly, but indirectly by experimental verification of deductions from the law.

(c) *Permittivity.* The force between two charges also depends on what separates them; its value is always reduced when an insulating material replaces a vacuum. To take this into account a medium is said to have *permittivity*, denoted by ϵ (epsilon) and ϵ is included in the denominator of the expression for Coulomb's law. A material with high permittivity is one which reduces appreciably the force between two charges compared with the vacuum value.

SI units, which are used in this book, are ' rationalized '. This means that the values of certain constants are adjusted so that π does not occur in formulae in frequent use, but it does occur in others. The formulae thus conveniently simplified usually refer to situations in which there is plane symmetry (later we shall see that uniform electric and magnetic fields are in this category) and where we would not logically expect π to occur: in an unrationalized system it does. In a rationalized system 4π or 2π appear if there is spherical or cylindrical symmetry respectively (certain non-uniform fields are examples).

Spherical symmetry is associated with point charges (p. 17) and the equation for Coulomb's law is rationalized by including 4π in the denominator. Hence

$$F = \frac{1}{4\pi\epsilon} \cdot \frac{Q_1 Q_2}{r^2}.$$

If F is in newtons, r in metres, Q_1 and Q_2 in coulombs (1 coulomb being the charge flowing per second through a conductor in which there is a steady current of 1 ampere) then the unit of ϵ is $C^2\,N^{-1}\,m^{-2}$ since $\epsilon = Q_1 Q_2/(4\pi F r^2)$.

The permittivity of a vacuum is denoted by ϵ_0 (epsilon nought) and is called the *permittivity of free space*. The numerical value of ϵ_0 is found experimentally by an indirect method (p. 41) which does not involve the difficult task of measuring the force between known ' point ' charges at a given separation in a vacuum. The result is

$$\epsilon_0 = 8.85 \times 10^{-12}\,C^2\,N^{-1}\,m^{-2}.$$

We can also write

$$1/(4\pi\epsilon_0) = 8.98 \times 10^9\,N\,m^2\,C^{-2}$$

and so

$$F \simeq 9 \times 10^9\,Q_1 Q_2/r^2.$$

The permittivity of air at s.t.p. is $1.0005\,\epsilon_0$ and we can usually take ϵ_0 as the value for air. A more widely used unit for permittivity is the *farad per metre* (F m^{-1}), as explained later (p. 34).

ELECTRIC FIELDS

Electric field strength

(a) *Definition.* A resultant force changes motion. Many everyday forces are pushes or pulls between bodies in contact. In other cases forces arise between bodies that are separated from one another. Electric, magnetic and gravitational effects involve such action-at-a-distance forces and to deal with them physicists find the idea of a *field of force* (or simply a *field*) useful. Fields of these three types have common features as well as important differences.

An electric field is a region where an electric charge experiences a force—just as a hayfield is a region in which hay is found. If a very small, positive point charge Q is placed at any point in an electric field and it experiences a force F, then the *field strength E* (also called the *E-field*) at that point is defined by the equation

$$E = \frac{F}{Q}.$$

In words, the magnitude of E is the force per unit charge and its direction is that of F (i.e. of the force which acts on a positive charge). Field strength E is thus a vector. Note that we refer to E as the force *per* unit charge and not as the force *on* unit charge. A finite charge (such as a unit charge) might affect the field by inducing charges on neighbouring bodies and so we must *imagine* a very small test charge $+Q$ to be placed at the point since we require to know E before $+Q$ was introduced into the field.

If F is in newtons (N) and Q is in coulombs (C) then the unit of E is the newton per coulomb (N C^{-1}). A commoner but equivalent unit is the *volt per metre* (V m^{-1}), as we shall see later.

(b) *E due to a point charge.* The magnitude of E due to an isolated positive point charge $+Q$, at a point P distance r away, in a medium of permittivity ϵ,

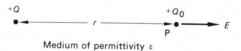

Fig. 1.5   Medium of permittivity ε

can be calculated by imagining a very small charge $+Q_0$ to be placed at P, Fig. 1.5. By Coulomb's law, the force F on Q_0 is

$$F = \frac{1}{4\pi\epsilon} \cdot \frac{QQ_0}{r^2}.$$

But E is the force per unit charge, that is

$$E = \frac{F}{Q_0}$$

$$= \frac{1}{4\pi\epsilon} \cdot \frac{Q}{r^2}.$$

10

E is directed away from $+Q$, as shown. If a point charge $-Q$ replaced $+Q$, E would be directed towards $-Q$ since unlike charges attract.

The above expression shows that E decreases with distance from the point charge according to an inverse square law. The field due to an isolated point charge is thus non-uniform but it has the same values at equal distances from the charge and so has spherical symmetry. In Fig. 1.6, if the magnitude of the field strength due to point charge $+Q$ is E at A, what is it at (i) B, (ii) C?

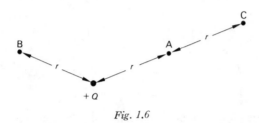

Fig. 1.6

(c) *Field strength and charge density.* So far as external effects are concerned an isolated spherical conductor having a charge Q uniformly distributed over its surface behaves like a point charge Q at its centre (p. 17). If r is the radius of the sphere, the field strength E at its surface is thus given by

$$E = \frac{1}{4\pi\epsilon} \cdot \frac{Q}{r^2}.$$

The charge per unit area of the surface of the conductor is called the *charge density σ* (sigma) and since a sphere has surface area $4\pi r^2$ we have $\sigma = Q/(4\pi r^2)$. Therefore $Q = 4\pi r^2 \sigma$ and so

$$E = \frac{\sigma}{\epsilon}.$$

This expression will be used later (p. 38). It has been derived by considering a sphere but it gives E at the surface of any charged conductor. Thus it can be seen to apply to a plane surface if the radius of the sphere, which does not appear in the expression, is allowed to tend to infinity.

Field lines

An electric field can be represented and so visualized by electric field lines. These are drawn so that (i) the field line at a point (or the tangent to it if it is curved) gives the direction of E at that point, i.e. the direction in which a positive charge would accelerate, and (ii) the number of lines per unit cross-section area is proportional to E. The field line is imaginary but the field it represents is real.

Electric field patterns similar to the magnetic field patterns given by iron filings can be obtained using tiny ' needles ' of an insulating substance such as semolina powder or grass seeds, suspended by previous stirring in fresh castor oil in a glass dish. An electric field is created by applying a high p.d. from a van de Graaff generator to metal electrodes dipping in the oil, Fig. 1.7. The powder or seed orientates itself to form different patterns according to the shape of the electrodes.

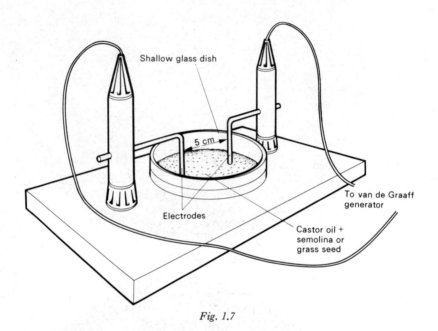

Fig. 1.7

Some electric field patterns are shown in Fig. 1.8. A uniform field is one in which E has the same magnitude and direction at all points, there is plane symmetry and the field lines are parallel and evenly spaced. In Fig. 1.8a the field is uniform between the plates away from the edges. When E varies in magnitude and direction with position, the field is non-uniform. In Fig. 1.8b the field is non-uniform but radial and there is spherical symmetry.

Electric field patterns are useful in designing electronic devices such as cathode-ray tubes. The engineer is often able to sketch intuitively the pattern for a given electrode arrangement and so predict the probable behaviour of the device.

Field lines are also referred to as ' lines of force ' and the term is appropriate when considering electric and gravitational fields because the field lines do indicate the direction in which a charge or mass experiences a force. This is not so in the magnetic case, as will be seen later, and therefore in general it is preferable to talk about field lines rather than lines of force.

Fig. 1.8a and b

Electric potential

Information about the field may be given by stating the *field strength* at any point; alternatively the *potential* can be quoted. Before discussing this idea some basic mechanics will be revised briefly.

(*a*) *Work and energy.* In science, work is done when the point of application of a force (or a component of it) undergoes a displacement in its own direction. The product of the force (or its component) F and the displacement s is taken as a measure of the work done W, i.e. $W = F \times s$. When F is in newtons and s in metres, W is in newton-metres or joules.

If a body A exerts a force on body B and work is done, a transfer of energy occurs *which is measured by the work done.* Thus, if we raise a mass m through a vertical height h, the work done W by the force we apply (i.e. by mg) is $W = mgh$ (assuming the earth's gravitational field strength g is constant). The energy transfer is mgh and we consider that the system gains and stores that amount of gravitational potential energy in its gravitational field. This energy is obtained from the conversion of chemical energy by our muscular activity. When the mass falls the system loses gravitational potential energy and, neglecting air resistance, there is a transfer of kinetic energy to the mass equal to the work done by gravity.

(*b*) *Meaning of potential.* A charge in an electric field experiences a force and if it moves work will, in general, be done. If a positive charge is moved from A to B in a direction opposite to that of the field E, Fig. 1.9*a*, an external agent

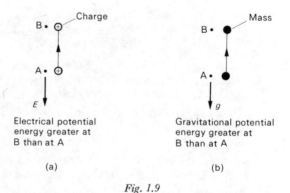

Electrical potential energy greater at B than at A

Gravitational potential energy greater at B than at A

(a)

(b)

Fig. 1.9

has to do work against the forces of the field and energy has to be supplied. As a result, the system (of the charge in the field) gains an amount of electrical potential energy equal to the work done. This is analogous to a mass being raised in the earth's gravitational field g, Fig. 1.9b. When the charge is allowed to return from B to A, work is done by the forces of the field and the electrical potential energy previously gained by the system is lost. If, for example, the motion is in a vacuum, an equivalent amount of kinetic energy is transferred to the charge.

In general, the potential energy associated with a charge at a point in an electric field depends on the location of the point and the magnitude of the charge (since the force acting depends on the latter, i.e. $F = QE$). Therefore if we state the magnitude of the charge we can describe an electric field in terms of the potential energies of that charge at different points. A unit positive charge is chosen and the change of potential energy which occurs when such a charge is moved from one point to another is called the change of *potential* of the field itself.

Hence the potential at B in Fig. 1.9a exceeds that at A by the work which must be done against the electric force to take unit positive charge from A to B. To be strictly accurate, however, as we were when we defined E (p. 10), we should refer to the work done *per* unit charge when a *very small* charge moves from one point to the other since the introduction of a unit charge would in general modify the field.

If for theoretical purposes we select as the zero of potential the potential at an infinite distance from any electric charges, then the *potential at a point in a field can be defined as the work done per unit positive charge moving from infinity to the point*, always assuming the charge does not affect the field. The choice of the zero of potential is purely arbitrary and whilst infinity may be a few hundred metres in some cases, in atomic physics where distances of 10^{-10} m are involved, it need only be a very small distance away from the charge responsible for the field.

Potential is a property of a *point* in a field and is a scalar since it deals with a quantity of work done or potential energy per unit charge. The symbol for potential is V and the unit a *joule per coulomb* (J C^{-1}) or *volt* (V).

Just as a mass moves from a point of higher gravitational potential to one of lower potential (i.e. to fall towards the earth's surface), so a positive charge is urged by an electric field to move from a point of higher electrical potential to one of lower potential. Negative charges move in the opposite direction if free to do so.

(*c*) *Potential and field strength compared.* When describing a field, potential is usually a more useful quantity than field strength because, being a scalar, it can be added directly when more than one field is concerned. Field strength is a vector and addition by the parallelogram law is more complex. Also, it is often more important to know what energy changes occur (rather than what forces act) when charges move in a field and these are readily calculated if potentials are known (see p. 21).

Equipotentials

All points in a field which have the same potential can be imagined as lying on a surface—called an *equipotential* surface. When a charge moves on such a surface no energy change occurs and no work is done. The force due to the field must therefore act at right angles to the equipotential surface at any point and so equipotential surfaces and field lines always intersect at right angles.

A field can therefore be represented pictorially by field lines and by equipotential surfaces (or lines in two dimensional diagrams). Equipotential surfaces for a point charge are concentric spheres (circles in two dimensions), Fig. 1.10*a*; there is spherical symmetry. The plane symmetry of a uniform field is seen in Fig. 1.10*b*. If equipotentials are drawn so that the change of potential from one to the next is constant, then the spacing will be closer where the field is stronger. To perform a certain amount of work in such regions a shorter distance need be travelled.

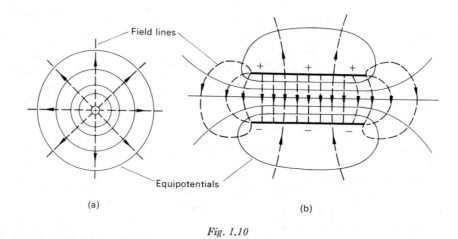

(a) (b)

Fig. 1.10

The surface of a conductor in electrostatics (i.e. one in which no current is flowing) must be an equipotential surface since any difference of potential would cause a redistribution of charge in the conductor until no field existed in it.

Potential due to a point charge

We wish to find the potential at A in the field of—and distant r from—an isolated point charge $+Q$ situated at O in a medium of permittivity ϵ, Fig. 1.11. Imagine a very small point charge $+Q_0$ is moved by an external agent from C, distance x from A, through a very small distance δx to B without affecting the field due to $+Q$.

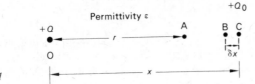

Fig. 1.11

Assuming the force F on Q_0 due to the field remains constant over δx, the work done δW by the external agent over δx against the force of the field is

$$\delta W = F(-\delta x).$$

The negative sign is inserted to show that the displacement δx is in the opposite direction to that in which F acts. By Coulomb's law

$$F = \frac{QQ_0}{4\pi\epsilon} \cdot \frac{1}{x^2}$$

$$\therefore \delta W = \frac{QQ_0}{4\pi\epsilon} \cdot \frac{(-\delta x)}{x^2}.$$

The total work done W in bringing Q_0 from infinity to A is

$$W = \frac{-QQ_0}{4\pi\epsilon} \int_\infty^r \frac{dx}{x^2} = \frac{-QQ_0}{4\pi\epsilon} \left[\frac{-1}{x}\right]_\infty^r$$

$$= \frac{QQ_0}{4\pi\epsilon} \cdot \frac{1}{r}.$$

The potential V at A is the work done per unit positive charge brought from infinity to A. Hence

$$V = \frac{W}{Q_0} = \frac{1}{4\pi\epsilon} \cdot \frac{Q}{r}.$$

What would be the analogous expression for the gravitational potential V at a distance r from a point mass M?

Potential due to a conducting sphere

(a) *An expression.* A charge $+Q$ on an isolated conducting sphere is uniformly distributed over its surface (due to the repulsion of like charges) and has a radial electric field pattern, Fig. 1.12a. The field at any point outside the sphere is exactly the same as if the whole charge were concentrated as a point charge $+Q$ at the centre of the sphere, Fig. 1.12b. (This can be shown to follow from Coulomb's law; a spherical mass similarly behaves as if its whole mass were concentrated at its centre.)

17

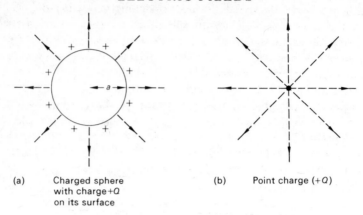

(a)   Charged sphere
with charge +Q
on its surface

(b)   Point charge (+Q)

Fig. 1.12

From the expression already obtained for a point charge, we can say that **the** potential V at a point P distance r from the centre of the sphere, is

$$V = \frac{1}{4\pi\epsilon} \cdot \frac{Q}{r}.$$

If the radius of the sphere is a, the potential V at its surface is

$$V = \frac{1}{4\pi\epsilon} \cdot \frac{Q}{a}.$$

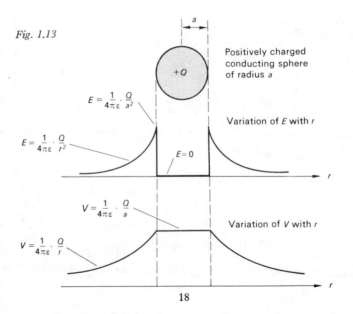

Fig. 1.13

Positively charged conducting sphere of radius a

$$E = \frac{1}{4\pi\varepsilon} \cdot \frac{Q}{a^2}$$

$$E = \frac{1}{4\pi\varepsilon} \cdot \frac{Q}{r^2}$$

$E = 0$

Variation of E with r

$$V = \frac{1}{4\pi\varepsilon} \cdot \frac{Q}{a}$$

$$V = \frac{1}{4\pi\varepsilon} \cdot \frac{Q}{r}$$

Variation of V with r

At all points inside the sphere the field strength is zero, otherwise field lines would link charges of opposite sign in the sphere and such a state of affairs is impossible under static conditions in a conductor. (This may also be shown to be a result of Coulomb's law.) It follows that no work is done when a charge is moved between any two points inside the sphere. The potential is thus the same at all points throughout the sphere and equal to that at the surface. As well as talking about the potential at a point in a field we also consider an insulated conductor to have a potential.

The variation of E and V at points outside and inside a positively charged conducting sphere should therefore be as shown in the graphs of Fig. 1.13. Outside the sphere E varies as $1/r^2$ and, as we have just seen, theory predicts that V varies as $1/r$.

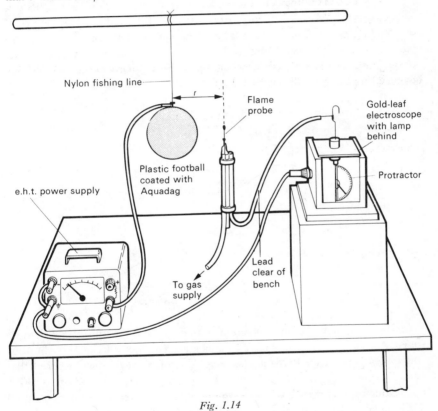

Fig. 1.14

(b) *Flame probe investigation.* The results for V outside may be investigated experimentally using the apparatus of Fig. 1.14. A probe,[1] in the form of a small gas flame at the point of a hypodermic needle, is connected to an electroscope

[1]The construction and action of the probe are described in Appendix 1, p. 528.

calibrated to measure potentials (see p. 55). The potential measured is that at the probe. The conducting sphere is charged to 1500 V from an e.h.t. power supply and the potentials (V) noted from the electroscope when the probe is at different distances (r) from the centre of the sphere. (In particular, at a distance of twice the radius from the centre of the sphere the potential should be 750 V.) The sphere should be ' isolated ' (as we shall see presently) by being well away from walls, bench tops, the experimenter etc., and the lead from the probe must also be clear of the bench. If the results confirm that $V \propto 1/r$ then this may be taken as indirect evidence of Coulomb's law since the $1/r$ law for potential is a consequence of a $1/r^2$ law for E.

(c) *Effect of neighbouring bodies.* The electric potential at a point in a field due to a charged body is not determined solely by the charge unless the body is ' isolated ' as we assume in theory. In practice it is affected by the presence of other bodies, charged or uncharged, and the surrounding material. Thus the potential at a point near a positively charged body increases when another positively charged body approaches and decreases when a negatively charged body is brought up.

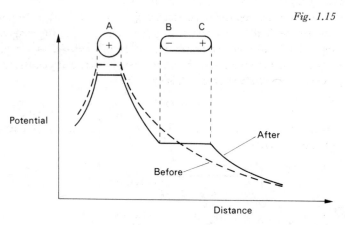

Fig. 1.15

In Fig. 1.15 the variation of potential with distance from a positively charged sphere A is shown before and after an uncharged conductor BC is brought near. Initially each point on BC is at the potential which previously existed at that point. Momentarily, therefore, B is at a higher potential than C and so, since BC is a conductor, electrons flow from C to B (i.e. from a lower to a higher potential) until the potential is the same all over BC—called the *potential of the conductor.* Electrostatic induction has occurred and the induced negative charge at B lowers the potential there, as well as at all points between A and B (including that of A). The induced positive charge at C raises the potential at C and at points beyond. The constant potential of BC lies between the original potentials at B and C.

Potential difference

The idea of potential applies not only to the electric fields produced in air or a vacuum by static charges (on an insulator or a conductor) but also to those in a wire having a battery or power supply across its ends and which cause charges to move as an electric current in the wire. The term *potential* is useful in both electrostatics and in current electricity.

However, there is an important difference between the two cases. In electrostatics when a charge moves in the direction of the field, the potential energy lost by the field-charge system can be regained if the charge is moved by an external agent in the opposite direction. In current electricity, energy lost by the electric field inside a conductor is irrecoverable since the heat produced cannot be converted back into other forms of energy by reversing the current.

In practice, especially in current electricity, we are usually concerned with the difference of potential or p.d. between two points in an electric field so that the zero of potential does not matter. The p.d. between two points is the *work done per unit charge* (or the energy change per unit charge) passing from one point to the other. The same symbol is used for p.d. as for potential, i.e. V, and the same unit, i.e. joule per coulomb ($J\ C^{-1}$) or volt (V).

If the p.d. between two points in an electric field is 10 volts then 10 joules of work are done per coulomb of charge moving from one point to the other and an energy change of 10 joules per coulomb occurs. In general if V is the p.d. (in V) between two points in an electric field, the energy change W occurring when a charge Q (in C) moves through the p.d. is given by

$$W = QV.$$

Two examples follow to show how, using this expression, energy changes in electric fields can be calculated.

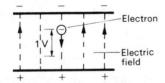

Fig. 1.16

First, suppose we wish to know the energy change when an electron ' falls ' through a p.d. of 1 V in an electric field in a vacuum, i.e. it travels between two points whose p.d.s differ by 1 V, Fig. 1.16. The electron is accelerated by the force acting on it due to the field and work is done. The energy change W is found from $W = QV$ where $Q = 1.6 \times 10^{-19}$ C (the electronic charge) and $V = 1.0\ V = 1.0\ J\ C^{-1}$

$$\therefore \quad W = (1.6 \times 10^{-19}\ C)(1.0\ J\ C^{-1})$$

$$= 1.6 \times 10^{-19}\ J.$$

This tiny amount of energy is called an *electron-volt* (eV) and is a unit of energy (not SI) much used in atomic physics.

Second, an example from current electricity. If a p.d. of 12 V maintains a current of 3.0 A through a resistor, the electrical energy W (from the electric field in the resistor) changed to heat per second is obtained from $W = QV = ItV$ (since $Q = It$) where $I = 3.0$ A $= 3.0$ C s^{-1}, $t = 1.0$ s and $V = 12$ V $= 12$ J C^{-1}.

$$\therefore \quad W = (3.0 \text{ C s}^{-1})(1.0 \text{ s})(12 \text{ J C}^{-1})$$
$$= 3.0 \times 1.0 \times 12 \quad \text{C s}^{-1} \times \text{s} \times \text{J C}^{-1}$$
$$= 36 \text{ J}.$$

Relation between E and V

Consider a charge $+Q$ at a point A in an electric field where the field strength is E, Fig. 1.17. The force F on Q is given by

$$F = EQ.$$

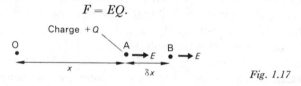

Fig. 1.17

If Q moves a very short distance δx from A to B in the direction of E, then (assuming E is constant over AB) the work done δW by the electric force on Q is

$$\delta W = \text{force} \times \text{distance}$$
$$= F \, \delta x$$
$$= EQ \, \delta x.$$

If the p.d. between B and A is δV, we have by the definition of p.d.

$$\delta V = \text{work done per unit charge}$$
$$= - \frac{\delta W}{Q} = - \frac{EQ \, \delta x}{Q}.$$

That is, $\qquad \delta V = -E \, \delta x.$

The negative sign is inserted to show that if displacements in the direction of E are taken to be positive, then when δx is positive, δV is negative, i.e. the potential decreases. On the other hand, if the charge is moved in a direction opposite to that of E, δx is negative and δV is positive, indicating an increase of potential as occurs in practice.

In the limit, as $\delta x \to 0$, E becomes the field strength at a point (A) and in calculus notation

$$E = - \frac{dV}{dx}.$$

dV/dx is called the *potential gradient* in the x-direction and so the field strength at a point equals the negative of the potential gradient there. Potential gradient is a vector and is measured in volts per metre $(V\,m^{-1})$. It follows that this is also a unit of E (as well as $N\,C^{-1}$ — an equivalent but less-used unit).

In a uniform field E is constant in magnitude and direction at all points, hence dV/dx is constant, i.e. the potential changes steadily with distance. The field near the centre of two parallel metal plates is uniform and if this is created by a p.d. V between plates of separation d, then

$$E = -\frac{V}{d}$$

where E is the field strength at any point in the *uniform* region (not at the edges). For example, if $V = 2.0 \times 10^3$ V and $d = 1.0$ cm $= 1.0 \times 10^{-2}$ m,

$$E = V/d \text{ (numerically)}$$

$$= (2.0 \times 10^3 \text{ V})/(1.0 \times 10^{-2} \text{ m})$$

$$= 2.0 \times 10^5 \text{ V m}^{-1} \text{ (or N C}^{-1}).$$

Gravitational analogy

Analogies exist between electric and gravitational fields.

(*a*) *Inverse square law of force.* Coulomb's law is similar in form to Newton's law of universal gravitation. Both are inverse square laws with $1/(4\pi\epsilon)$ in the electric case corresponding to the gravitational constant G. The main difference is that whilst electric forces can be attractive or repulsive, gravitational forces are always attractive. Two types of charge are known but there is only one type of matter. By comparison with electric forces, gravitational forces are extremely weak (p. 30).

Coulomb's law	*Newton's law*

$$F = \frac{1}{4\pi\epsilon} \cdot \frac{Q_1 Q_2}{r^2} \qquad\qquad F = G \cdot \frac{m_1 m_2}{r^2}$$

(*b*) *Field strength.* The field strength at a point in a gravitational field is defined as the force acting per unit mass placed at the point. Thus if a mass m in kilograms experiences a force F in newtons at a certain point in the earth's field, the strength of the field at that point will be F/m in newtons per kilogram. This is also the acceleration a the mass would have in metres per second squared if it fell freely under gravity at this point (since $F = ma$). The gravitational field strength and the acceleration due to gravity at a point thus have the same value (i.e. F/m) and the same symbol, g, is used for both. At the earth's surface $g = 9.8$ N kg^{-1} = 9.8 m s^{-2} (vertically downwards).

ELECTRIC FIELDS

Electric field strength
(at distance r from point charge Q)

$$E = \frac{1}{4\pi\epsilon} \cdot \frac{Q}{r^2} \text{ (in N C}^{-1})$$

Gravitational field strength
(at distance r from point mass m)

$$g = G \cdot \frac{m}{r^2} \text{ (in N kg}^{-1})$$

Expressing forces in terms of field strengths we also have

Electric force
(on a charge Q)
$$F = QE$$

Gravitational force
(on a mass m)
$$F = mg$$

(c) *Field lines and equipotentials.* These can also be drawn to represent gravitational fields but such fields are so weak, even near massive bodies, that there is no method of plotting field lines similar to those used for electric (and magnetic) fields. Field lines for the earth are directed towards its centre and the field is spherically symmetrical. Over a small part of the earth's surface the field can be considered uniform, the lines being vertical, parallel and evenly spaced. Figs 1.18a and b represent uniform electric and gravitational fields.

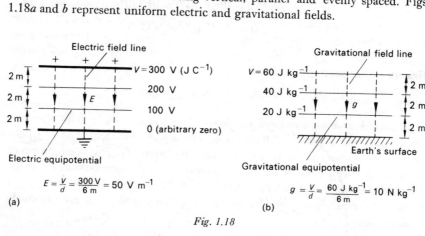

Fig. 1.18

(d) *Potential and p.d.* Electric potentials and p.d.s are measured in joules per coulomb (J C^{-1}) or volts; gravitational potentials and p.d.s are measured in joules per kilogram (J kg^{-1}). If the p.d. between two points in the earth's gravitational field is 20 J kg^{-1}, the work done and the change of potential energy will be 20 J when 1 kg moves from one point to the other. In general if V is the p.d. between two points, the energy change W which occurs when a mass m moves from one point to the other is given by $W = mV$. From this expression the energy required to send a rocket from one point in space to another can be calculated (see p. 27).

Electrical p.d. and energy change
$$W = QV$$

Gravitational p.d. and energy change
$$W = mV$$

24

As a mass moves away from the earth the potential energy of the earth-mass system increases, transfer of energy from some other source being necessary. If infinity is taken as the zero of gravitational potential (i.e. a point well out in space where no more energy is needed for the mass to move further away from the earth) then the potential energy of the system will have a negative value except when the mass is at infinity. At every point in the earth's field the potential is therefore negative (see expression below), a fact which is characteristic of fields that exert attractive forces. Fields giving rise to repulsive forces cause positive potentials.

Electric potential
(at distance r from point charge Q)

$$V = \frac{1}{4\pi\epsilon} \cdot \frac{Q}{r}$$

Gravitational potential
(at distance r from point mass m)

$$V = -G \cdot \frac{m}{r}$$

Fig. 1.19a shows the variation of gravitational potential V with distance r from the centre of a spherical mass or the variation of electrical potential V with distance r from the centre of a negatively charged sphere, i.e. it applies to an attractive force field. Fig. 1.19b shows the variation of electrical potential V with distance r from the centre of a positively charged field, i.e. it refers to a repulsive force field.

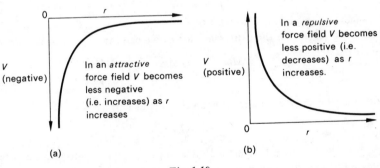

Fig. 1.19

Field treatment of space travel

The motion of spacecraft and satellites in the earth's gravitational field is often considered in terms of ' forces '. As an alternative, a ' field ' treatment may be given which involves working out energy changes from potential differences.

(a) *Satellite orbits.* The gravitational potential V at a distance r from the centre of the earth is $-GM/r$ where M is the earth's mass. Potential equals the potential energy per unit mass (or charge, in the electrical case, see p. 15) and so a satellite of mass m at a distance r from the centre of the earth has potential energy

$E_p = -GMm/r$. If the satellite is moving with speed v in a circular orbit of radius r, Fig. 1.20, its kinetic energy $E_k = \frac{1}{2} mv^2$. Hence the total energy E of the satellite in orbit is given by

$$E = E_p + E_k = -GMm/r + \tfrac{1}{2}mv^2.$$

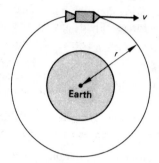

Fig. 1.20

From the expression for centripetal force and the law of gravitation we have

$$\frac{mv^2}{r} = \frac{GMm}{r^2}$$

$$\therefore \quad mv^2 = \frac{GMm}{r}$$

Substituting for mv^2 in the expression for E, we can say

$$E = -\frac{GMm}{r} + \tfrac{1}{2} \cdot \frac{GMm}{r} = -\tfrac{1}{2} \cdot \frac{GMm}{r}.$$

The potential energy E' of a mass m on the earth's surface is

$$E' = -GMm/R$$

where R is the radius of the earth. Hence to put a satellite into circular orbit of radius r it must be given the extra amount of energy $(E - E')$ where

$$E - E' = -\tfrac{1}{2} \cdot \frac{GMm}{r} - \frac{(-GMm)}{R}$$

$$= GMm \left(\frac{1}{R} - \frac{1}{2r} \right).$$

If the orbit is close to the earth's surface (e.g. at a height of 100–200 km), $r \simeq R$ and

$$E - E' = \tfrac{1}{2} \cdot \frac{GMm}{R}.$$

If this extra energy is all given in the form of kinetic energy ($\frac{1}{2}mu^2$) at lift-off, then

$$\tfrac{1}{2}mu^2 = \tfrac{1}{2}\frac{GMm}{R}$$

$$\therefore u^2 = \frac{GM}{R}.$$

Taking $G = 6.7 \times 10^{-11}$ N m^2 kg^{-2}, $M = 6.0 \times 10^{24}$ kg and $R = 6.4 \times 10^6$ m, and substituting in the expression for u^2, we get $u = 8.0$ km s^{-1}. The speed the satellite must be given at launching to go into a circular orbit close to the earth's surface will be greater than 8.0 km s^{-1} since air resistance has been neglected.

In practice circular orbits are seldom attained: most are elliptical.

(b) *Speed of escape*. Some values are given in Table 1.1 of the gravitational potential ($-GM/r$) at different distances r from the centre of the earth.

Table 1.1

r ($\times 10^6$ m)	6.4	6.6	20	40	400	∞
$-GM/r$ ($\times 10^6$ J kg^{-1})	-63	-61	-20	-10	-1.0	0.00

As r increases, the potential becomes less negative, i.e. it increases, and the potential difference between any two values of r equals the change of potential energy when a mass of 1 kg moves from one point to the other. For example, if a spacecraft travels from the earth's surface ($r = 6.4 \times 10^6$ m) to the moon ($r = 400 \times 10^6$ m), the gain of potential energy per kg is $(63 - 1) \times 10^6 = 62 \times 10^6$ J kg^{-1}. The spacecraft loses this amount of kinetic energy and so the energy needed per kg to get the craft from the earth's surface to the distance of the moon is 62×10^6 J kg^{-1}.

To escape to 'infinity' 63×10^6 J kg^{-1} is required, and if this is given to a spacecraft of mass m at lift-off as kinetic energy, we can say

$$\tfrac{1}{2}mu^2 = 63 \times 10^6 \, m$$

where u is the 'escape' speed from the surface of the earth. Hence

$$u \simeq 11 \text{ km s}^{-1}.$$

Potential and field strength calculations

1. *Find (a) the potential and (b) the field strength at points A and B, Fig. 1.21, due to two small spheres X and Y, 1.0 m apart in air and carrying charges of $+ 2.0 \times 10^{-8}$ C and -2.0×10^{-8} C respectively. Assume the permittivity of air $= \epsilon_0$ and $1/(4\pi\epsilon_0) = 9.0 \times 10^9$ N m^2 C^{-2}.*

Fig. 1.21

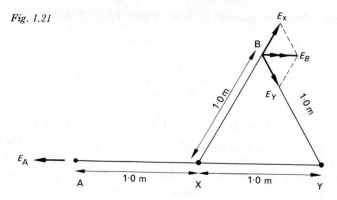

(*a*) Potential at A due to X

$$= \frac{1}{4\pi\epsilon_0} \cdot \frac{Q}{r}$$

$$= \frac{(9.0 \times 10^9 \, \text{N m}^2 \, \text{C}^{-2}) \times (+2.0 \times 10^{-8} \, \text{C})}{1.0 \, \text{m}}$$

$$= \frac{9.0 \times 10^9 \times 2.0 \times 10^{-8}}{1.0} \quad \frac{\text{N m}^2 \, \text{C}^{-2} \, \text{C}}{\text{m}}$$

$$= + 1.8 \times 10^2 \, \text{N m C}^{-1} \quad (1 \, \text{N m} = 1 \, \text{J})$$

$$= + 1.8 \times 10^2 \, \text{V} \quad (1 \, \text{J C}^{-1} = 1 \, \text{V}).$$

Similarly the potential at A due to Y

$$= \frac{9.0 \times 10^9 \times (-2.0 \times 10^{-8})}{2.0} \, \text{V}$$

$$= -0.90 \times 10^2 \, \text{V}.$$

Potential is a scalar quantity and is added algebraically, therefore the potential V_A at A due to X and Y is

$$V_A = (1.8 - 0.90) \times 10^2 \, \text{V}$$

$$= 0.90 \times 10^2 \, \text{V} = 90 \, \text{V}.$$

Since B is equidistant from equal and opposite charges, the potential V_B at B due to X and Y is

$$V_B = (1.8 - 1.8) \times 10^2 \, \text{V} = 0.$$

28

ELECTRIC FIELDS

(*b*) Field strength at A due to X

$$= \frac{1}{4\pi\epsilon_0} \cdot \frac{Q}{r^2}$$

$$= \frac{9.0 \times 10^9 \times 2.0 \times 10^{-8}}{1.0^2} \quad \frac{\text{N m}^2\,\text{C}^{-2}\,\text{C}}{\text{m}^2}$$

$$= 1.8 \times 10^2\,\text{N C}^{-1}\,(\text{or V m}^{-1})\text{ towards the left.}$$

Similarly the field strength at A due to Y

$$= 0.45 \times 10^2\,\text{N C}^{-1}\,(\text{or V m}^{-1})\text{ towards the right.}$$

The resultant field strength E_A at A due to X and Y is

$$E_A = (1.8 - 0.45) \times 10^2\,\text{N C}^{-1}\,(\text{or V m}^{-1})$$

$$= 1.3(5) \times 10^2\,\text{N C}^{-1}\,(\text{or V m}^{-1})\text{ towards the left.}$$

At B, the field strengths due to X and Y have the directions shown and the resultant field strength E_B is their vector sum. Hence

$$E_B = E_X \cos 60 + E_Y \cos 60$$

$$= E_X \quad (\text{since } E_X = E_Y \text{ and } \cos 60 = \tfrac{1}{2})$$

$$= \frac{9.0 \times 10^9 \times 2.0 \times 10^{-8}}{1.0^2}$$

$$= 1.8 \times 10^2\,\text{N C}^{-1}\,(\text{or V m}^{-1})\text{ towards the right.}$$

2. *Two large horizontal, parallel metal plates are 2.0 cm apart in vacuo and the upper is maintained at a positive potential relative to the lower so that the field strength between them is 2.5 × 10⁵ V m⁻¹. (a) What is the p.d. between the plates? (b) If an electron of charge 1.6 × 10⁻¹⁹ C and mass 9.1 × 10⁻³¹ kg is liberated from rest at the lower plate, what is its speed on reaching the upper plate?*

(*a*) If E is the field strength (assumed uniform) and V is the p.d. between two plates distance d apart, we have from $E = V/d$ (p. 23) that

$$V = Ed$$

$$= (2.5 \times 10^5\,\text{V m}^{-1}) \times (2.0 \times 10^{-2}\,\text{m})$$

$$= 5.0 \times 10^3\,\text{V.}$$

(*b*) The energy change (i.e. work done) W which occurs when a charge Q moves through a p.d. of V in an electric field is given by $W = QV$. There is a transfer of electrical potential energy from the field to kinetic energy of the electron. We have

$$QV = \tfrac{1}{2}mv^2$$

29

where v is the required speed and m is the mass of the electron. Therefore

$$\tfrac{1}{2}mv^2 = QV$$

$$\therefore \quad v = \sqrt{\frac{2QV}{m}}$$

$$= \sqrt{\frac{(2 \times 1.6 \times 10^{-19}\,\text{C}) \times (5.0 \times 10^3\,\text{V})}{(9.1 \times 10^{-31}\,\text{kg})}}$$

$$= \sqrt{\frac{2 \times 1.6 \times 10^{-19} \times 5.0 \times 10^3}{9.1 \times 10^{-31}}} \quad \frac{\text{C J C}^{-1}}{\text{kg}} \quad (1\,\text{V} = 1\,\text{J C}^{-1})$$

$$= \sqrt{\frac{16}{9.1} \times 10^{15}\,\text{m}^2\,\text{s}^{-2}} \quad (1\,\text{J} = 1\,\text{N m} = 1\,\text{kg m s}^{-2}\,\text{m})$$

$$= 4.2 \times 10^7\,\text{m s}^{-1}.$$

3. *Compare the electrical and gravitational forces between a proton of charge $+e$ and mass M at a distance r from an electron of charge $-e$ and mass m (e.g. as in a hydrogen atom), given that*

$$e = 1.6 \times 10^{-19}\,\text{C}$$
$$m = 9.1 \times 10^{-31}\,\text{kg}$$
$$M = 1.7 \times 10^{-27}\,\text{kg}$$
$$1/(4\pi\epsilon_0) = 9.0 \times 10^9\,\text{N m}^2\,\text{C}^{-2}$$
$$G = 6.7 \times 10^{-11}\,\text{N m}^2\,\text{kg}^{-2}.$$

$$\text{Electrical attraction} = \frac{1}{4\pi\epsilon_0} \cdot \frac{Q_1 Q_2}{r^2} = \frac{1}{4\pi\epsilon_0} \cdot \frac{e^2}{r^2}.$$

$$\text{Gravitational attraction} = \frac{GmM}{r^2}.$$

Hence,

$$\frac{\text{electrical attraction}}{\text{gravitational attraction}} = \frac{1}{4\pi\epsilon_0} \cdot \frac{e^2}{GmM}$$

$$= \frac{(9.0 \times 10^9\,\text{N m}^2\,\text{C}^{-2})}{(6.7 \times 10^{-11}\,\text{N m}^2\,\text{kg}^{-2})} \times \frac{(1.6 \times 10^{-19}\,\text{C})^2}{(9.1 \times 10^{-31}\,\text{kg}) \times (1.7 \times 10^{-27}\,\text{kg})}$$
$$\simeq 10^{39} : 1.$$

The gravitational force is insignificant compared with the electrical force.

ELECTRIC FIELDS

QUESTIONS

1. An isolated conducting spherical shell of radius 10 cm in vacuo carries a positive charge of 1.0×10^{-7} C. Calculate (a) the electric field intensity, and (b) the potential, at a point on the surface of the conductor. Sketch a graph to show how one of these quantities varies with distance along a radius from the centre to a point well outside the spherical shell. Point out the main features of the graph. (Permittivity of free space $= 8.9 \times 10^{-12}$ F m^{-1}.) (J.M.B.)

2. Considering a hydrogen atom to consist of a proton and an electron at an average separation of 0.50×10^{-10} m, find (a) the electrical potential at 0.50×10^{-10} m from the proton, (b) the potential energy of the electron, and (c) the energy required to remove the electron from the atom assuming the electron is at rest. (Charge on proton $= +1.6 \times 10^{-19}$ C; charge on electron $= -1.6 \times 10^{-19}$ C; $1/(4\pi\epsilon_0) = 9.0 \times 10^9$ N m^2 C^{-2}). Note. In practice other factors as well as the potential energy have to be considered when calculating the energy needed to remove an electron from an atom. Suggest one.

3. Four infinite conducting plates A, B, C and D, of negligible thickness, are arranged parallel to one another so that the distance between adjacent plates is 2.0 cm. The outer plates A and D are earthed and the inner ones B and C are maintained at potentials of $+20$ V and $+60$ V respectively. Draw a graph showing how the potential between the plates varies as a function of position along a line perpendicular to the plates. What is the magnitude and direction of the electric field between each pair of adjacent plates? (O. and C. part qn.)

4. Many electrical insulators are ionic compounds consisting of arrays of dipoles, i.e. pairs of equal but oppositely charged ions a small distance apart. The properties of the dipoles account for the fact that such compounds have very strong internal electric fields but almost zero external fields.

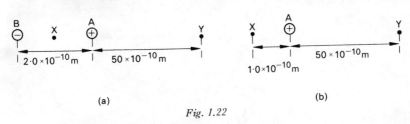

(a)   (b)

Fig. 1.22

In Fig. 1.22a, A and B form an ionic dipole with charges of $+1.6 \times 10^{-19}$ C and -1.6×10^{-19} C respectively and at a separation of 2.0×10^{-10} m. What is the field strength due to the dipole at (a) X, mid-way between A and B, and (b) Y, 50×10^{-10} m to the right of A? Find the ratio of these two field strengths. $(1/(4\pi\epsilon_0) = 9 \times 10^9$ N m^2 C^{-2}).

In Fig. 1.22b A is a single charge of $+1.6 \times 10^{-19}$ C. What is the field at (c) X, and (d) Y? Again, find the ratio.

Compare the two ratios. Does the external field due to a dipole decrease more rapidly with distance than that of a single charge?

31

5. The diagram, Fig. 1.23, shows a pair of flat, wide conducting plates. They are parallel and are connected to a steady potential difference of 2000 V. An oil drop between the plates moves from D_1 to D_2 along a straight line at right angles to the plates and between their centres.

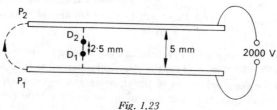

Fig. 1.23

(*i*) If the drop carries five electron charges, each 1.6×10^{-19} C, and moves 2.5 mm (half the distance between the plates), how much electrical energy is transformed? Give the unit of energy appropriate to your answer.

(*ii*) State what you can about the amount of electrical energy transformed if the drop were to move along the curved path P_1P_2 from one plate to the other outside the plates.

Explain your statement.

(*O. and C. Nuffield*)

6. An electric field is maintained between two parallel metal plates by a p.d. of 1000 V. A charge of 10^{-15} C is moved by the forces of the field from one plate to the other.

(*i*) How much electrical energy is transferred from the field-charge system? (*ii*) How much work is done by the electrical force? (*iii*) If the plates are 20 mm apart what is the value of the electrical force assuming it is constant, i.e. the field between the plates is uniform? (*iv*) What is the field strength in N C⁻¹? (*v*) What is the field strength in V m⁻¹?

2 Capacitors

Capacitance

(a) *Definition.* Charge given to an isolated conductor may be thought of as being ' stored ' on it. The amount it will take depends on the electric field thereby created at the surface of the conductor. If this is too great there is breakdown in the insulation of the surrounding medium, resulting in sparking and discharge of the conductor.

In terms of potential we can say that the smaller the change of potential of a conductor when a certain charge is transferred to it, the more charge it can ' store ' before breakdown occurs. The change in potential due to a given charge depends on the size of the conductor, the material surrounding it and the proximity of other conductors.

The idea that an insulated conductor in a particular situation has a certain *capacitance*, or charge-storing ability, is useful and is defined as follows: if the potential of an insulated conductor changes by V when given a charge Q, the *capacitance* C of the conductor is

$$C = \frac{Q}{V}.$$

In words, *capacitance equals the charge required to cause unit change in the potential of a conductor.* If Q is in coulombs (C) and V in volts (V) then the unit of C is a coulomb per volt (C V^{-1}) or a *farad* (F). This is a very large unit and the microfarad (1 μF = 10^{-6} F) or the picofarad (1 pF = 10^{-12} F) is generally used.

Experiment (see p. 36) and theory both show that for an insulated conductor in a given situation, $V \propto Q$, i.e. C is a constant.

(b) *Capacitance of an isolated conducting sphere.* The potential V of such a sphere of radius a, in a medium of permittivity ϵ and having a charge Q, is given by

$$V = \frac{1}{4\pi\epsilon} \cdot \frac{Q}{a}. \tag{p. 18}$$

33

But
$$C = \frac{Q}{V}$$

$$\therefore \quad C = 4\pi\epsilon \cdot a$$

The capacitance of a sphere is therefore proportional to its radius. We also see from this expression that if C is in farads and a in metres then ϵ can be expressed in farads per metre (F m^{-1}), since $\epsilon = C/(4\pi a)$.

(c) *Practical zero of potential*. The theoretical zero of potential is taken, as we have seen, as the potential of points at infinity. However, when making actual measurements this is an impracticable zero and the potential of the earth (itself a conductor) is adopted as the practical zero.

When a charged conductor X of small capacitance and an uncharged conductor Y of large capacitance are connected, they end up with the same potential which is less than X's was originally and X loses most of its charge to Y. There is charge flow until the potentials of X and Y are the same.

If a is the radius of the earth, its capacitance C is

$$C = 4\pi\epsilon \cdot a$$

$$= (1/9 \times 10^{-9} \text{ F m}^{-1}) \times (6 \times 10^6 \text{ m})$$

$$= 7 \times 10^{-4} \text{ F}$$

$$= 700 \ \mu\text{F}.$$

This is a large capacitance compared with that of other conductors used in electrostatics. Therefore when a charged conductor is 'earthed'—say by touching it (the human body conducts)—it loses most of its charge to the earth, i.e. is discharged and acquires earth potential, i.e. zero potential. The earth has such a large capacitance that any change in its potential due to the loss or gain of charge, because of connection to another conductor, is negligible. It thus provides a satisfactory practical zero of potential.

Capacitors

A capacitor is designed to store electric charge and basically consists of two conductors, such as a pair of parallel metal plates, separated by an insulator. The symbol for a capacitor is ⊣⊢ and the conductors are usually referred to as 'plates' whatever their form.

(a) *Action of a capacitor*. A capacitor is readily 'charged' by applying a p.d. across the plates from a battery or other power supply. Insight into the process can be obtained from the circuit of Fig. 2.1 using a 500-μF electrolytic capacitor (see p. 44).

When the circuit is completed, equal-sized flicks occur on the meters indicating momentary but identical current flow through each. The current directions

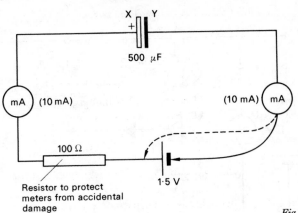

Fig. 2.1

round the circuit are the same, which suggests that as much charge, in the form of a current pulse, leaves one plate of the capacitor as enters the other. Thus if charge $+Q$ has flowed on to X, an equal charge $+Q$ has flowed off Y.

In terms of electron flow we can say that when the supply is connected to the capacitor, electrons flow from the negative of the supply on to plate Y and from plate X towards the positive of the supply at the same rate. The positive and negative charges that appear on X and Y respectively oppose further electron flow and as these charges build up, the p.d. between X and Y increases until it equals the p.d. of the supply (1.5 V). Electron flow then stops. There is never a steady current and no permanent meter deflections.

We say the capacitor has charge Q, meaning that one plate has charge $+Q$ and the other charge $-Q$; as much charge flows off one plate as flows on to the other, Fig. 2.2.

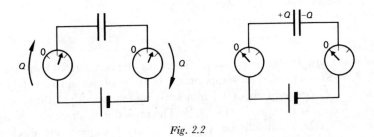

Fig. 2.2

Making the connection shown by the dotted line causes the meters to again give equal, momentary deflections but in the opposite direction to that indicated previously. Electrons flow back from Y to X until the positive charge on X is neutralized. This momentary pulse of current thus discharges the capacitor, leaving zero charge on its plates.

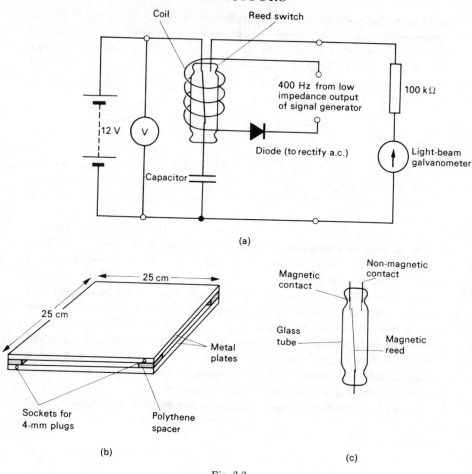

Fig. 2.3

(*b*) *Factors affecting the charge stored.* These may be investigated by the circuit of Fig. 2.3*a* using a parallel plate capacitor in the form of two square metal plates each of side 25 cm, kept about 1.5 mm apart by four small polythene spacers (5 mm × 5 mm) at the corners, Fig. 2.3*b*. The capacitor is alternately charged, usually from a 12-V smooth d.c. supply (e.g. dry batteries), and discharged through a sensitive light-beam galvanometer 400 times a second by a *reed switch*.

A reed switch is shown in Fig. 2.3*c* and when *rectified* a.c. passes through the coil surrounding it, the reed and magnetic contact become oppositely magnetized and attract on the conducting half-cycle. On the non-conducting half-cycle the reed, no longer magnetized, springs back to the non-magnetic contact. The number of charge-discharge actions per second equals the frequency of the a.c. supply to the coil. If this is high enough current pulses follow one another so rapidly that the galvanometer deflection is steady and represents the average current I through it. When Q is the charge stored on the capacitor and released on discharge through the galvanometer and when f is the switching frequency (i.e. that of the a.c. supply) then

$$I = \text{charge passing per second}$$

$$= Qf.$$

If f is constant then $I \propto Q$. The 100–kΩ protective resistor prevents excessive current pulses when the reed switch contacts close; the capacitor will not discharge completely if the resistor is too large. How could you check that discharge is complete?

If the charging p.d. V is varied, we find that the charge Q on the capacitor, i.e. the galvanometer reading I, is directly proportional to V.

When V is kept constant and the separation d of the plates varied (using small insulating spacers at the corners of the plates), it can be inferred that Q is inversely proportional to d if d is not too large. (A graph of Q against $1/d$ does not pass through the origin because of the capacitance between the upper plate and the bench or other nearby objects.)

If V and d are fixed and the area of overlap A of the plates is varied, we can conclude that Q is directly proportional to A. Finally, inserting a solid slab of insulator (e.g. Perspex) in the space between the plates causes Q to increase (if V, d and A are fixed).

This investigation may also be performed using a d.c. amplifier electrometer (Appendix 2, p. 529).

Summarizing the results, we have

$$Q \propto V, \; Q \propto 1/d, \; Q \propto A$$

Hence
$$Q \propto \frac{VA}{d}.$$

The capacitance C of a capacitor is defined (as for an insulated conductor) as the charge stored per unit p.d. between its plates. That is,

$$C = \frac{Q}{V}$$

Hence
$$C \propto \frac{A}{d} = \text{constant} \cdot \frac{A}{d}.$$

It will be seen shortly the capacitance of a parallel plate capacitor is given by

$$C = \frac{\epsilon A}{d}.$$

CAPACITORS

The constant of proportionality thus being ϵ, the permittivity of the insulating medium (also called the *dielectric*) between the plates.

Measurement of capacitance

A reed switch is used as in the circuit of Fig. 2.3a (p. 36) but for capacitances of a few microfarads the light-beam galvanometer is replaced by a milliammeter (1 mA) and a protective resistor of 220 Ω is suitable. For smaller capacitances a microammeter (100 μA) with a 2-kΩ resistor is required. In both cases the reed switch can be operated from a 50-Hz supply.

The capacitance C is found from

$$C = \frac{Q}{V} = \frac{I}{fV} \quad \text{(since } I = Qf\text{)}$$

where I and V are the two meter readings and f is the frequency of the supply.

The effect of connecting capacitors in series and in parallel may be investigated.

Parallel plate capacitor

An expression for the capacitance of a parallel plate capacitor is required.

Consider a capacitor with plates of common area A, separated by a medium of thickness d and permittivity ϵ, Fig. 2.4. If one plate has charge $+Q$ and the other $-Q$, the charge density σ is Q/A. Assuming the field between the plates

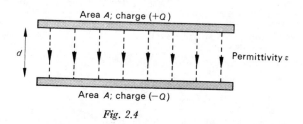

Area A; charge $(+Q)$

d

Permittivity ε

Area A; charge $(-Q)$

Fig. 2.4

is uniform, the field strength E is the same at all points and is given by

$$E = \frac{\sigma}{\epsilon} = \frac{Q}{A\epsilon}.$$

(see p. 11)

Further, if V is the p.d. between the plates then

$$E = \frac{V}{d}.$$

(see p. 23)

Therefore,
$$\frac{Q}{A\epsilon} = \frac{V}{d}.$$

Hence
$$\frac{Q}{V} = \frac{A\epsilon}{d}$$

$$\therefore \quad C = \frac{A\epsilon}{d} \quad \text{(since } Q = VC\text{)}.$$

C will be in farads if A is in m², d in m and ϵ in F m^{-1}. It is worth noting that since the field between the plates is uniform, there is plane symmetry and π does not appear in the formula for C. It is an example of the simplification of a common formula by the use of rationalized units (p. 9). In practice, the expression is not strictly true due to non-uniformity of the field at the edges of the plates.

Large capacitance values are obtained by having ' plates ' with a large overlapping area that are very close together and are separated by a dielectric with high permittivity.

Permittivity

Earlier we saw that the force between two charges depended on the intervening medium and the idea of permittivity ϵ was introduced to take this into account. We will now consider this idea further, especially in relation to capacitors.

(a) *Relative permittivity or dielectric constant* ϵ_r. Experiment shows that inserting an insulator or dielectric between the plates of a capacitor increases its capacitance (p. 37). If C_0 is the capacitance of a capacitor when a vacuum separates its plates and C is the capacitance of the same capacitor with a dielectric filling the space between the plates, the relative permittivity ϵ_r of the dielectric is defined by

$$\epsilon_r = \frac{C}{C_0}.$$

Taking a parallel plate capacitor as an example we have

$$\epsilon_r = \frac{C}{C_0} = \frac{\epsilon A/d}{\epsilon_0 A/d} = \frac{\epsilon}{\epsilon_0}$$

where ϵ is the permittivity of the dielectric and ϵ_0 is that of a vacuum (i.e. of free space). The expression for the capacitance of a parallel plate capacitor with a dielectric of relative permittivity ϵ_r can therefore be written as

$$C = \frac{A\epsilon_r\epsilon_0}{d}.$$

CAPACITORS

Relative permittivity has no units, unlike ϵ and ϵ_0 which have; it is a pure number without dimensions. For air at atmospheric pressure $\epsilon_r = 1.0005$, which is near enough 1 and so for most purposes $\epsilon_{air} = \epsilon_0$. Table 2.1 gives some values of ϵ_r.

Table 2.1

Dielectric	Relative permittivity ϵ_r
Vacuum	1.0000
Air at s.t.p.	1.0005
Polythene	2.3
Perspex	2.6
Paper (waxed)	2.7
Mica	7
Water (pure)	80
Barium titanate	1200

The near impossibility of removing all the impurities dissolved in water makes it unsuitable in practice as a dielectric.

(b) *Action of a dielectric.* The molecules of a dielectric which is between the plates of a charged capacitor are in an electric field. The positive nuclei are urged in the direction of the field and the negative electrons in the opposite direction. As a result the molecules become distorted or *polarized* by the field with one end having an excess of positive charge and the other of negative charge. Electric dipoles are thus formed, Fig. 2.5a.

Fig. 2.5

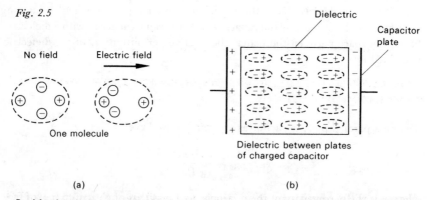

(a)   (b)

Inside the dielectric the positive and negative ends of adjacent dipoles cancel each other's effects but at the surfaces of the dielectric unneutralized charges appear which have opposite signs to those on the plates, Fig. 2.5b. Thus the positive potential of the positive plate is reduced by the negative charge at one surface of the dielectric and the negative potential of the negative plate is made less negative by the positive charge at the other surface of the dielectric. The p.d. between the plates is thereby less than it would otherwise be and so more charge is required before the p.d. between the plates equals the applied (i.e. charging) p.d. The capacitance is thus increased.

Some molecules, called *polar* molecules, are polarized even in the absence of an electric field and consequently they increase the capacitance even more than does a dielectric with *non-polar* molecules. The high relative permittivity of water arises from the H_2O molecule being polar.

(*c*) *Measurement of* ϵ_0. The circuit is the same as that for investigating the factors affecting the charge stored by a capacitor (p. 36) and is shown again in Fig. 2.6. As before, a capacitor in the form of a large pair of parallel metal plates is alternately charged from a smooth d.c. supply (0–12 V) and discharged through a light-beam galvanometer (on its most sensitive range, e.g. × 1) about 400 times a second by a vibrating reed switch.

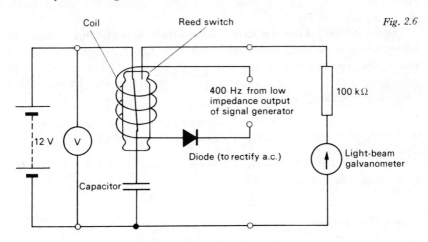

Coil   Reed switch   Fig. 2.6

400 Hz from low impedance output of signal generator

100 kΩ

12 V   V

Diode (to rectify a.c.)

Light-beam galvanometer

Capacitor

The capacitance C of a parallel plate air capacitor having a plate overlap area A and plate separation d is given by

$$C = \frac{\epsilon_{air}A}{d}$$

$$\therefore \quad \epsilon_{air} = \frac{Cd}{A}$$

where ϵ_{air} is the permittivity of air which, to a good approximation, is ϵ_0. If Q is the charge stored on the capacitor when the applied p.d. is V (measured on the voltmeter) then since $Q = VC$ we have

$$\epsilon_0 = \frac{Qd}{AV}.$$

CAPACITORS

The steady current I recorded by the galvanometer is Qf, where f is the switching frequency, provided the capacitor is fully charged and discharged during each contact of the reed switch. Hence from $I = Qf$ we get

$$\epsilon_0 = \frac{Id}{fVA}.$$

Knowing the current sensitivity of the galvanometer (marked on the instrument or supplied by the manufacturer, and typically 20–25 mm μA^{-1}), I can be found in amperes. The switching frequency f is obtained from the dial reading on the signal generator and may be checked using a CRO and a low voltage 50-Hz mains supply (see p. 436). The area A of one of the plates is readily measured as is the separation d if the top capacitor plate rests on four small insulating spacers (each about 1.5 mm thick) at the corners of the bottom plate.

Very small charges are involved and, to prevent leakage, the connection from the top plate to the reed switch should not touch anything. The accepted value of ϵ_0 is 8.85×10^{-12} F m^{-1}.

The permittivity of different materials can also be found by this method if the experiment is done with the material completely filling the space between the plates. The permittivity ϵ of the material is then $\epsilon = I_1 d/(fVA)$ where I_1 is the galvanometer current. The ratio of the galvanometer currents with and without the material between the plates gives the relative permittivity ϵ_r. That is,

$$\epsilon_r = \frac{\epsilon}{\epsilon_0} = \frac{I_1 d/(fVA)}{Id/(fVA)} = \frac{I_1}{I}.$$

Both ϵ_0 and ϵ_r may be determined in a similar manner using a d.c. amplifier electrometer (Appendix 2, p. 529).

Types of capacitor

Capacitors are used in electric circuits for various purposes, as we will see later. Different types have different dielectrics. The choice of type depends on the value of capacitance and stability (i.e. ability to keep the same value with age, temperature change, etc.) needed and on the frequency of any alternating current (a.c.) that will flow in the capacitor (this affects the power loss).

For every dielectric material there is a certain potential gradient at which it breaks down and a spark passes. The working p.d. of a capacitor is thus determined by the thickness of the dielectric. Liquid and gaseous dielectrics recover when the applied p.d. is reduced below the breakdown value; only some solid dielectrics do.

(a) *Paper, plastic, ceramic and mica capacitors.* Waxed paper, plastics (e.g. polystyrene), ceramics (e.g. talc with barium titanate added) and mica (which occurs naturally and splits into very thin sheets of uniform thickness) are all

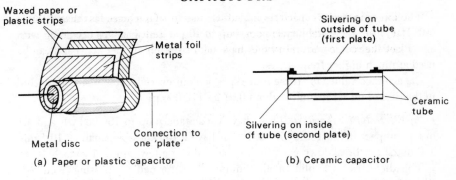

(a) Paper or plastic capacitor   (b) Ceramic capacitor

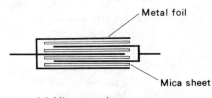

Fig. 2.7

(c) Mica capacitor

used as dielectrics. Typical constructions are shown in Fig. 2.7 and actual capacitors in Fig. 2.8.

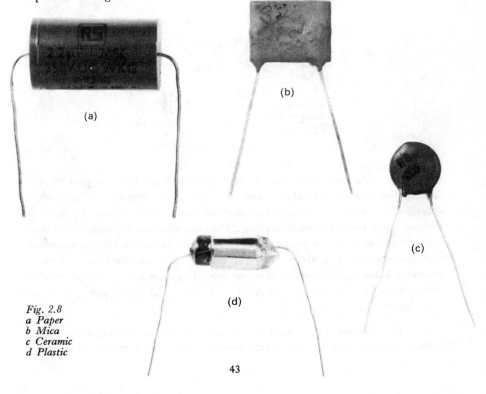

Fig. 2.8
a Paper
b Mica
c Ceramic
d Plastic

The losses in paper capacitors limit their use to frequencies less than 1 MHz (10^6 Hz); also, their stability is poor—up to 10 per cent changes occurring with age. Plastic, ceramic and mica types have better stability (1 per cent) and can be used at much higher frequencies.

Capacitance values for these four types seldom exceed a few microfarads and in the case of mica the limit is about 0.01 μF (10 000 pF).

(b) *Electrolytic capacitors.* These have capacitances up to 100 000 μF and are quite compact because the dielectric can have a thickness as small as 10^{-4} mm and not suffer breakdown even for applied p.d.s of a few hundred volts.

The dielectric is a film of aluminium oxide formed by passing a current through a strip of paper soaked with aluminium borate solution, separating two aluminium foil electrodes, Fig. 2.9a. The oxide forms on the anode, which acts as one plate of the capacitor. The borate solution, being an electrolyte, is the other; connection to it is made via the cathode (the other piece of aluminium foil).

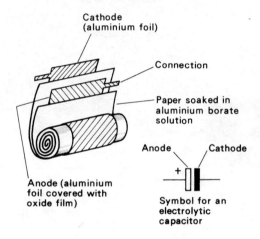

Cathode
(aluminium foil)

Connection

Paper soaked in
aluminium borate
solution

Anode   Cathode

Anode (aluminium
foil covered with
oxide film)

Symbol for an
electrolytic
capacitor

Fig. 2.9a

In use, a ' leakage ' current of the order of 1 mA must always pass through the capacitor, in the correct direction, to maintain the dielectric. The anode is therefore coloured red to show that it must be at a higher potential than the cathode when the capacitor is in circuit. Care must be taken to ensure that there is direct current (d.c.) in the circuit and that the capacitor is correctly connected.

Some electrolytic capacitors are shown in Fig. 2.9b; the case is often of aluminium and acts as the negative terminal. They are not used in a.c. circuits where the frequency exceeds about 10 kHz. Their stability is poor (10–20 per cent) but in many cases this does not matter.

(c) *Air capacitors.* Air is used as the dielectric in variable capacitors. These consist of two sets of parallel metal plates, one set is fixed and the other moves

Fig. 2.9b.

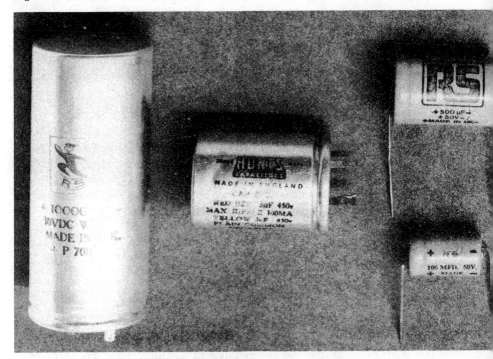

on a spindle within—but not touching—the fixed set, Fig. 2.10. The interleaved area varies, thus changing the capacitance.

Losses in an air dielectric are very small at all frequencies. However, relatively large thicknesses are needed because breakdown occurs at a potential gradient which is small compared with those of other dielectrics. Variable capacitors are used to tune radio and TV receivers.

Moving plates

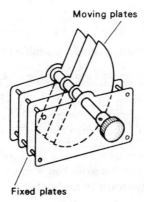

Fixed plates

Fig. 2.10

Capacitor networks

A network of capacitors has a combined or equivalent capacitance which can be calculated.

(a) *Capacitors in parallel.* In Fig. 2.11a the three capacitors of capacitance C_1, C_2 and C_3 are in parallel. *The applied p.d. V is the same across each but the charges are different* and are given by

$$Q_1 = VC_1 \qquad Q_2 = VC_2 \qquad Q_3 = VC_3.$$

The total charge Q on the three capacitors is

$$Q = Q_1 + Q_2 + Q_3$$
$$Q = V(C_1 + C_2 + C_3).$$

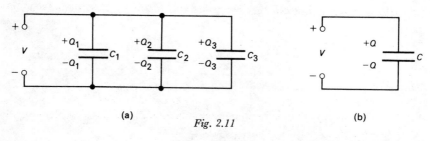

(a)

(b)

Fig. 2.11

If C is the capacitance of the single equivalent capacitor, it would have charge Q when the p.d. across it is V, Fig. 2.11b. Hence

$$Q = VC$$
$$\therefore C = C_1 + C_2 + C_3.$$

The combined capacitance of capacitors of 1 μF, 2 μF and 3 μF in parallel is 6 μF.

It should be noted that the charges on capacitors in parallel are in the ratio of their capacitances, i.e.

$$Q_1 : Q_2 : Q_3 = C_1 : C_2 : C_3.$$

The expression for capacitors in parallel is similar to that for resistors in series.

(b) *Capacitors in series.* The capacitors in Fig. 2.12a are in series and have capacitances C_1, C_2 and C_3. Suppose a p.d. V applied across the combination causes the motion of charge from plate Y to plate A so that a charge $+Q$ appears on A and an equal but opposite charge $-Q$ appears on Y. This charge $-Q$ will induce a charge $+Q$ on plate X if the plates are large and close together. The plates X and M and the connection between them form an insulated conductor whose net charge must be zero and so $+Q$ on X induces a charge $-Q$ on M. In turn this charge induces $+Q$ on L and so on.

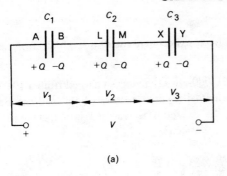

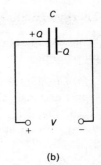

(a)

(b)

Fig. 2.12

Capacitors in series thus *all have the same charge* and the p.d.s across each are given by

$$V_1 = \frac{Q}{C_1} \qquad V_2 = \frac{Q}{C_2} \qquad V_3 = \frac{Q}{C_3}.$$

The total p.d. V across the network is

$$V = V_1 + V_2 + V_3$$

$$\therefore \; V = \frac{Q}{C_1} + \frac{Q}{C_2} + \frac{Q}{C_3}$$

$$= Q\left(\frac{1}{C_1} + \frac{1}{C_2} + \frac{1}{C_3}\right).$$

If C is the capacitance of the single equivalent capacitor, it would have charge Q when the p.d. across it is V, Fig. 2.12*b*.

$$V = \frac{Q}{C}.$$

Therefore

$$\frac{Q}{C} = \left(\frac{1}{C_1} + \frac{1}{C_2} + \frac{1}{C_3}\right)$$

$$\therefore \; \frac{1}{C} = \frac{1}{C_1} + \frac{1}{C_2} + \frac{1}{C_3}.$$

The combined capacitance of capacitors of 1 μF, 2 μF and 3 μF in series is 6/11 μF, i.e. the equivalent capacitance is less than the smallest capacitance.

The expression for capacitors in series is similar to that for resistors in parallel.

Notes. 1. For capacitors in parallel the p.d. across each is the same.

2. For capacitors in series each has the same charge.

Energy of a charged capacitor

Some basic facts can be established from experiments with a large electrolytic capacitor before the theory is developed.

(a) *Discharge through a motor.* If a 10 000-μF capacitor is charged from a 10-V supply and then discharged through a small electric motor, a light load can be raised, Fig. 2.13. Energy stored in the capacitor is changed into mechanical energy. Are any other forms of energy produced?

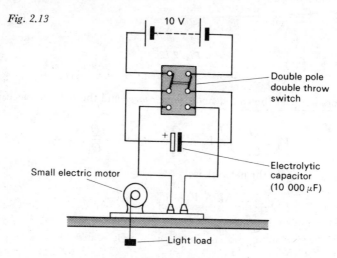

Fig. 2.13

Double pole double throw switch

Electrolytic capacitor (10 000 μF)

Small electric motor

Light load

(b) *Discharge through lamps.* Using the circuit of Fig. 2.14a a 10 000-μF capacitor is charged from a 3-V supply then discharged through a 2.5-V, 0.3-A lamp, and the brightness is noted. If the procedure is repeated with two lamps in series (or in parallel) the brightness of each is much less than before. With a

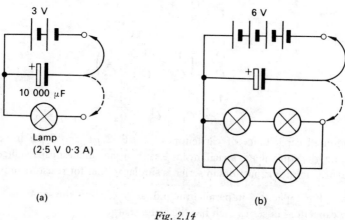

Fig. 2.14

6-V charging supply, two lamps in series flash brighter and longer than one did at 3 V. However, *four* lamps arranged as in Fig. 2.14*b* light up to about the same brightness and for about the same time as one lamp did on 3 V. How many times more energy is stored in the capacitor at 6 V than at 3 V? What value of charging p.d. is needed to light *nine* lamps similarly to one lamp at 3 V?

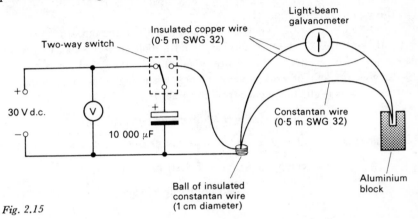

Fig. 2.15

(*c*) *Discharge through a heating coil.* The capacitor in Fig. 2.15 is charged and then discharged through the resistance coil made from 2 m of constantan wire (SWG 32). The resulting temperature rise in the coil is detected by a copper-constantan thermocouple and indicated by a light-beam galvanometer on its most sensitive range ($\times$ 1). One discharge at 20 V gives roughly the same deflection as four successive discharges at 10 V.

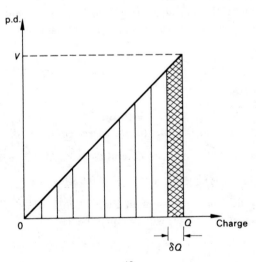

Fig. 2.16

CAPACITORS

From this experiment and the previous one we can say that doubling the p.d. to which a capacitor is charged quadruples the energy stored, i.e. if the p.d. is V then the energy is proportional to V^2. Can you say why?

(*d*) *Graphical treatment.* The charge on a capacitor is directly proportional to the p.d. across it ($Q = VC$). A graph of p.d. against charge is therefore a straight line through the origin, Fig. 2.16.

Suppose a capacitor has capacitance C and that when the p.d. is V the charge is Q. If the capacitor starts to discharge and initially a very small charge δQ passes from the negative to the positive plate, then by the definition of p.d. the resulting energy loss (or work done) is $V . \delta Q$—assuming δQ is so small that the decrease in V is negligible. Hence from the graph,

$$\text{energy loss} = \text{area of shaded strip.}$$

If the capacitance discharges completely so that Q and V fall to zero, we have

$$\text{total energy loss} = \text{area of all strips}$$

$$= \text{area of triangle below graph}$$

$$= \tfrac{1}{2}QV.$$

This is the energy stored in the capacitor. We can also write, since $Q = VC$, that

$$\text{total energy} = \tfrac{1}{2}V^2C = \tfrac{1}{2}Q^2/C.$$

The energy stored is thus proportional to V^2, as suggested by the previous experiments. If Q is in coulombs, V in volts and C in farads, the energy is in joules. How much energy is stored in a 10 000-μF capacitor charged to 30 V? The energy of a charged capacitor is considered to be stored as electrical energy in the field in the medium between the plates.

The expression $\tfrac{1}{2}QV$ is analogous to $\tfrac{1}{2}Fe$ for the energy stored in a wire that obeys Hooke's law, when a tension F causes an extension e.

Capacitor calculations

1. *In the circuit of Fig. 2.17a,* $C_1 = 2$ μF, $C_2 = C_3 = 0.5$ μF *and E is a 6-V battery. For each capacitor calculate (a) the charge on it, and (b) the p.d. across it.*

The combined capacitance C_4, of C_2 and C_3 in parallel is given by

$$C_4 = C_2 + C_3 = 0.5 \ \mu\text{F} + 0.5 \ \mu\text{F} = 1 \ \mu\text{F.}$$

The circuit may now be redrawn as in Fig. 2.17b in which C_1 and C_4 are in series. Their charges Q_1 and Q_4 will be equal, hence

$$Q_1 = Q_4 = V_1 C_1 = V_4 C_4$$

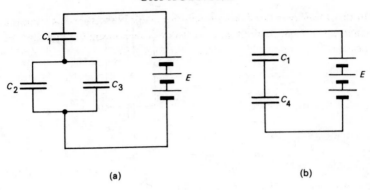

(a)   (b)

Fig. 2.17

where V_1 and V_4 are the p.d.s across C_1 and C_4 respectively. Therefore

$$\frac{V_1}{V_4} = \frac{C_4}{C_1} = \frac{1}{2}.$$

But $\qquad V_1 + V_4 = 6$

$\qquad\qquad \therefore \quad V_1 = 2\text{ V and } V_4 = 4\text{ V}.$

Also $\qquad Q_1 = V_1 C_1 = (2\text{ V}) \times (2 \times 10^{-6}\text{ F}) = 4 \times 10^{-6}\text{ C}$

$\qquad\qquad \therefore \quad Q_4 = 4 \times 10^{-6}\text{ C} = 4\ \mu\text{C}.$

The p.d. across the combined capacitance C_4 equals that across each of C_2 and C_3. Hence

$$V_2 = V_3 = V_4 = 4\text{ V}.$$

Now $\qquad Q_2 = V_2 C_2 \text{ and } Q_3 = V_3 C_3$

$$\therefore \quad \frac{Q_2}{Q_3} = \frac{C_2}{C_3} = \frac{0.5}{0.5} \quad (\text{since } V_2 = V_3)$$

$$= 1.$$

But $\qquad Q_2 + Q_3 = Q_4 = 4 \times 10^{-6}\text{ C}$

$\qquad\qquad \therefore \quad Q_2 = Q_3 = 2 \times 10^{-6}\text{ C} = 2\ \mu\text{C}.$

Therefore $\qquad Q_1 = 4\ \mu\text{C} \qquad Q_2 = Q_3 = 2\ \mu\text{C}$

$$V_1 = 2\text{ V} \qquad V_2 = V_3 = 4\text{ V}.$$

2. *A 10-μF capacitor is charged from a 30-V supply and then connected across an uncharged 50-μF capacitor. Calculate (a) the final p.d. across the combination, and (b) the initial and final energies, Fig. 2.18a and b.*

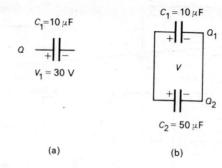

Fig. 2.18

(a)   (b)

(*a*) Initial charge Q on $C_1 = V_1 C_1 = (30 \text{ V}) \times (10 \times 10^{-6} \text{ F})$

$$= 3.0 \times 10^{-4} \text{ C} = 300 \ \mu\text{C}.$$

When C_1 and C_2 are connected (in parallel) the charge Q is shared in the ratio of their capacitances. That is,

$$\frac{Q_1}{Q_2} = \frac{C_1}{C_2} = \frac{10}{50} = \frac{1}{5}.$$

But $Q = Q_1 + Q_2 = 300 \ \mu\text{C}$

$$\therefore \quad Q_1 = 50 \ \mu\text{C and } Q_2 = 250 \ \mu\text{C}.$$

The final, common p.d. V is given by

$$V = \frac{Q_1}{C_1} \left(\text{or } \frac{Q_2}{C_2}\right) = \frac{50 \times 10^{-6} \text{ C}}{10 \times 10^{-6} \text{ F}} = 5.0 \text{ V}.$$

(*b*) Initial energy $= \frac{1}{2} Q V_1 = \frac{1}{2} \times 3.0 \times 10^{-4} \times 30 \text{ J}$

$$= 4.5 \times 10^{-3} \text{ J}.$$

Final energy $= \frac{1}{2} Q_1 V + \frac{1}{2} Q_2 V = \frac{1}{2} V (Q_1 + Q_2) = \frac{1}{2} Q V$

$$= \frac{1}{2} \times 3.0 \times 10^{-4} \times 5.0 \text{ J}$$

$$= 0.75 \times 10^{-3} \text{ J}.$$

Note. The apparent loss of energy is due to the production of heat when charge flows in the wires connecting the capacitors.

Discharge of a capacitor

Many electronic circuits involve capacitors charging and discharging through resistors. The way in which these processes occur is typical of other growth and decay effects in physics (e.g. radioactive decay), chemistry, biology and economics. All may be described either graphically or mathematically (using calculus) by the same equation. Here we shall consider mainly the decay of charge on a capacitor.

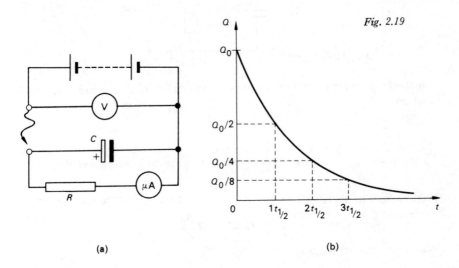

Fig. 2.19

(a)

(b)

(*a*) *Exponential decay curve.* First, using the circuit of Fig. 2.19*a* it can be shown that the larger the values of C and R, the slower the discharge.

To obtain a decay curve a convenient value for C is 500 μF and for R 100 kΩ. The capacitor is charged to about 10 V and then readings taken on the micro-ammeter (100 μA) at 10-second intervals during discharge.

The decay curve of charge Q against t, Fig. 2.19*b*, will have the same shape as the decay curve of I against t since $Q = VC$ (where Q is the charge on the capacitor and V the p.d. across it at time t) and $V = IR$ (V causes I through R), therefore $Q = IRC$. But R and C are constant and so $Q \propto I$.

Examination of the decay curve shows that it decreases by the same fraction in successive equal time intervals. Thus if it falls from Q_0 to $Q_0/2$ in time $t_{\frac{1}{2}}$, it will also fall from $Q_0/2$ to $Q_0/4$ in the next time interval $t_{\frac{1}{2}}$ and so on. Time $t_{\frac{1}{2}}$ is therefore the time for the charge to fall by half and is known as the *half-life* of the decay process. For the values of R and C used, $t_{\frac{1}{2}} \simeq 35$ s. In general the time for the charge to fall by *any* fraction (not just a half) is constant and this is a characteristic of any quantity which decays by an *exponential* law. The graph in Fig. 2.19*b* is an exponential decay curve.

An alternative arrangement for obtaining a decay curve using a d.c. amplifier is given in Appendix 2 on page 529.

(*b*) *Calculus treatment.* If at a certain time t during the discharge of a capacitor of capacitance C through a resistance R, Fig. 2.20, the p.d. across the capacitor is V and the charge on it is Q, then

$$Q = VC.$$

53

CAPACITORS

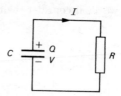

Fig. 2.20

The discharge current I at time t equals the rate of loss of charge and in calculus notation

$$I = -\frac{dQ}{dt}.$$

The negative sign shows that Q decreases as t increases. We also have that

$$V = IR.$$

From these three equations

$$Q = -CR\frac{dQ}{dt}.$$

Rearranging and integrating

$$\int_0^t \frac{dt}{CR} = -\int_{Q_0}^Q \frac{dQ}{Q} = \int_Q^{Q_0} \frac{dQ}{Q}$$

where Q_0 is the initial charge. Therefore

$$\frac{t}{CR} = \log_e \left(\frac{Q_0}{Q}\right)$$

$$\therefore \quad Q = Q_0 e^{-t/CR}.$$

This is the equation of the exponential decay curve.

When $Q = Q_0/2$, $t = t_{\frac{1}{2}}$ and

$$\frac{Q_0}{2} = Q_0 e^{-t_{\frac{1}{2}}/CR}$$

$$\therefore \quad \tfrac{1}{2} = e^{-t_{\frac{1}{2}}/CR}$$

$$\therefore \quad e^{t_{\frac{1}{2}}/CR} = 2$$

$$\therefore \quad t_{\frac{1}{2}} = CR \log_e 2 = 0.693\, CR.$$

In the circuit of Fig. 2.19a, if $C = 500\ \mu\text{F} = 500 \times 10^{-6}\ \text{F} = 5 \times 10^{-4}\ \text{F}$ and $R = 100\ \text{k}\Omega = 10^5\ \Omega$, then $CR = 50\ \text{s}$

$$\therefore \quad t_{\frac{1}{2}} = 34.7\ \text{s}.$$

It can be seen that CR has units of time by expressing C and R in terms of the units from which they are derived. Thus

$$C\,(\text{farads}) = \frac{Q\,(\text{coulombs})}{V\,(\text{volts})} = \frac{It\,(\text{amperes} \times \text{seconds})}{V\,(\text{volts})}$$

$$R\,(\text{ohms}) = \frac{V\,(\text{volts})}{I\,(\text{amperes})}$$

$$\therefore\; F \times \Omega = \left(\frac{A \times s}{V}\right) \times \left(\frac{V}{A}\right) = s.$$

The product CR is called the *time constant* of the circuit; it is an important and useful quantity to know. When $t = CR$ we have

$$Q = Q_0\,e^{-t/CR} = \frac{Q_0}{e}.$$

Now e, the natural base of logs, equals 2.7 (approximately) and so

$$Q = 0.37\,Q_0.$$

That is, CR is the time for Q to fall to 37 per cent of its initial value.

When a capacitor charges up through a resistor, the charge (and p.d.) grows exponentially. A circuit for viewing the charge and discharge curves on a CRO is given in Fig. 2.21.

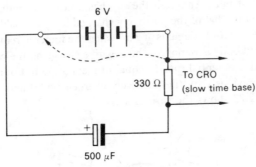

6 V

330 Ω

To CRO
(slow time base)

+

500 μF

Fig. 2.21

Electroscopes

The *gold-leaf* electroscope, Fig. 2.22a, was used in most of the early work on electrostatics. The *Braun* electroscope, Fig. 2.22b, is a more robust instrument.

Basically an electroscope is a capacitor, the leaf (or pointer) and rod to which it is attached forming one plate and the case the other. When a p.d. is applied to the ' plates ', a small charge flows on to the leaf and an equal but opposite charge is induced on the inside of the case. The forces of the resulting electric field between the ' plates' deflect the leaf, and the deflection is a measure of the p.d.

CAPACITORS

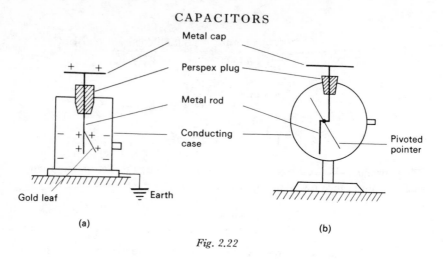

Metal cap

Perspex plug

Metal rod

Conducting case

Gold leaf   Earth

Pivoted pointer

(a)   (b)

Fig. 2.22

between leaf etc. and case. If the case is earthed, as it will be if it is made of wood and stands on a wooden bench, then the *electroscope records potential*.

The last point can be shown by charging by induction an electroscope with an earthed case. Thus when a positively charged body is brought near the electroscope cap, the electric field due to the charge raises the potential of the cap and leaf and deflection of the latter occurs. If the cap is now earthed by touching it with the finger, the leaf falls. Although the cap has a negative (induced) charge, its effect is cancelled by that of the positively charged body so making the potential of the cap and leaf zero. It remains at zero when the finger is removed. When the positively charged body is removed, the negative charge on the cap gives the cap and leaf a negative potential (nearly equal in magnitude but opposite in sign to that produced previously by the positively charged body) and there is again a deflection. Fig. 2.23 shows the various stages.

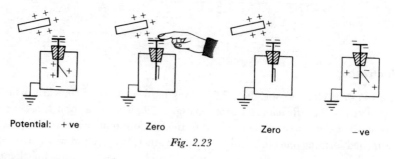

Potential: + ve   Zero   Zero   − ve

Fig. 2.23

If the case stands on an insulator and the cap is given a charge, what happens if the cap and case are (*i*) connected, Fig. 2.24*a*, (*ii*) not connected, Fig. 2.24*b*?

56

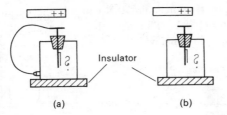

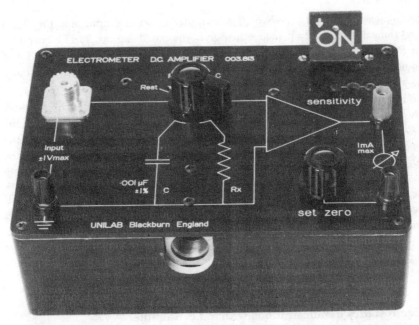

Fig. 2.24

The d.c. amplifier as an electrometer

An electrometer measures p.d.s, including those produced electrostatically. An electroscope with a scale for reading deflections is an example of such an instrument.

Fig. 2.25

The d.c. amplifier, one make of which is shown in Fig. 2.25, can also be used as an electrometer. It consists of an amplifier whose stages are *directly coupled* (hence d.c.) and are not joined by capacitors as in a normal amplifier (see p. 449). Basically a d.c. amplifier acts as a very high resistance voltmeter (about $10^{13} \, \Omega$) but it can be adapted to measure current and charge. In all cases, however, a p.d. of up to 1 V (for the instrument shown) has to be applied to the input terminals and this controls the output current which is recorded on a meter (1 mA). The meter has to be calibrated to read the input p.d. directly—as described below.

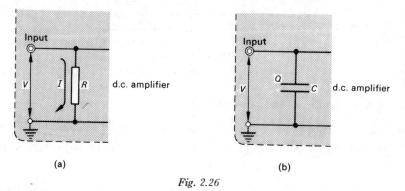

(a)   (b)

Fig. 2.26

To measure current, a resistor R is connected across the input as in Fig. 2.26a and the calibrated output meter then gives the p.d. V across R. The current I in R will be V/R. If $R = 10^{11}\ \Omega$, a meter reading corresponding to a 1 V input indicates that $I = V/R = 1/10^{11} = 10^{-11}$ A. (The much higher input resistance of the amplifier—about $10^{13}\ \Omega$—in parallel with R can be ignored.)

The measurement of charge requires a known capacitor C across the input as in Fig. 2.26b. The p.d. V across C, i.e. the input p.d., is again shown by the calibrated output meter. The unknown charge Q on the capacitor (which is responsible for the p.d.) is obtained from $Q = VC$. For example, if $C = 10^{-8}$ F (0.01 μF) then when the output meter gives a reading of 1 mA, the input p.d. $V = 1$ V and so $Q = 10^{-8}$ C, the input capacitance of the amplifier being neglected. Almost complete transfer of charge to the amplifier capacitor C will occur if the capacitance of the object bringing the charge to the input terminal is small compared with that of the amplifier capacitor. Also, the input resistance of the amplifier is so high that leakage of charge from the amplifier capacitor through the instrument is negligible. There is, therefore, ample time to make a reading since the p.d. across the capacitor remains constant for a long time.

Electrostatics demonstrations with a d.c. amplifier

(*a*) *Calibration*. After connecting a suitable meter (1 mA), the amplifier is switched on and the INPUT SWITCH is set to ' rest ', then the SET ZERO control is adjusted for zero deflection of the meter.

The amplifier in Fig. 2.25, is calibrated by setting the INPUT SWITCH to ' R ', connecting the circuit of Fig. 2.27 to give exactly a 1-V input, adjusting the SENSITIVITY control with a screw-driver till the 1-mA meter reads full scale and then removing the calibrating circuit.

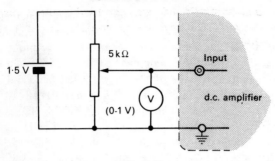

Fig. 2.27

Once calibrated, the electrical zero on the meter can be moved anywhere on the scale using the SET ZERO control.

(b) *Use as electrometer.* A capacitor has to be connected across the input to suit the charge to be measured. Its capacitance must be such that the charge does not cause the p.d. across it to exceed 1 V (10^{-7} F is often satisfactory).

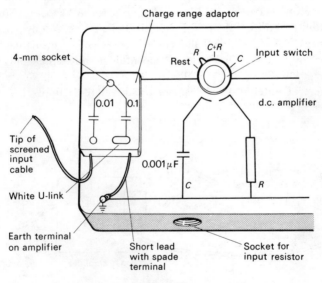

Fig. 2.28

In the instrument of Fig. 2.25, a charge range adaptor (consisting of two capacitors) is plugged directly into the input socket (with the engraved panel uppermost) and the spade terminal on the short lead is joined to the earth terminal on the amplifier, Fig. 2.28. Charges can be applied either to the tip of the screened input cable or directly to a small brass disc (2 cm in diameter) plugged into the 4-mm socket on the adaptor. To measure charges up to 10^{-7} C the right-hand and centre sockets are linked by the white U link (as shown) and the INPUT SWITCH on the amplifier set to ' R ' (any input resistor being unscrewed from the front of the amplifier). This connects a 0.1-μF (10^{-7}-F) capacitor across the input. For the 10^{-8} C range, the left-hand and centre sockets are linked and the INPUT SWITCH left at ' R '; the 0.01-μF (10^{-8}-F) capacitor in the adaptor is now in circuit. The most sensitive range of 10^{-9} C is obtained by removing the U link and setting the INPUT SWITCH to ' C ', thereby connecting a 0.001-μF (10^{-9}-F) capacitor in the amplifier itself across the input. Setting the INPUT SWITCH to ' rest ' discharges the capacitor in use: this is often necessary between readings.

(*c*) *Positive and negative charges.* The SET ZERO control on the amplifier should be altered to give a centre-zero on the meter; deflections due to charges of different sign will then be clearly indicated. If a cellulose acetate (or Perspex) strip, charged positively by rubbing, is brought near to or drawn firmly across the small brass input disc, a deflection occurs which is opposite to that obtained when a negatively charged polythene strip is used.

(*d*) ' *Spooning* ' *charge.* It can be shown that charge is a *quantity* of something that can be measured and passed from place to place. A ' spoonful ' of positive charge is transferred from the positive terminal of an e.h.t. supply (e.g. 1 kV) to the amplifier using say a 3-cm diameter metal sphere on an insulating handle as the ' spoon ', Fig. 2.29. Each additional ' spoonful ' produces the same increase of deflection.

The effect of different-sized ' spoons ' and different supply p.d.s may be investigated.

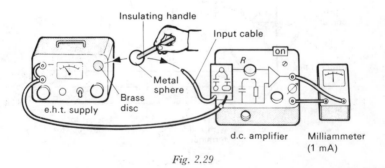

Fig. 2.29

(*e*) *Electrostatic induction.* Two metal spheres (3 cm in diameter) on insulated stands, initially touching each other, are used, Fig. 2.30. A charged strip is brought near one sphere, the spheres are separated and the strip removed. The charges on the spheres are then tested by touching each in turn on the small disc plugged into the input socket on the amplifier (meter set to centre-zero).

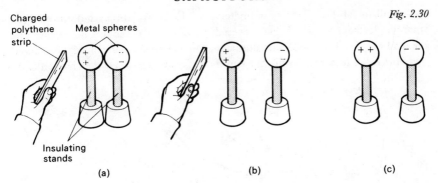

Fig. 2.30

(*f*) *Distribution of charge on conductors.* The charge per unit area of the surface of a conductor, i.e. the *charge density*, is greatest on the most highly convex parts. This may be shown by transferring charge with a proof-plane (a small metal disc on an insulating handle) from different parts of various conductors on insulated stands, Fig. 2.31. The pear-shaped conductor (for which the 10^{-8} C range may be necessary) and the can may be charged four or five times by an electrophorus; the disc can be the metal plate of a charged electrophorus.

The charge acquired by the proof-plane is proportional to the charge density on the surface touched since the proof-plane in effect becomes part of the charged surface. Although the charge density on the surface of a conductor may vary from place to place, the potential is the same at all points.

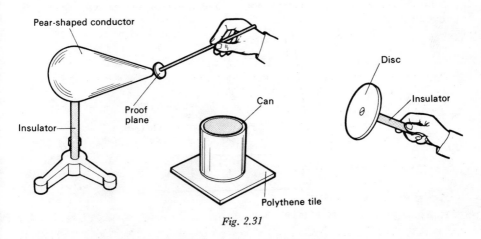

Fig. 2.31

(*g*) *Faraday's ice-pail.* Faraday investigated electrostatic induction using, by chance, a pail he had for storing ice. His experiment may be repeated with a metal vessel plugged into the input socket of the amplifier (or charge range adaptor), Fig. 2.32.

CAPACITORS

A small metal disc (2 cm in diameter) on a long insulating handle is charged from an electrophorus and then lowered into the can without touching it. The meter deflects and the deflection does not change with the position of the charged body provided it is well within the can, nor does it alter when the charged body

Fig. 2.32

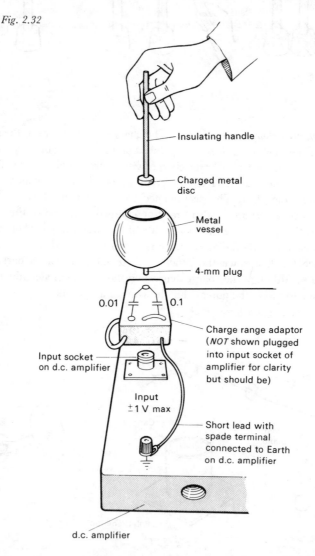

Insulating handle

Charged metal disc

Metal vessel

4-mm plug

0.01   0.1

Charge range adaptor (*NOT* shown plugged into input socket of amplifier for clarity but should be)

Input socket on d.c. amplifier

Input ±1 V max

Short lead with spade terminal connected to Earth on d.c. amplifier

d.c. amplifier

touches the inside of the can and is then removed from it. The disc is found to be completely discharged. (*Note.* The insulating handle on the disc should be long to minimize 'hand capacitance' effects and it is advisable to discharge it before use by quickly passing it several times through a flame.)

The conclusions to be drawn are:

(*i*) a charged body enclosed in a hollow conductor induces an equal charge of opposite sign on the inside of the conductor and an equal charge of the same sign as its own on the outside, Fig. 2.33, and

(*ii*) inside a hollow conductor the net charge is zero (as may have been discovered from investigating the charge density inside the can in (*f*) above); any excess charge resides on the *outside* surface.

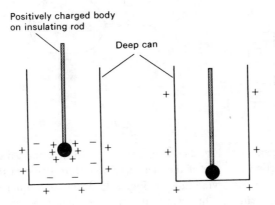

Fig. 2.33

The experiment also shows that all the charge on a conductor can be transferred to a hollow can if it touches the can *well down inside*; only part would be transferred if the outside were touched.

(*h*) *Conservation of charge.* The ' ice-pail ' is used as in (*g*). A strip of cellulose acetate and one of polythene, both first discharged in a flame, are held inside the ' ice-pail ' and rubbed together. There should be little deflection (meter centre-zero) until one or other of the strips is removed, when approximately equal and opposite deflections are obtained.

(*i*) *Charge sharing.* One of the capacitors in the amplifier should be charged from a 1-V supply to give a full-scale meter deflection. If another capacitor of the same value is then connected externally across the input and earth sockets, the meter reading is halved. The effect on charge sharing of connecting a larger and a smaller capacitor can also be tried.

Action at points

The point of a pin is highly curved and if the charge density is sufficiently great, an intense electric field arises near the point. Background ionizing radiation (p. 499) produces electrons in the air which are accelerated by the intense field and cause ionization of the air by collision (p. 406). Ions having the same sign as the charge on the pin are strongly repelled from it to create an electric ' wind '.

CAPACITORS

Ions with charges of opposite sign to that on the pin are attracted to the point and neutralize its charge. The net result of the ' action at points ' is the apparent loss of charge from the pin to the surrounding air by what is termed a *point* or *corona discharge*.

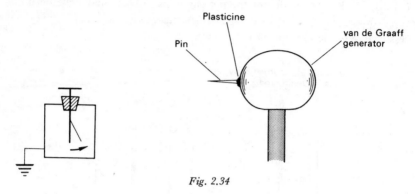

Fig. 2.34

The arrangement of Fig. 2.34 may be used to show a point discharge from a pin and collection of some of the charge by an electroscope in the path of the ' wind '. The electric ' windmill ' of Fig. 2.35 provides another demonstration of the electric ' wind '. The ' windmill ' revolves rapidly when connected to a high potential due to the ' wind ' streaming away from each point. Why should apparatus working at high p.d.s not have sharply curved parts?

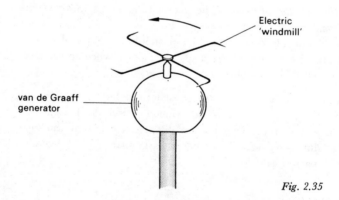

Fig. 2.35

As well as losing charge, a sharp point can also collect it. Thus, when a negatively charged polythene strip is held close to the point of a pin taped to the cap of an electroscope, Fig. 2.36, a deflection is obtained which remains on removing the polythene. Electrostatic induction occurs as shown, then due to 'action at points ' a positive ion ' wind ' is created, neutralizing part of the charge on the strip. At the same time negative ions in the air are attracted to the point, neu-

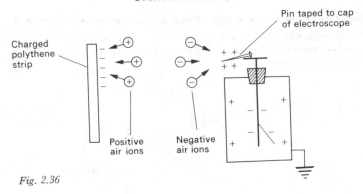

Fig. 2.36

tralizing its induced positive charge. Consequently the electroscope gains negative charge, apparently collected by the point of the pin.

van de Graaff generator

(*a*) *Action.* Large generators of this type are used to develop p.d.s of up to 14 million volts for accelerating atomic particles in nuclear physics research. In the generator of Fig. 2.37 beams of protons (or deuterons) from a source in the dome at the top are accelerated down the column and cause nuclear reactions when they hit ' targets ' of different materials at the bottom.

Fig. 2.37

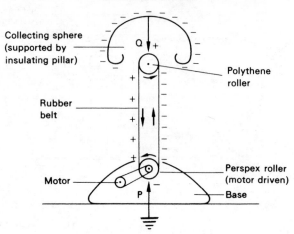

Fig. 2.38

A simplified school version is shown in Fig. 2.38. Initially a positive charge is produced on the motor-driven Perspex roller by friction between it and the rubber belt. This charge induces a negative charge on the earthed comb of metal points P (by drawing electrons from earth) which is then sprayed off by ' action at points ' on to the outside of the belt and carried upwards. The comb of metal points Q is connected to the inside of an insulated metal sphere and when the negative charge on the upward moving belt reaches Q, positive charge is induced in Q due to the repulsion of negative charge to the surface of the sphere. By ' action at points ' Q has thus apparently drawn off negative charge from the belt on to the sphere. The belt becomes positively charged as it moves down due to Q spraying charge on to it and because of friction between the belt and the polythene roller.

A large force is set up as a result of repulsion between the negative charge on the collecting sphere and the negative charge on the upward moving belt. The motor driving the roller therefore has to do work against this repulsion.

(*b*) *Estimating the p.d.* School-type generators can produce p.d.s of up to a few hundred thousand volts, the insulation of the surrounding air being an important limiting factor.

A rough idea of the p.d. can be obtained by finding the greatest separation of the collecting and discharging spheres for a spark to pass between them. Under dry, dust-free conditions at s.t.p., spark discharges 1 cm, 7 cm and 12 cm long occur between two *rounded* electrodes when the p.d.s are roughly 30 kV, 250 kV and 750 kV respectively.

Another method, not requiring the assumption of the above information, involves connecting a microammeter (100 μA) in series with the collecting sphere

66

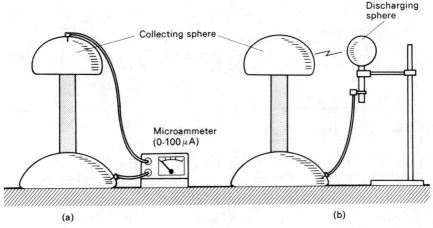

Fig. 2.39

and base of the generator and measuring the steady current I which passes with the generator working at a certain speed, Fig. 2.39a. The meter is then removed and the discharging sphere brought up until regular sparking *just* occurs between it and the collecting sphere, Fig. 2.39b, with the generator running at the same speed as before. The number of sparks passing per second is then found with a stop watch.

If Q is the charge which builds up on the collecting sphere and passes per spark and t is the interval between sparks, then the discharge may be regarded as equivalent to a steady current of value Q/t. Assuming that this current is the same as the current I recorded by the microammeter, we can say $I = Q/t$. Suppose there were two sparks per second, then $t = 0.5$ s. Also taking $I = 10$ μA $= 1.0 \times 10^{-5}$ A, then

$$Q = It = (1.0 \times 10^{-5} \text{ A}) \times (0.5 \text{ s}) = 5.0 \times 10^{-6} \text{ C}.$$

But $Q = VC$ where V is the potential of the collecting sphere and C is its capacitance. For an isolated sphere (which the collecting ' sphere ' is not), $C = 4\pi\epsilon_0 a$ where a is the radius of the sphere (p. 34). If $a = 13$ cm $= 13 \times 10^{-2}$ m then $C = (4\pi \times 9.0 \times 10^{-12} \text{ F m}^{-1}) \times (13 \times 10^{-2} \text{ m})$ since $\epsilon_0 \simeq 9.0 \times 10^{-12}$ F m^{-1}. Therefore $C \simeq 1.5 \times 10^{-11}$ F

$$\therefore \quad V = \frac{Q}{C} = \frac{5.0 \times 10^{-6} \text{ C}}{1.5 \times 10^{-11} \text{ F}}$$

$$\simeq 3.3 \times 10^5 \text{ V} \quad (330 \text{ kV}).$$

CAPACITORS

QUESTIONS

1. (a) In the circuit of Fig. 2.40a what is the p.d. across each capacitor? What is the total charge stored? What is the capacitance of the single capacitor which would store the same charge as the two capacitors together?

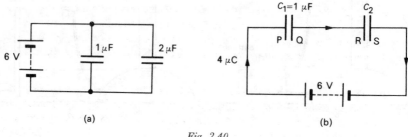

Fig. 2.40

(b) In Fig. 2.40b if a charge of 4 μC flows from the 6-V battery to plate P of the 1-μF capacitor, what charge flows from (i) Q to R, and (ii) S to the battery? What is the p.d. across (iii) C_1, and (iv) C_2? What is the capacitance of (v) C_2, and (vi) the single capacitor which is equivalent to C_1 and C_2 in series, and what charge would it store?

2. (a) What is the capacitance of a small metal sphere of radius 1 cm (take $4\pi\epsilon_0 = 10^{-10}$ F m^{-1})? (b) What charge is stored on it when it briefly touches the 1-kV terminal of an e.h.t. power supply? (c) If the sphere now shares its charge with a 0.001-μF capacitor, what is the ratio of the charge on the capacitor to that on the sphere? (d) What, in effect, will be the charge on (i) the capacitor, and (ii) the sphere? (e) If the capacitor had had a capacitance of 1 pF (10^{-12} F) how would the charge have been shared?

3. Describe an experiment to compare the capacitances of two capacitors.

A capacitor is charged through a large series resistance. Sketch a curve of the increase in potential difference across the capacitor with time.

Under what circumstances is it necessary to know the capacitance of an electroscope or electrostatic voltmeter?

A steady potential of 100 V is maintained across the combination of a capacitor of capacitance 5.0 μF in series with one of 2.0 μF. Calculate (a) the potential difference across each, (b) the charge on each, and (c) the energy stored in each capacitor.

(A.E.B.)

4. (a) A parallel plate capacitor consists of two square plates each of side 25 cm, 3.0 mm apart. If a p.d. of 200 V is applied, calculate the charge on the plates with (i) air, and (ii) paper of relative permittivity 2.5, filling the space between them. ($\epsilon_0 = 8.9 \times 10^{-12}$ F m^{-1}.)

(b) A tubular 0.10-μF capacitor is to be made from a sheet of plastic 2.0 cm wide and 1.0 $\times$ 10^{-3} cm thick rolled between metal foil of the same width. What length of plastic is required if its relative permittivity is 2.4 and $\epsilon_0 = 8.9 \times 10^{-12}$ F m^{-1}?

5. Define *electric field strength*.

How would you demonstrate experimentally that (*a*) a charge given to a hollow conductor is confined to the outside surface, and (*b*) the potential is constant over the surfaces of the charged conductor?

A fine layer of silver is deposited on each side of a sheet of mica 2.4 cm^2 in area to form a capacitor of capacitance 1.2×10^{-4} μF. If the dielectric constant (relative permittivity) of the mica is 6.0 find the thickness of the sheet of mica. If the insulation of the mica breaks down when subjected to an electric field strength of 4.0×10^5 V cm^{-1} and it is desirable in practice never to exceed one-half of this field strength, what is the maximum working voltage of the capacitor? ($\epsilon_0 = 8.9 \times 10^{-12}$ F m^{-1}) (*L.*)

6. A 100-μF capacitor is charged from a supply of 1000 V, disconnected from the supply and then connected across an uncharged 50-μF capacitor. Calculate the energy stored initially and finally in the two capacitors. What conclusion do you draw from a comparison of these results? (*J.M.B.*)

7. (*a*) Explain how field strength is related to potential difference in an electrostatic field.

The field strength E close to a plane conductor in free space, carrying charge density σ, is σ/ϵ_0. Use this information to derive an expression for the capacitance of a parallel plate capacitor in free space where ϵ_0 is the permittivity of free space.

(*b*) A 2-μF capacitor is required which will work at 1000 V d.c. but the only capacitors available are each 2-μF with a working voltage of 400 V d.c. Describe how such capacitors could be combined to give an equivalent capacitor of the required rating. (*J.M.B.Eng. Sc.*)

8. Define *electric potential* and *capacitance of an insulated conductor*. Explain carefully, with reference to your definitions, why the presence of a similar earthed conductor near and parallel to an isolated sheet of metal considerably increases the capacitance of the sheet.

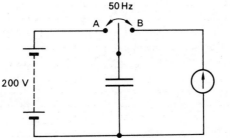

Fig. 2.41

Fig. 2.41 shows a capacitor in a circuit in which a vibrating switch first charges it up to 200 V and then discharges it through a galvanometer, repeating the sequence 50 times a second. The mean current registered by the galvanometer is 11.0 μA. Use this information to find the capacitance of the capacitor in farads.

CAPACITORS

The capacitor in fact consists of two flat square metal sheets, 50 cm × 50 cm placed parallel one over the other and 2.0 mm apart. Calculate the magnitude of ϵ_0, the permittivity of free space (or air), and state the units in which it is measured.

If a capacitor is to be accurately measured with such a circuit it is essential that the part of the circuit between B and the galvanometer shall be well insulated from the part between A and the battery. What must be the minimum resistance of any possible leakage path between these parts if the error introduced is not to exceed 1 per cent in measuring this capacitor? (S.)

9. Discuss the essential differences between current and static electricity.

Describe and explain the working of a device which can supply a current of the order of 10^{-6} A and maintain a potential difference of the order of 10^6 V.

A high voltage generator continuously supplies charge to a spherical dome of radius 15 cm, which gives 10 sparks each minute to a nearby earthed sphere. While the generator is operating at the same rate, the dome is connected to earth through a galvanometer, which indicates a steady current of 2.0×10^{-6}A. Calculate

(a) the charge passed in each spark,

(b) the maximum potential difference between the dome and earth,

(c) the distance from the surface of the dome to the surface of the nearby earthed sphere.

Make the simplifying assumptions that an isolated spherical dome of 1 cm radius has a capacitance of 10^{-12} F, and that in this system the mean potential gradient needed for sparks to pass is 4×10^4 V cm^{-1}. (C.)

10. What is meant by *the electrostatic potential at a point*?

Illustrate, by a suitably-labelled sketch in each case, how it is possible to set up electrostatic systems in which (a) a conductor at earth potential carries a net positive charge, (b) a conductor at earth potential has regions of both negative and positive charge, and (c) a conductor with no net charge is at a positive potential with respect to earth.

How would you verify experimentally that both negative and positive charges are present in case (b)?

A radioactive source emits beta particles (electrons) at a substantially constant rate of 3.7×10^4 per second. If the source, which is a metal sphere 1 mm in diameter, is electrically insulated, how long will it take for its potential to rise by 1 V, assuming 90 per cent of the beta particles emitted escape from the source? The capacitance in SI units of an isolated spherical conductor of radius r is $4\pi\epsilon_0 r$.

(Charge on the electron = 1.6×10^{-19} C; $\epsilon_0 = 8.9 \times 10^{-12}$ F m^{-1}.)

(O. and C.)

3 Magnetic fields

Importance of electromagnetism

The discovery in 1819 by Professor Oersted at the University of Copenhagen that an electric current is accompanied by magnetic effects saw the birth of electromagnetism. Since then, the subject has been an extremely fruitful field of study for the imagination of scientists and the creative genius of engineers.

In science, Faraday announced the discovery of electromagnetic induction in 1831 and developed the idea of a ' field ' for dealing with such action-at-a-distance effects. Clerk Maxwell put Faraday's field ideas into mathematical form and predicted electromagnetic waves. Einstein pondered over the mysteries of electromagnetism, especially the need for relative motion, and was led to the theory of relativity.

In electrical engineering the dynamo, the motor and the transformer were the outcome of Faraday's work. Edison was responsible for opening the first power station in 1882, designed to supply electricity to domestic consumers in New York. Shortly afterwards Tesla invented the induction motor which does not require brushes (to supply current to its rotor) and which is now the indispensable servant of industry. In 1887 Hertz produced radio-type electromagnetic waves and then Marconi, contrary to all theoretical predictions, transmitted them across the Atlantic. And so the remarkable story continues with, for example, the development today of the linear motor which, unlike conventional rotary motors, is flat and eliminates the energy losses occurring when rotary motion is converted to the often required linear motion.

Modern civilization is heavily indebted to electromagnetism, the topic of this chapter and the next.

Fields due to magnets

The magnetic properties of a magnet appear to originate at certain regions in the magnet which we call the *poles*: in a bar magnet these are near the ends. Experiments show that (*i*) magnetic poles are of two kinds, (*ii*) like poles repel each other and unlike poles attract, (*iii*) poles always seem to occur in equal and opposite pairs, and (*iv*) when no other magnet is near, a freely-suspended magnet sets so that the line joining its poles (i.e. its magnetic axis) is approximately parallel to the earth's north-south axis.

The last fact suggests that the earth itself behaves like a large permanent magnet and it makes it appropriate to call the pole of a magnet which points (more or less) towards the earth's geographical North Pole, the *north pole* of the magnet and the other the *south pole*. What kind of magnetic pole must be near the earth's geographical North Pole?

The space surrounding a magnet where a magnetic force is experienced is called a *magnetic field*. The direction of a magnetic field at a point is taken as the direction of the force that acts on a north magnetic pole there. A magnetic field can be represented by magnetic *field lines* drawn so that (*i*) the line (or the tangent to it if it is curved) gives the direction of the field at that point and (*ii*) the number of lines per unit cross-section area is an indication of the ' strength ' of the field. Arrows on the lines show the direction of the field and since a north pole is repelled by the north pole of a magnet and attracted by the south, the arrows point away from north poles and towards south poles.

Field lines can be obtained quickly with iron filings or accurately plotted using a small compass (i.e. a pivoted magnet). Some typical field patterns are shown in Fig. 3.1. The field round a bar magnet varies in strength and direction from point to point, i.e. is non-uniform. Locally the earth's magnetic field is uniform; the lines are parallel, equally spaced and point north.

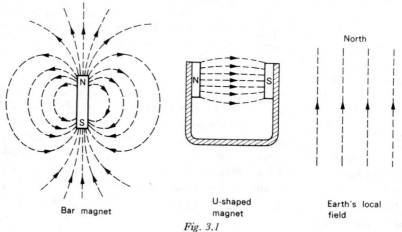

| Bar magnet | U-shaped magnet | Earth's local field |

Fig. 3.1

Earth's magnetic field

The magnitude and direction of the earth's field varies with position over the earth's surface and it also seems to be changing gradually with time. The pattern of field lines is similar to that which would be given if there was a strong bar magnet at the centre of the earth, Fig. 3.2a. At present there is no generally accepted theory of the earth's magnetism but it may be caused by electric currents circulating in its core due to convection currents arising from radioactive heating inside the earth.

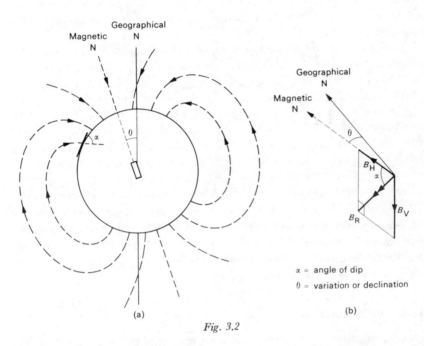

α = angle of dip

θ = variation or declination

(a)   (b)

Fig. 3.2

In Britain the earth's field is inclined downwards at an angle of about 70° to the horizontal—called the *angle of dip* α. At the magnetic poles (which are near the geographical poles but whose positions change slowly) it is vertical and $\alpha = 90°$. At the magnetic equator it is parallel to the earth's surface, i.e. horizontal and $\alpha = 0°$.

It is convenient to resolve the earth's ' field strength ' B_R into horizontal and vertical components, B_H and B_V respectively. We then have from Fig. 3.2b that

$$B_H = B_R \cos \alpha$$

and

$$B_V = B_R \sin \alpha.$$

Also

$$\tan \alpha = \frac{B_V}{B_H}.$$

Instruments such as compass needles whose motion is confined to a horizontal plane are affected by B_H only.

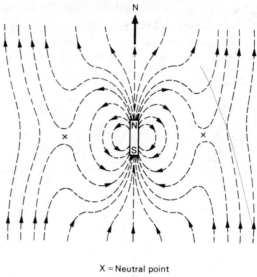

X = Neutral point

Fig. 3.2c

A *neutral point* is a place where two magnetic fields are equal and opposite and the resultant force is zero. Two such points are shown in Fig. 3.2c in the combined field due to the earth and a bar magnet with its N pole pointing N. Where will they be when the N pole of the magnet points S?

Fields due to currents

A conductor carrying an electric current is surrounded by a magnetic field. The field patterns due to conductors of different shapes can be obtained as for a magnet using iron filings or a plotting compass.

The lines due to a straight wire are circles, concentric with the wire, Fig. 3.3a. The *right-hand screw rule* is a useful aid for predicting the direction of the field knowing the direction of the current. It states that *if a right-handed screw moves forward in the direction of the current (conventional), then the direction of rotation of the screw gives the direction of the lines*. Figs. 3.3b and c illustrate the rule; in b the current is flowing out of the paper and the dot in the centre of the wire is the point of an approaching arrow; in c the current is flowing into the paper and the cross is the tail of a receding arrow.

The field pattern due to a current in a plane circular coil is shown in Fig. 3.4a, and one due to a current in a long cylindrical coil—called a *solenoid*—in Fig. 3.4b. In both cases the field direction is again given by the right-hand screw rule. A solenoid produces a field similar to that of a bar magnet; in Fig. 3.4b the left-hand end behaves like the north pole of a bar magnet and the right-hand end

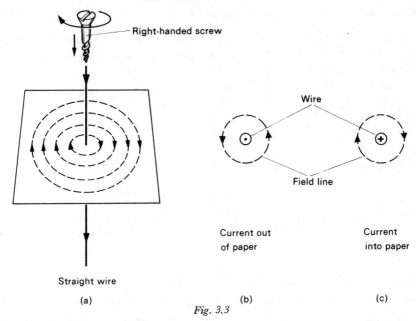

Right-handed screw

Wire

Field line

Current out
of paper

Current
into paper

Straight wire

(a)

(b)

(c)

Fig. 3.3

like the south pole. In general, the field produced by any magnet can also be produced by a current in a suitably-shaped conductor. It is this fact which leads us to believe that *all* magnetic effects are due to electric currents, which in the case of permanent magnets arise from the motion of electrons within the atoms themselves.

In brief, we can say that a static electric charge gives rise to an electric field, whilst a moving electric charge creates a magnetic field as well as an electric field.

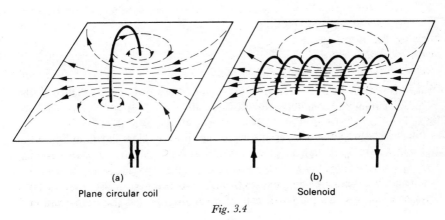

(a)

Plane circular coil

(b)

Solenoid

Fig. 3.4

Force on a current in a magnetic field

A magnet in a uniform magnetic field experiences two equal but opposite parallel forces, i.e. a couple, which gives it an angular acceleration and—provided there is damping—it ultimately comes to rest with its axis parallel to the field, Fig. 3.5. In a non-uniform field, the poles of the magnet are acted on by unequal forces and these may cause translational as well as rotational acceleration. A current-carrying conductor behaves like a magnet and so it is not surprising to find that it too experiences a force in a magnetic field. In the next section we will see how this force can be used to define and measure the strength of a magnetic field. But first some basic experimental facts.

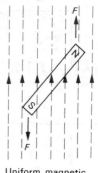

Fig. 3.5

Uniform magnetic field

(*a*) *Fleming's left-hand rule.* Using the apparatus of Fig. 3.6 it can be observed that when current passes through the ' bridge ' wire, it shoots along the ' rails ' if the magnetic field of the U-shaped magnet is perpendicular to it (as shown). There is no motion if the field is parallel to it. From such experiments we conclude

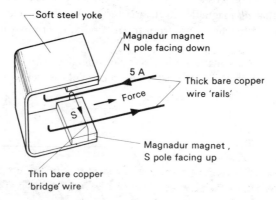

Soft steel yoke

Magnadur magnet N pole facing down

5 A

Force

Thick bare copper wire 'rails'

Magnadur magnet , S pole facing up

Thin bare copper 'bridge' wire

Fig. 3.6

that the force on a current-carrying conductor (*i*) is always perpendicular to the plane containing the conductor and the direction of the field in which it is placed, and (*ii*) is greatest when the conductor is at right angles to the field.

The facts about the relative directions of current, field and force are summarized by Fleming's left-hand (or motor) rule. It states that *if the thumb and first two fingers of the left hand are held each at right angles to the other, with the First finger pointing in the direction of the Field and the seCond finger in the direction of the Current, then the Thumb predicts the direction of the Thrust or force*, Fig. 3.7.

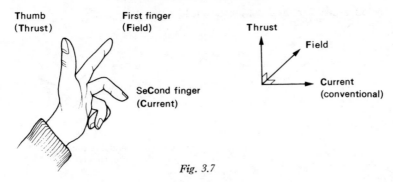

Thumb
(Thrust)

First finger
(Field)

Thrust

Field

SeCond finger
(Current)

Current
(conventional)

Fig. 3.7

(*b*) *Factors affecting the force.* The effect of the current, the length of the conductor and the strength and orientation of the magnetic field may be investigated

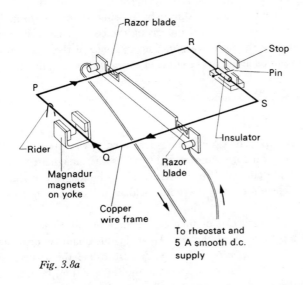

Razor blade

R

Stop

Pin

P

S

Rider

Q

Insulator

Magnadur
magnets
on yoke

Razor
blade

Copper
wire frame

To rheostat and
5 A smooth d.c.
supply

Fig. 3.8a

by passing current through one arm PQ of a copper wire frame balanced on two razor blades via which the current enters and leaves, Fig. 3.8*a*. The ends of the wire in the arm RS of the frame are held together by an insulator and no current flows in RS. If the magnet is arranged so that PQ rises when current passes, the balance can be restored and the force measured by placing a suitable rider (a short length of wire) on PQ.

To see how the force depends on the *current* (measured by an ammeter which strictly speaking should be calibrated using a current balance, see p. 93), it is convenient to find the currents required to balance one, two and three identical riders on PQ. The effect on the force of doubling and trebling the *length* of conductor carrying the same current in the same magnetic field can be found by

77

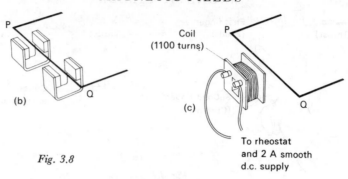

(b)

Coil
(1100 turns)

(c)

To rheostat
and 2 A smooth
d.c. supply

Fig. 3.8

placing a second and a third, equally strong magnet beside the first, Fig. 3.8*b*. If the magnetic field is now produced by passing a current through a coil of many turns (e.g. 1100), Fig. 3.8*c*, increasing the coil current can reasonably be expected to increase the *field strength*. The response of the force on PQ to this may then be observed and also the effect of the *angle* between the direction of the field and PQ.

Magnetic flux density

(*a*) *Definition of B.* The experiments described in the previous section show that the force F on a wire lying at right angles to a magnetic field is directly proportional to the current I in the wire and to the length l of the wire in the field. It also depends on the magnetic field, a fact we can use to define the strength of the field.

Electric field strength E is defined as the force per unit charge; gravitational field strength g is force per unit mass. An analogous quantity for magnetic fields is the *flux density* or *magnetic induction*, B (also called the *B-field*), defined as *the force acting per unit current length*, i.e. the force acting per unit length on a conductor which carries unit current and is at right angles to the direction of the magnetic field. In symbols, B is defined by the equation

$$B = \frac{F}{Il}.$$

Thus if $F = 1$ newton when $I = 1$ ampere and $l = 1$ metre then $B = 1$ newton per ampere metre, i.e. the SI unit of B is the newton per ampere metre ($\text{N A}^{-1}\,\text{m}^{-1}$), which is given the special name of 1 *tesla* (T).

B is a vector whose direction at any point is that of the field line at the point. Its magnitude may be represented pictorially by the number of field lines passing through unit area; the greater this is, the greater the value of B.

(b) Expression for force on a current. Rearranging the equation defining B we see that the force F on a conductor of length l, carrying a current I and lying at right angles to a magnetic field of flux density B, is given by

$$F = BIl.$$

If the conductor and field are not at right angles, but make an angle θ with one another, Fig. 3.9, the expression becomes $BIl \sin \theta$. When $\theta = 90°$, $\sin \theta = 1$ and $F = BIl$ as before. If $\theta = 0$, the conductor and field are parallel and, as experiment confirms, $F = 0$.

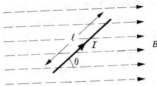

Fig. 3.9

(c) Measuring B using a current balance. The copper wire frame used previously to investigate the factors affecting the force on a current in a magnetic field (p. 77) is a very simple ' current balance ' (see p. 93) and enables us to measure the flux density of a field. For example, suppose a rider of mass 0.084 g is required to counterpoise the frame when arm PQ, of length 25 cm and carrying a current of 1.2 A, is inside and in series with a flat, wide solenoid, Fig. 3.10. We then have

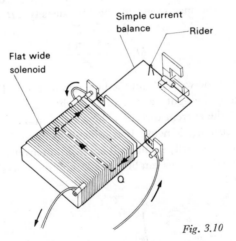

Fig. 3.10

$F = $ force on arm PQ

$\quad = $ weight of rider = mass of rider $\times g$

$\quad = (0.084 \times 10^{-3}\,\text{kg}) \times (10\,\text{N kg}^{-1})$

$\quad = 0.084 \times 10^{-2}\,\text{N}$

$$I = 1.2 \text{ A}$$

$$l = 0.25 \text{ m.}$$

The flux density B (assumed uniform) inside the solenoid is given by

$$B = \frac{F}{Il} = \frac{8.4 \times 10^{-4} \text{ N}}{1.2 \text{ A} \times 0.25 \text{ m}}$$

$$= 2.8 \times 10^{-3} \text{ T.}$$

The field in the gap between two Magnadur magnets on a mild steel yoke (see Fig. 3.8a) is about ten times stronger than this, whilst in the earth's magnetic field in Britain, $B_R \simeq 5.3 \times 10^{-5}$ T and $B_H \simeq 1.8 \times 10^{-5}$ T. A flux density of 1 T would be produced by a fairly strong magnet.

Suggest (i) a source of error in the above measurement of B and (ii) how the result could be made more reliable.

The Biot-Savart law

In the previous section we saw how B could be found experimentally; it can also be calculated using the Biot-Savart law.

The calculation involves considering a conductor as consisting of a number of very short lengths, each of which contributes to the total field at any point. Biot and Savart *stated* that for a very short length δl of conductor, carrying a steady current I, the magnitude of the flux density δB at a point P distant r from δl is

$$\delta B \propto \frac{I \, \delta l \sin \theta}{r^2}$$

where θ is the angle between δl and the line joining it to P, Fig. 3.11. The direction of δB, given by the right-hand rule, is at right angles to the plane containing δl and P and, for the current direction shown, acts into the paper. The product $I \, \delta l$ is called a 'current element'. It is difficult to prove Biot and Savart's law directly for an infinitesimally small conductor, i.e. a current element, but experimentally verifiable deductions can be made from it for 'life-sized' conductors as we shall see shortly.

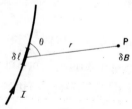

Fig. 3.11

MAGNETIC FIELDS

Before the Biot-Savart law can be used in calculations it has to be expressed as an equation and a constant of proportionality introduced. Now δB depends on one other factor besides I, δl, θ and r, and that is the medium through which the magnetic force acts. This factor can be incorporated in the Biot-Savart law if we regard the constant of proportionality as a property of the medium. The constant is called the *permeability* of the medium and is denoted by μ (mu).

The permeability of a vacuum is denoted by μ_0 (mu nought), its value is *defined* (from the definition of the ampere, p. 93) to be $4\pi \times 10^{-7}$ and its unit is the *henry per metre* (H m^{-1}), as we shall see later (p. 141). Air and most other materials except ferromagnetics have nearly the same permeability as a vacuum. (Note that whilst the value of ϵ_0—the permittivity of free space—is found by experiment, that for μ_0 is defined.)

Rationalization (see p. 9) is achieved by introducing 4π in the denominator. The Biot-Savart equation for a current element becomes

$$\delta B = \frac{\mu_0 I \, \delta l \sin \theta}{4\pi r^2}.$$

It is an inverse square relation.

Calculation of flux density

In most cases the calculation requires the use of calculus but for a circular coil the value of B at the centre can be found fairly simply.

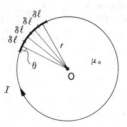

Fig. 3.12

(i) *Circular coil.* Suppose the coil is in air, has radius r, carries a steady current I and is considered to consist of current elements of length δl. Each element is at distance r from the centre O of the coil and is at right angles to the line joining it to O, i.e. $\theta = 90°$, Fig. 3.12. At O the total flux density B is the sum of the flux densities δB due to all the elements. That is

$$B = \sum \frac{\mu_0 I \, \delta l \sin \theta}{4\pi r^2} = \frac{\mu_0 I \sin \theta}{4\pi r^2} \sum \delta l.$$

But $\sum \delta l$ = total length of the coil = $2\pi r$ and $\sin \theta = \sin 90 = 1$. Hence

$$B = \frac{\mu_0 I \, 2\pi r}{4\pi r^2} = \frac{\mu_0 I}{2r}.$$

If the coil has N turns each of radius r

$$B = \frac{\mu_0 N I}{2r}.$$

(*ii*) *Very long straight wire.* It can be shown (see Appendix 3, p. 531) that the value of B at a perpendicular distance a from a very long straight wire carrying a current I, Fig. 3.13, is

$$B = \frac{\mu_0 I}{2\pi a}.$$

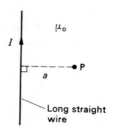

Fig. 3.13

In this case the field is non-uniform, there is cylindrical symmetry (Fig. 3.3a) and 2π appears quite appropriately. If $I = 10$ A and $a = 1.0$ cm $= 0.010$ m, then $B = 2.0 \times 10^{-4}$ T.

(*iii*) *Very long solenoid.* If the solenoid has N turns, length l and carries a current I, the flux density B at a point O on the axis near the centre of the solenoid, Fig. 3.14, is found to be given by

$$B = \frac{\mu_0 N I}{l}$$

$$= \mu_0 n I$$

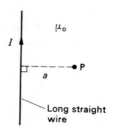

Fig. 3.14

where $n = N/l$ = number of turns per unit length. B thus equals μ_0 multiplied by the *ampere-turns per metre*. The field is fairly uniform over the cross-section of the solenoid for most of its length (Fig. 3.4b) and the flux density is given by the above expression. The absence of π from it may be noted and is what we expect from the plane symmetry of a uniform field using a rationalized system.

For a solenoid of 400 turns, 25 cm long, carrying a current of 5.0 A we have
$n = 400/0.25 = 1600$ turns per metre, $I = 5.0$ A and so $B \simeq 1.0 \times 10^{-2}$ T.

At P, a point at the end of a long solenoid,

$$B = \frac{\mu_0 n I}{2}.$$

That is, B at a point at the end of the solenoid's axis is half the value at the centre.

These expressions are strictly true only for an infinitely long solenoid: in practice they are accurate enough for most purposes if the length of the solenoid is at least ten times its diameter.

(iv) *Helmholtz coils.* Two flat coaxial coils having the same radius r, the same number of turns N and separated by a distance r, Fig. 3.15, produce an almost uniform field over some distance near their common axis midway between them when the same current I flows round each in the same direction. It is given by

$$B = \frac{8\mu_0 NI}{5\sqrt{5}\ r}$$

$$\simeq 0.72 \frac{\mu_0 NI}{r}.$$

Helmholtz coils are used in a school laboratory measurement of e/m for electrons (p. 385).

Fig. 3.15

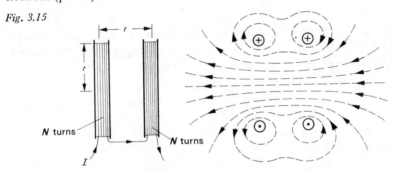

N turns

N turns

I

Magnetic field measurements

Measurements of B for different current-carrying conductors can be made and compared with the values and variations predicted by the expressions derived from the Biot and Savart law. Whilst a simple current balance of the type used previously (Fig. 3.10, p. 79) is satisfactory for defining B, it is not sensitive enough for showing how B varies from one point to another in a field. However, two other devices are convenient. One is the search coil, used with alternating fields produced by alternating currents and the other is the Hall probe for steady fields produced by direct currents.

(*a*) *Search coil and a.c.* A typical search coil has 5000 turns of wire and an external diameter of no more than 1.5 cm so that it samples the field over a small area. If it is connected to the Y-plates of a CRO (time base off), the length of the vertical trace is proportional to the maximum value of the alternating flux density, or any such component, *acting along the axis of the search coil at right angles to its face*. (The technique uses an effect called *mutual induction* which will be considered in the next chapter, p. 107. The CRO is in fact being used as a voltmeter to measure an e.m.f. induced in the search coil by the alternating magnetic field. The sensitivity of the method increases with the frequency of the a.c. in the conductor producing the field and 10 kHz is a convenient frequency to use.) Fig. 3.16 shows two types of mounting for a search coil.

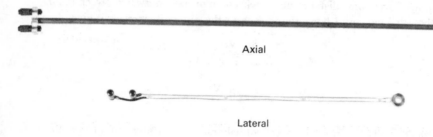

Axial

Lateral

Fig. 3.16 Search coils

(*b*) *Hall probe and d.c.* The Hall probe consists of a slice of semiconducting material on the end of a rod, Fig. 3.17, connected to a light-beam galvanometer via a circuit box, Fig. 3.19*b*. When a battery in the box is switched on, current is supplied to the slice. The ' balance ' control is adjusted for zero deflection on the galvanometer and the probe is then inserted in the field to be measured. The change in the galvanometer reading is proportional to the flux density (or its component) *at right angles to the slice*. The action of the probe depends on the Hall effect to be discussed shortly (p. 89).

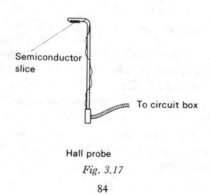

Semiconductor
slice

To circuit box

Hall probe

Fig. 3.17

(c) *Investigations.* If actual values of B are required, the search coil or Hall probe would have to be calibrated (How?). For studying comparisons and relationships this is not necessary.

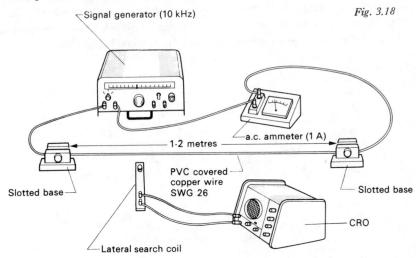

Fig. 3.18

A long straight wire may be investigated using the arrangement in Fig. 3.18. What is the direction of B near the wire? Is B the same all along the wire at a constant distance from it? Is B directly proportional to the current in the wire? Does B vary as $1/a$ (where a is the perpendicular distance from the wire), as theory suggests?

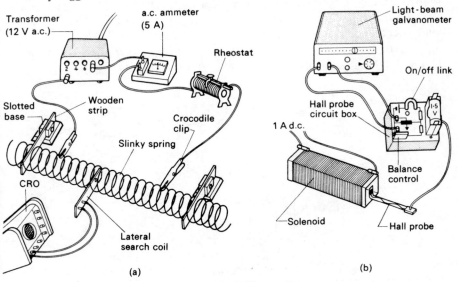

(a)   (b)

Fig. 3.19

MAGNETIC FIELDS

Using a Slinky spring as a solenoid, Fig. 3.19*a*, find if the field is uniform across the width of the solenoid but zero outside. Is *B* constant along most of the length of the solenoid? Can we accept the theoretical prediction that for a long solenoid *B* at the end is half that at the centre? How does *B* at the centre alter if the solenoid is stretched, the current remaining constant? Answers to some of these questions may also be found with the apparatus of Fig. 3.19*b*.

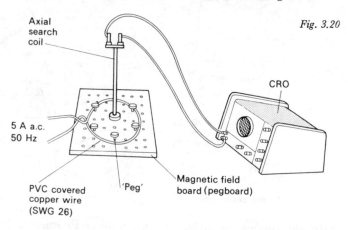

Fig. 3.20

A 'magnetic field board' like that in Fig. 3.20 enables circular coils to be studied. Do the direction and magnitude of *B* vary over the plane of the coil? When the current and radius are constant how does *B* at a given point vary with the number of turns? With a certain number of turns and current, is *B* at the centre inversely proportional to the radius of the coil? The field along the axis should vary with distance *x* along the axis from the centre as $r^2/(x^2 + r^2)^{3/2}$ where *r* is the coil radius. Does it?

For a compact coil of, say, 240 turns how do the values of *B* compare at P and Q, Fig. 3.21? Is there a relationship between *B* and *x* along the axis of the coil?

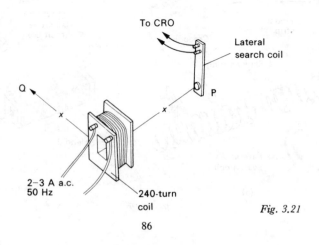

Fig. 3.21

86

Force on a charge in a magnetic field

An electric current is regarded as a drift of charges and so it is reasonable to assume that the force experienced by a current-carrying conductor in a magnetic field is the resultant of the forces acting on the charges constituting the current. Accepting this, an expression for the force on a single charged particle moving in a magnetic field can be derived.

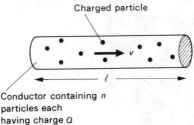

Charged particle

Conductor containing n
particles each
having charge Q

Fig. 3.22

Consider a length l of conductor containing n charged particles each of charge Q and average drift velocity v, Fig. 3.22. Each particle will take the same time t to travel distance l, and so

$$v = l/t.$$

The total charge passing through any cross-section of the conductor in time t is nQ. Therefore the current I is given by

$$I = \text{total charge passing a section/time}$$

$$= nQ/t$$

$$= nQv/l \quad (\text{since } t = l/v).$$

If the conductor makes an angle θ with a uniform magnetic field of flux density B, then

$$\text{force on conductor} = BIl \sin \theta \qquad (\text{p. 79})$$

$$= B\left(\frac{nQv}{l}\right) l \sin \theta$$

$$= BnQv \sin \theta.$$

Hence the force on one charged particle is $BQv\sin \theta$. If the charge moves at right angles to the field, $\theta = 90°$, $\sin \theta = 1$ and the force F is given by

$$F = BQv.$$

If B is in teslas (T), Q in coulombs (C) and v in metres per second (m s^{-1}), then F is in newtons (N). The direction of the force is given by Fleming's left-hand rule (remembering that a negative charge moving one way is equivalent to conventional current flowing in the opposite direction) and is at right angles both to the field and to the direction of motion.

This force is also responsible for the deflection of a beam of charged particles travelling through a magnetic field in a vacuum, e.g. electrons in a cathode-ray tube. Since it is always at right angles to the path of the beam it only changes the direction of motion but not the speed. When the particles enter a uniform field at right angles, they are deflected into a circular path; for other angles (except 0°) they describe a helix. The circular and helical paths of electrons in a fine beam tube (p. 386) are shown in Figs. 3.23a and b.

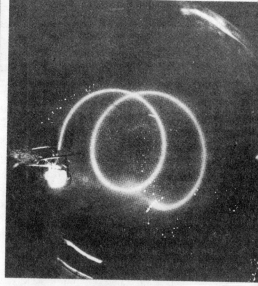

Fig. 3.23a and b

Charged particles thus tend to become 'trapped' in magnetic fields and their existence in certain regions round the earth, known as the van Allen radiation belts, is due to the earth's magnetic field. The behaviour of charged particles in magnetic fields is of considerable importance in atomic physics.

Hall effect

A current-carrying conductor in a magnetic field has a small p.d. across its sides, in a direction at right angles to the field. In the slab of conductor shown in Fig. 3.24a, the p.d., called the *Hall p.d.*, appears across XY.

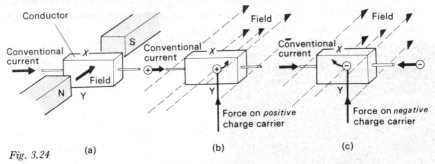

Fig. 3.24   (a)   (b)   (c)

(*a*) *Explanation.* The effect was discovered in 1879 by Hall and can be attributed to the forces experienced by the charge carriers in the conductor. These act at right angles to the directions of the magnetic field and the current (and are given by Fleming's left-hand rule) and cause the charge carriers to be pushed sideways, thus increasing their concentration towards one side of the conductor. As a result, a p.d. (and an electric field) is produced across the conductor.

With the magnetic field and conventional current having the directions shown in Figs. 3.24*b* and *c*, the forces act upwards and side X of the conductor develops the same sign as the charge carriers, i.e. positive in (*b*) and negative in (*c*). The Hall p.d. therefore reveals the sign of the charge carriers (more exactly, the majority charge carriers) in the conductor. In metals such as copper and aluminium they are negative; in semiconductors they may be positive or negative.

(*b*) *Demonstration of Hall effect.* Germanium, a semiconductor, is used because the effect is very much greater with it than with metals (for reasons given later); p.d.s of about 0.1 V are obtainable. Pure germanium has very few ' free ' electrons for current carrying but if it is ' doped ' with a small proportion of atoms of another element (e.g. antimony) which provides extra electrons, *n*-type germanium is formed with enhanced conduction properties. *p*-type germanium is ' doped ' with atoms (e.g. those of indium) which cause the germanium to behave as if it had positive charge carriers.

In the circuit of Fig. 3.25 a current of about 50 mA is passed through the Hall slice (a piece of doped germanium) via the current terminals P and Q. If the Hall p.d. connections X and Y on the slice are not directly opposite each other, a p.d. will exist between them even in the absence of a magnetic field and there will be a current through the microammeter. In this case the ' balance ' control should be altered to zero the meter and bring X and Y to the same potential. If a Magnadur magnet is then placed face downwards over the slice, the microammeter should indicate a Hall p.d., the sign of which for *p*-type germanium will

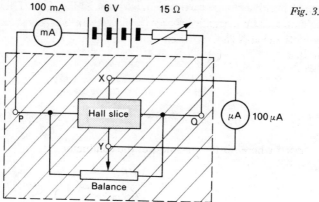

Fig. 3.25

be opposite to that for *n*-type. Application of the left-hand rule enables the sign of the charge carriers to be determined.

(*c*) *Expression for Hall p.d.* The Hall p.d. and electric field build up until the repulsive electrical force they cause on a charge carrier drifting through the conductor is equal and opposite to the deflecting magnetic force on it and no further charge displacement occurs.

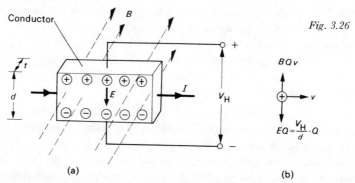

Fig. 3.26

If V_H is the Hall p.d. across the width d of the conductor and E is the final value of the electric field strength, then since electric field strength = potential gradient, we have

$$E = V_H/d.$$

If the charge carriers each have charge Q and drift velocity v then

electric force on each drifting charge = $EQ = V_H Q/d$

magnetic force on each drifting charge = $BQv.$

At equilibrium, $\qquad\qquad V_H Q/d = BQv$

$$\therefore \quad V_H = Bvd.$$

90

The case of positive charge carriers is illustrated in Fig. 3.26. Draw the corresponding diagrams for a negative charge carrier. It can be shown[1] for a conductor of cross-sectional area A carrying current I and having n charge carriers per unit volume, that

$$I = nAQv.$$

Substituting for v we get $\qquad V_H = \dfrac{BId}{nAQ}.$

But $A = d \times t$ where t is the thickness of the conductor

$$\therefore \quad V_H = \frac{BI}{nQt}.$$

The expression shows that V_H is greatest in materials for which n is small, i.e. semiconductors, since those have a smaller number of charge carriers per unit volume than metals.

(d) *Number of charge carriers per atom.* If V_H is measured for, say, a thin strip of aluminium foil carrying a large current in a strong magnetic field, n may be found. Thus if $B = 1.4$ T, $I = 10$ A, $t = 0.060$ mm $= 6.0 \times 10^{-5}$ m and $V_H = 10$ μV $= 10 \times 10^{-6}$ V, then taking $Q = $ electronic charge $= 1.6 \times 10^{-19}$ C we get

$$n = \frac{BI}{V_H Qt}$$

$$= \frac{1.4 \times 10}{10 \times 10^{-6} \times 1.6 \times 10^{-19} \times 6.0 \times 10^{-5}}$$

$$= 1.4 \times 10^{29} \text{ electrons per metre}^3.$$

A calculation using the Avogadro constant (6.0×10^{23} atoms mol^{-1}), the relative atomic mass and the density of aluminium (27 and 2.7 g cm^{-3} respectively) shows that there are 6.0×10^{28} aluminium atoms per cubic metre. Check this. The number of ' free ' (conduction) electrons per atom must therefore be between 2 and 3 ($[1.4 \times 10^{29}]/[6.0 \times 10^{28}] = 2.3$). Chemical evidence suggests that aluminium is trivalent, i.e. has three ' outer ' electrons (which participate in chemical reactions). Hall effect measurements do not conflict with this conclusion.

The ampere and current balances

(a) *Force between currents.* The ampere is the basic electrical unit of the SI system and has to be defined, like any other unit, so that it is accurately reproducible. This is achieved by basing the definition on the force between two long,

[1]See *Advanced Physics: Materials and Mechanics*, p. 58.

straight, parallel current-carrying conductors, i.e. by using the magnetic effect, and making measurements with a current balance.

The forces between currents may be demonstrated with two long narrow strips of aluminium foil arranged as in Fig. 3.27. When the currents flow in opposite directions the strips repel and when in the same direction they attract. In brief, *unlike currents repel, like currents attract.*

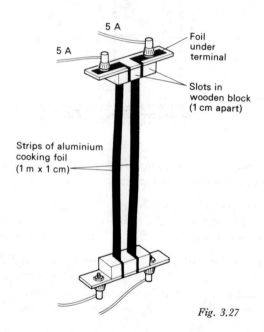

Fig. 3.27

To derive an expression for the force, consider two long, straight, parallel conductors, distance a apart in air, carrying currents I_1 and I_2 respectively, Fig. 3.28. The magnetic field at the right-hand conductor due to the current I_1 in the left-hand one is directed into the paper and its flux density B_1 is given by

$$B_1 = \frac{\mu_0 I_1}{2\pi a}.$$

(p. 82)

The force F acting on length l of the right-hand conductor (carrying current I_2) is therefore

$$F = B_1 I_2 l$$
$$= \frac{\mu_0 I_1 I_2 l}{2\pi a}.$$

The left-hand conductor experiences an equal and opposite force due to being in the field of the right-hand conductor.

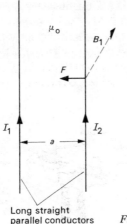

Long straight
parallel conductors   *Fig. 3.28*

(*b*) *The ampere and* μ_0. The definition of the ampere is based on the previous expression and may be stated as follows:

The ampere is the constant current which, flowing in two infinitely long, straight, parallel conductors of negligible circular cross-section, placed in a vacuum 1 metre apart, produces between them a force of 2 × 10⁻⁷ newton per metre of their length.

Previously it was defined in terms of the chemical effect, and the choice of a force of 2×10^{-7} newton was made in the magnetic effect definition to keep the value of the new unit as near as possible to that of the old one.

Once the ampere has been defined, the value of μ_0 follows. Thus we have from the definition,

$$I_1 = I_2 = 1\,\text{A}$$

$$l = a = 1\,\text{m}$$

$$F = 2 \times 10^{-7}\,\text{N}.$$

Substituting in $F = \mu_0 I_1 I_2 l/(2\pi a)$, we get

$$2 \times 10^{-7} = \mu_0 \times 1 \times 1 \times 1/(2\pi \times 1)$$

$$\therefore \quad \mu_0 = 4\pi \times 10^{-7}\ \text{H m}^{-1}.$$

This is the value given previously.

(*c*) *Current balances and absolute measurement of current.* A current balance enables a current to be measured by weighing and uses the fact that adjacent current-carrying conductors, be they straight wires as considered in the definition of the ampere or coils, exert forces on each other. The measurement, when reduced to its essentials, involves finding a mass, a length and a time and is said to be an *absolute* one; no electrical quantities as such have to be measured, only mechanical ones. (Basically, a force is determined from the acceleration it produces in a known mass and has dimensions MLT^{-2}.) Current balances are used primarily to calibrate more convenient types of current-measuring instruments.

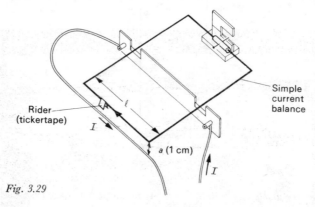

Rider (tickertape)

Simple current balance

Fig. 3.29

A simple arrangement is shown in Fig. 3.29 in which the current balance met earlier (p. 79) measures the current I (about 5 A) in a long, straight wire, close and parallel to its current-carrying arm of length l and in series with it. You should be able to show that $I^2 = mga/(2 \times 10^{-7} l)$, where m is the mass of the rider needed to counterpoise the balance. The accuracy of the result is poor because the force involved is very small. A larger force and greater accuracy are obtained if a wide, flat solenoid is used (see Fig. 3.10, p. 79) and I calculated from $I^2 = mg/(\mu_0 nl)$ where n is the number of turns per metre on the solenoid.

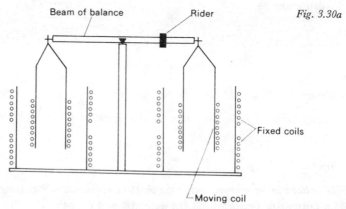

Beam of balance   Rider   *Fig. 3.30a*

Fixed coils

Moving coil

The principle of a practical current balance such as is used in standardizing laboratories like the National Physical Laboratory (NPL) is shown in Fig. 3.30a. The six coils carry the same current and are connected in series so that the forces on the two movable coils tip the balance to the same side. A rider of known mass is moved on the beam of the balance to restore equilibrium and enables the forces between the coils to be found. Knowing this and the value of B (from the dimensions of the coils), the current can be calculated to an accuracy of a few parts in a million. The NPL current balance is shown in Fig. 3.30*b*.

Fig. 3.30b

Couple on a coil in a magnetic field

Current-carrying coils in magnetic fields are essential components of electric motors and meters of various kinds.

(*a*) *Expression for couple.* Consider a rectangular coil PQRS of N turns pivoted so that it can rotate about a vertical axis YY¹ which is at right angles to a *uniform* magnetic field of flux density B, Fig. 3.31*a*. Let the normal to the plane of the coil make an angle θ with the field, Fig. 3.31*b*.

When current I flows in the coil each side experiences a force (since all make some angle with B), acting perpendicularly to the plane containing the side and the direction of the field. The forces on the top and bottom (horizontal) sides are

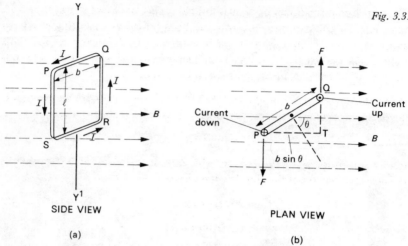

SIDE VIEW

(a)

PLAN VIEW

(b)

parallel to YY^1 and for the current direction shown, they lengthen the coil. The forces on the vertical sides, each of length l, are equal and opposite and have value F where

$$F = BIlN.$$

Whatever the position of the coil, its vertical sides are at right angles to B and so F remains constant. The forces constitute a couple whose moment (or torque) C is given by

$C =$ one force $\times$ perpendicular distance between lines of action of the forces

$\quad = F \times PT = F \times b \sin PQT \quad (b = \text{breadth of coil})$

$\quad = Fb \sin \theta$

$\quad = BIlN \times b \sin \theta$

$\quad = BIAN \sin \theta$

where $A = $ area of face of coil $= l \times b$. The couple causes angular acceleration of the coil which rotates until its plane is perpendicular to the field (i.e. $\theta = 0$) and then $C = 0$.

Circular coil

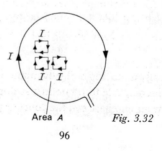

Area A *Fig. 3.32*

The expression for C can be shown to hold for a coil of *any* shape of area A. Thus a circular coil carrying current I can be regarded as consisting of a large number of tiny rectangular coils, each with current I flowing in the same direction as in the circular coil, Fig. 3.32. The forces on all sides of the tiny coils cancel except where they lie on the circular coil itself.

(*b*) *Electromagnetic moment of a coil.* From the above expression we see that C depends on, among other things, I, A and N. It is convenient to write

$$m = IAN$$

where m is a property of the coil and the current it carries, and is called the *electromagnetic moment* of the coil. We then have

$$C = Bm \sin \theta$$

or

$$m = C/(B \sin \theta).$$

If $B = 1$ T and $\theta = 90°$ then $m = C$ and we can define the *electromagnetic moment* of a coil as *the moment of the couple (or torque) acting on it when it lies with its plane parallel to a magnetic field of unit flux density*, Fig. 3.33.

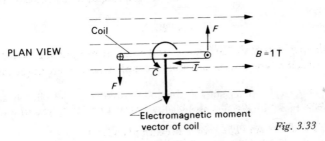

Fig. 3.33

m is a vector whose direction is taken to be that of the flux density created along the axis of the coil by the current in it. We can therefore regard the couple which acts on a current-carrying coil as trying to align the electromagnetic moment of the coil with the direction of the flux density of the applied magnetic field.

Moving coil galvanometers

A galvanometer detects (or measures, if its scale is calibrated) small currents passing through it or small p.d.s across it: the addition of a shunt or multiplier converts it to an ammeter or a voltmeter. Most d.c. meters are of the moving coil type.

(*a*) *Construction.* Basically a moving coil galvanometer consists of a coil of fine, insulated copper wire which is able to rotate in a strong magnetic field. The field is produced in the narrow air gap between the concave pole pieces of a permanent magnet and a fixed soft iron cylinder and is *radial*, i.e. the field lines in the gap appear to radiate from the central axis of the cylinder and are always parallel to the plane of the coil, Fig. 3.34.

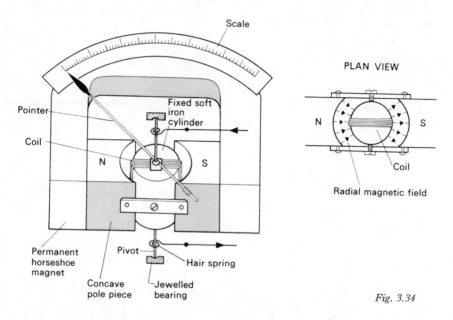

Fig. 3.34

In the *pointer-type* meter shown, the coil is pivoted on jewelled bearings and its rotation is resisted by hair springs above and below it. The springs also lead the current in and out of the coil. In another common type of construction called *taut-ribbon suspension*, the coil is suspended and controlled not by pivots, jewelled bearings and hair springs but by two gold alloy ribbons held taut by springs above and below it. The ribbons conduct the current to and from the coil.

In the most sensitive galvanometers a small concave mirror (instead of a pointer) is fixed to the coil, as shown in the taut-ribbon instrument of Fig. 3.35a. The mirror throws an image of an illuminated hair line on to a scale via a return mirror, Fig. 3.35b and the light beam acts as a weightless pointer whose effective length is the distance from the coil to the scale. The angular deflection of the coil is magnified twice by this optical system. Why? A modern *light-beam* galvanometer with a millimetre scale is shown in Fig. 3.35c.

(b) *Theory.* The magnetic field is *radial* and so the plane of the coil is parallel to it, whatever the deflection. The forces acting on the vertical sides are therefore always perpendicular to the sides and the deflecting couple consequently has a maximum value for all positions of the coil (since $\theta = 90°$ and sin 90 = 1). If the air gap is of constant width, the flux density B of the field is also nearly constant

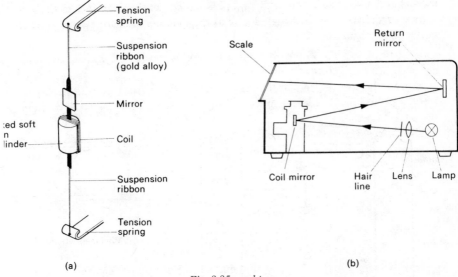

Fig. 3.35a and b.

and the moment C of the deflecting couple due to current I in the coil is given by

$$C = BIAN$$

where A is the mean area of the coil and N is the number of turns on it.

The coil rotates until the resisting couple C^1 due to the suspension is equal and opposite to C. If the deflection is then α and k is the moment of the couple needed to produce unit angular deflection (in newton metres per radian) of the suspension, we have

$$C^1 = k\alpha$$

(assuming Hooke's law holds for the suspension). Hence, since at equilibrium $C = C^1$

$$BIAN = k\alpha.$$

B, A, N and k are constants for a given meter and so

$$\alpha \propto I.$$

The use of a radial field and a uniform air gap thus results in the deflection α of the coil being directly proportional to the current I in it; the galvanometer scale is therefore linear.

(c) *Sensitivity*. The *current sensitivity* of a galvanometer is defined as the *deflection per unit current* and equals α/I. It follows from above that

$$\frac{\alpha}{I} = \frac{BAN}{k}.$$

Maximum current sensitivity therefore requires

(*i*) *B* to be large in the air gap, i.e. the permanent magnet should be strong and the air gap narrow.

(*ii*) *A* to be as large as possible but if the coil is too large it swings about its deflected position before a reading can be taken.

(*iii*) *N* to be large but not at the expense of having to use a wide air gap.

(*iv*) *k* to be small but if the opposition of the suspension is too weak readings again take time.

The *voltage sensitivity* is defined as α/V where α is the deflection when the p.d. across the galvanometer is V. If its resistance is R then since $I = V/R$, we have

$$\frac{\alpha}{I} = \frac{BAN}{k}$$

$$\therefore \frac{\alpha}{V} = \frac{BAN}{kR} \ .$$

Sensitivities are usually expressed in mm per μA or mm per μV, this being the deflection produced on a mm scale by 1 μA or 1 μV.

(*d*) *Choice of galvanometer.* When choosing a galvanometer for a particular task the resistance of the rest of the circuit has to be considered. It can be shown that we should aim at transferring maximum power to the meter, and this occurs when the resistance of the meter equals the resistance of the rest of the circuit. If the latter is low, as it often is in potentiometer, Wheatstone bridge and thermocouple circuits, then a low resistance galvanometer is required. Such instruments have high voltage sensitivity. For high resistance circuits, galvanometers should have a high resistance and high current sensitivity.

Typical data for the general purpose galvanometer shown in Fig. 3.35*c* is given in Table 3.1 for the ' direct ' position of the range switch.

Table 3.1

Galvanometer resistance Ω	Sensitivities		
	Current mm/μA	Voltage mm/μV	Charge mm/μC
14	25	1.8	75

(*e*) *Care and use of light-beam galvanometers.* Light-beam galvanometers should be handled with care. When not in use, being moved or when making connections, the range switch should be set at ' short ' (see p. 117). The zero is usually set either at the end or the centre of the scale by turning the range switch to ' × 0.001 ' and adjusting the set-zero control.

Fig. 3.35c

On ' direct ' the instrument is most sensitive and the coil swings freely. It is used in this position for ballistic work to measure charge (see p. 144). The sensitivity is about the same on ' × 1 ' but internal resistors (shunts) are then connected across the coil so that it is ' critically damped '. This means the coil reaches its steady deflection in the minimum of time, i.e. the movement is ' dead-beat ' (see p. 117). The coil is critically damped for all settings of the range switch (except ' direct ') unless the resistance of the external circuit is low compared with the critical damping resistance of the galvanometer (an external resistance of 120 Ω for the instrument of Fig. 3.35c).

The ' × 0.1 ' range is 10 times less sensitive than the ' × 1 ' setting. It is advisable to start on the ' × 0.001 ' range and increase the sensitivity as required.

QUESTIONS

1. In which of the four cases shown in Fig. 3.36 will the arm XY of the current balance experience a tilting force and will it be upwards or downwards?

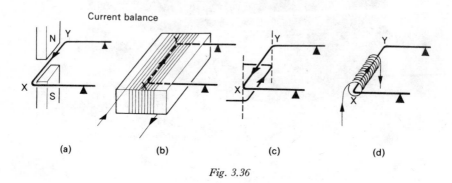

Fig. 3.36

2. A current of 5 A flows in a straight wire in a uniform flux density of 2×10^{-3} T. Calculate the force per unit length on the wire if it is (*i*) perpendicular to the field (*ii*) inclined at 30° to it.

3. A long straight vertical wire carrying 5.0 A in a downward direction passes through a horizontal board on which lines of force may be plotted. Draw a diagram of the lines of force to be expected and deduce a value for the horizontal component of the earth's magnetic field if there is a neutral point 5.0 cm from the wire. What would be the resultant horizontal field at a point an equal distance from the wire on a line through the wire at right angles to that joining the wire to the neutral point?
Permeability of free space $(\mu_0) = 4\pi \times 10^{-7}$ H m^{-1}. (*J.M.B.*)

4. What current must be passed through a flat circular coil of 10 turns and radius 5.0 cm to produce a flux density of 2.0×10^{-4} T at its centre? ($\mu_0 = 4\pi \times 10^{-7}$ H m^{-1})

5. Define the *magnitude* and *direction of a magnetic field*.
Describe experiments which would enable you to investigate how the magnetic field at the centre of a flat coil varies (*a*) with the current flowing in the coil, and (*b*) with the radius of the coil.
A flat circular coil of wire of 20 turns and of radius 10.0 cm is placed with its plane vertical and at 45° to the magnetic meridian. Calculate the current in the coil if a compass needle, free to move in a horizontal plane, points in the east-west direction when placed at the centre of the coil.
(Horizontal component of the earth's magnetic flux density = 2.0×10^{-5} T; $\mu_0 = 4\pi \times 10^{-7}$ H m^{-1}.) (*A.E.B.*)

6. Define the *ampere*.
Two long vertical wires, set in a plane at right angles to the magnetic meridian, carry equal currents flowing in opposite directions. Draw a diagram showing the

pattern, in a horizontal plane, of the magnetic flux due to the currents alone—that is, neglecting for the moment the earth's magnetic field.

Next, taking into account the earth's magnetic field, discuss the various situations that can give rise to neutral points in the plane of the diagram.

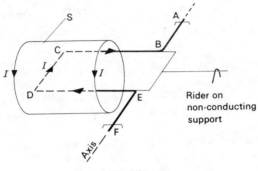

Fig. 3.37

Fig. 3.37 shows a simple form of current balance. The 'long' solenoid S, which has 2000 turns per metre, is in series with the horizontal rectangular copper loop ABCDEF, where BC = 10 cm and CD = 3.0 cm. The loop, which is freely pivoted on the axis AF, goes well inside the solenoid, and CD is perpendicular to the axis of the solenoid. When the current is switched on, a rider of mass 0.20 g placed 5.0 cm from the axis is needed to restore equilibrium. Calculate the value of the current, I.

$(\mu_0 = 4\pi \times 10^{-7}$ H m^{-1}; $g = 10$ N kg^{-1}) (O.)

7. Draw a diagram to show the magnetic field (magnetizing force) due to a long solenoid carrying a current. Write down an equation for the flux density at its centre and at its ends, explaining your units. Describe an experiment you could perform to verify that the ratio of these flux densities is as predicted by your formulae.

A metal wire 10 m long lies east-west on a wooden table. What p.d. would have to be applied to the ends of the wire, and in what direction, in order to make the wire rise from the surface? Assume that the electrical connections to the wire cause no appreciable restraint.

(Density of the metal = 1.0×10^4 kg m^{-3}, resistivity of the metal = 2.0×10^{-8} Ω m, horizontal component of earth's field = 1.8×10^{-5} T, $g = 9.8$ N kg^{-1}.) (C.)

8. An electron of charge e and mass m describes a circular path of radius r when it is projected with velocity v into a uniform magnetic field of flux density B. Derive an expression for the frequency of revolution.

How many orbits per second are made by an electron in a fine beam tube (Fig. 3.23a, p. 88,) if the flux density in the field due to the Helmholtz coils is 1.0×10^{-3} T? ($e = 1.6 \times 10^{-19}$ C; $m = 9.1 \times 10^{-31}$ kg.)

MAGNETIC FIELDS

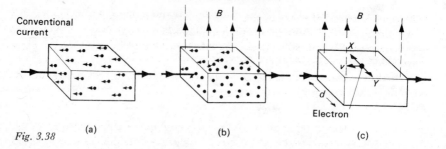

Fig. 3.38 (a) (b) (c)

9. (a) In Fig. 3.38a electrons are shown drifting from right to left through a block of conductor. A flux density is applied as in Fig. 3.38b and the concentration of electrons increases near the front edge of the block. Why? If the current had consisted of positive charges drifting from right to left would they have been pushed towards the front?

(b) An electric field and a p.d. (the Hall p.d.) are created across the block and soon attain maximum values. Why?

(c) Fig. 3.38c shows an electron (charge e, mass m) drifting with velocity v near the middle of the conductor (width d) when the electric field and Hall p.d. have their maximum values, say E and V_H respectively. It is undeflected because two equal and opposite forces X and Y act on it. What are they? Derive an equation relating them (let B be the magnetic flux density).

(d) Hence calculate v if $V_H = 8$ μV, $B = 0.4$ T and $d = 2$ cm.

10. Calculate the magnetic flux density at a point 2.0 cm from a long straight wire carrying a current of 10 A. Hence calculate the force which would be exerted on a 50-cm length of another straight wire, parallel to the first and 2.0 cm away from it, if this second wire carried a current of 20 A. ($\mu_0 = 4\pi \times 10^{-7}$ H m^{-1}.)

(S.)

11. Two long parallel wires in air with axes 50 cm apart carry currents of 100 A in opposite directions. Find (a) the magnetic field strength on the axis of one wire, due to the current in the other, and (b) the approximate value in newtons of the force per metre length on each wire.

Show clearly on a diagram the distribution of magnetic field around the wires and the direction of the force on one of them. Ignore the presence of the earth's magnetic field. ($\mu_0 = 4\pi \times 10^{-7}$ H m^{-1}.)

(L. part qn.)

12. A rectangular coil 10 cm $\times$ 2.0 cm consisting of 100 turns is suspended vertically from the middle of a short side in a radial magnetic flux density of 2.0×10^{-2} T and supplied with current from a 25-V d.c. supply. Give a diagram of the arrangement showing the directions of the current, magnetic flux density (magnetic induction) and forces. If the resistance of the coil is 100 Ω, calculate the deflecting torque on the suspension.

(J.M.B.)

13. Describe a simple experiment which demonstrates that a force is experienced by a current-carrying conductor in a magnetic field. State the factors that determine (a) the magnitude of the force, and (b) its direction.

A rectangular coil of wire carrying a steady current is pivoted on an axis which is at right angles to a radial magnetic field. Obtain an expression for the torque experienced by the coil, and explain the relevance of this result to the design of moving coil galvanometers.

A moving coil galvanometer has a resistance of 25 Ω and gives a full-scale deflection when carrying a current of 4.4×10^{-6} A. What current will give a full-scale deflection when the galvanometer is shunted by a 0.10-Ω resistance?   (O.)

14. Describe with the aid of diagrams the structure and mode of action of a moving coil galvanometer having a linear scale and suitable for measuring small currents. If the coil is rectangular, derive an expression for the deflecting couple acting upon it when a current flows in it, and hence obtain an expression for the current sensitivity (defined as the deflection per unit current).

If the coil of a moving coil galvanometer having 10 turns and of resistance 4.0 Ω is removed and replaced by a second coil having 100 turns and of resistance 160 Ω calculate

(a) the factor by which the current sensitivity changes, and

(b) the factor by which the voltage sensitivity changes.

Assume that all other features remain unaltered.   (J.M.B.)

4 Electromagnetic induction

Inducing e.m.f.s

An electric current creates a magnetic field, the reverse effect of producing electricity by magnetism was discovered independently in 1831 by Faraday in England and Henry in America and is called electromagnetic induction. Induced e.m.f.s can be generated in two ways.

(*a*) *By relative movement* (the dynamo effect). If a bar magnet is moved in and out of a stationary coil of wire connected to a centre-zero galvanometer, Fig. 4.1, a small current is recorded *during the motion* but not at other times. Movement of the coil towards or away from the stationary magnet has the same results. Relative motion between magnet and coil is necessary.

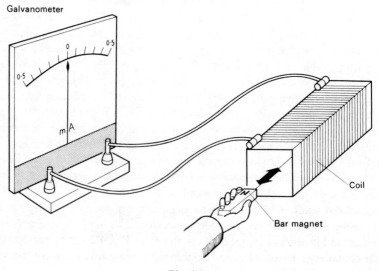

Fig. 4.1

ELECTROMAGNETIC INDUCTION

Observation shows that the direction of the induced current depends on the direction of relative motion. Also the magnitude of the current increases with the speed of motion, the number of turns on the coil and the strength of the magnet.

Although it is current we detect in this demonstration, an e.m.f. must be induced in the coil to cause the current. The induced e.m.f. is the more basic quantity and is always present even when the coil is not in a complete circuit. The value of the induced current depends on the resistance of the circuit as well as on the induced e.m.f.

We will be concerned here only with e.m.f.s induced in conductors but they can be produced in any medium—even a vacuum, where they play a basic role in the electromagnetic theory of radiation.

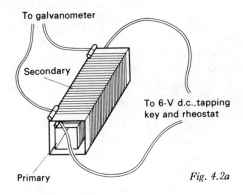

To galvanometer

Secondary

To 6-V d.c.,tapping key and rheostat

Primary

Fig. 4.2a

(*b*) *By changing a magnetic field* (the transformer effect). In this case two coils are arranged one inside the other as in Fig. 4.2*a*. One coil, called the *primary*, is in series with a 6-V d.c. supply, a tapping key and a rheostat. The other, called the *secondary*, is connected to a galvanometer. Switching the current on or off in the primary causes a pulse of e.m.f. and current to be induced in the secondary. Varying the primary current by quickly altering the value of the rheostat has the same effect. Electromagnetic induction thus occurs when there is any *change* in the primary current and so also in the magnetic field it produces.

Cases of electromagnetic induction in which current changes in one circuit cause induced e.m.f.s in a neighbouring circuit, not connected to the first, are examples of *mutual induction*—the transformer principle. (Our use of search coils when investigating magnetic fields due to various current-carrying conductors in the previous chapter depended on it.)

The induced e.m.f. is increased by having a soft iron rod in the coils or better still, by using coils wound on a complete iron ring. The iron ring and coils used by Faraday in his original experiment are shown in Fig. 4.2*b*. It is worth noting that the secondary current is in one direction when the primary current increases and in the opposite direction when it decreases.

107

Fig. 4.2b and c

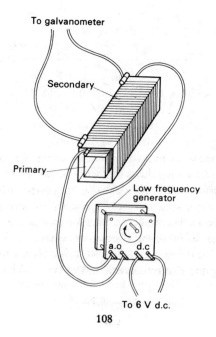

To galvanometer

Secondary

Primary

Low frequency
generator

a.o   d.c

To 6 V d.c.

An alternating current is continually increasing and decreasing first in one direction and then in the opposite direction. The magnetic field accompanying it changes similarly and if a.c. is applied to the primary coil we would expect an induced e.m.f. (also alternating?) to be induced in the secondary. A simple arrangement for investigating the effect is shown in Fig. 4.2c; very low frequency a.c. is produced when the generator is hand-operated.

Magnetic flux

Electromagnetic induction is one of those action-at-a-distance effects whose mechanism is not revealed to our senses. Consequently we have to invent a conceptual model or theory which enables us to picture and to ' explain ' (i.e. describe in terms of the model) what might be happening.

Faraday suggested a model based on magnetic field lines. He proposed that an e.m.f. is induced in a conductor *either* when there is a change in the number of lines ' linking ' it (i.e. passing through it) *or* when it ' cuts ' across field lines. The two statements often come to the same thing, as can be seen by considering Fig. 4.3 (and later, the experiment described on p. 113). Thus if the coil moves towards the magnet from X to Y, the number of lines ' linking ' or ' threading ' it increases from three to five; alternatively we can say it ' cuts ' two lines in moving from X to Y.

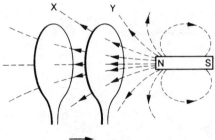

Fig. 4.3

Before expressing these ideas mathematically in the next section, we require some way of deciding how many lines link a coil or are cut by it. In the previous chapter (p. 72) it was suggested that magnetic field lines could be drawn so that the number per unit cross-sectional area represents the magnitude B of the flux density. It would then be reasonable to take the product, $B \times A$, as a measure of the number of lines linking a coil of cross-sectional area A.

This product is called the *magnetic flux* (Φ). It is a more useful concept than B alone for quantifying our electromagnetic induction model and it is defined by the equation

$$\Phi = BA$$

where B is the flux density acting at right angles to and over an area A, Fig. 4.4a. In words, *the magnetic flux through a small plane surface is the product of the flux density normal to the surface and the area of the surface.* (Why a small surface?) If $B = 1$ tesla (T) and $A = 1$ square metre (m²) then Φ is defined to be 1 weber (Wb).

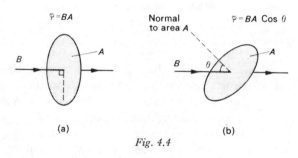

(a) (b)

Fig. 4.4

Since $B = \Phi/A$, the reason for calling B the flux density will now be evident. Another unit of B is therefore the weber per square metre (Wb m⁻²), i.e. 1 T = 1 Wb m⁻².

In general, if the normal to the area A makes an angle θ with B, Fig. 4.4b, then the flux is $BA \cos \theta$.

If Φ is the flux through the cross-sectional area A of a coil of N turns, the total flux through it, called the *flux-linkage*, is $N\Phi$ since the same flux Φ links each of the N turns.

Although the term *flux* suggests that something flows along the field lines, this is not so, but references to the flux ' entering ', ' passing through ' or ' leaving ' a coil are in common use.

Faraday's law

(a) *Statement.* The law states that the *induced e.m.f. is directly proportional to the rate of change of flux-linkage or rate of flux-cutting.* In calculus notation it can be written

$$\mathscr{E} \propto \frac{d}{dt}(N\Phi)$$

or

$$\mathscr{E} = \text{constant} \times \frac{d}{dt}(N\Phi)$$

where $\mathscr{E}$ is the induced e.m.f. and $d(N\Phi)/dt$ is the rate of change of flux-linkage or the rate of flux-cutting. The law is found to be true for the dynamo and the transformer types of induction.

If instead of defining the weber in terms of the tesla (as we did in the previous section), we now redefine it as *the magnetic flux which induces in a one-turn coil an*

ELECTROMAGNETIC INDUCTION

e.m.f. of 1 volt when the flux is reduced to zero in 1 second, then the constant of proportionality in the above equation is 1. That is, if $d(N\Phi) = 1$ Wb when $\mathscr{E} = 1$ V and $dt = 1$ s, then we can write $1 = \text{constant} \times 1/1$. Hence

$$\mathscr{E} = \frac{d}{dt}(N\Phi)$$

where $\mathscr{E}$ is in volts, $d\Phi/dt$ in webers per second and N is a number of turns if we are considering a coil.

(b) *Numerical example.* Suppose a single-turn coil of cross-sectional area 5.0 cm² is at right angles to a flux density of 2.0×10^{-2} T, which is then reduced steadily to zero in 10 s. The flux-linkage change $d(N\Phi) = $ (number of turns) × (change in B) × (area of coil) $= (1) \times (2.0 \times 10^{-2}\text{ T}) \times (5.0 \times 10^{-4}\text{ m}^2) = 1.0 \times 10^{-5}$ Wb. The change occurs in time $dt = 10$ s, hence the e.m.f. $\mathscr{E}$ induced in the coil is given by

$$\mathscr{E} = \frac{d}{dt}(N\Phi) = \frac{1.0 \times 10^{-5}}{10}\frac{\text{Wb}}{\text{s}}$$

$$= 1.0 \times 10^{-6}\text{ V}.$$

If the coil had 5000 turns, the *flux-linkage* would be 5000 times as great, (i.e. $N\,d\Phi = 5.0 \times 10^{-2}$ Wb), hence

$$\mathscr{E} = 5000 \times 10^{-6} = 5.0 \times 10^{-3}\text{ V}.$$

If the normal to the plane of this coil made an angle of 60° (instead of 0°) with the field then $N\,d\Phi = 5.0 \times 10^{-2}\cos 60 = 2.5 \times 10^{-2}$ Wb and

$$\mathscr{E} = 2.5 \times 10^{-3}\text{ V}.$$

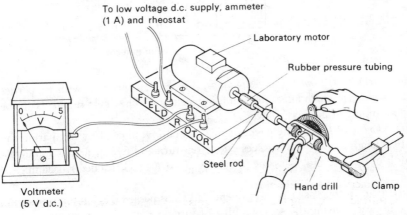

Fig. 4.5

(c) *Experimental test of the law.* Two methods will be outlined, one for each type of induction.

(i) *Using a motor as a dynamo.* The arrangement of Fig. 4.5 uses a laboratory motor (12 V) with separate field and rotor terminals. When the hand drill is turned the rotor coil rotates and ' cuts ' the flux due to the direct current (about 1 A) in the field coil.

To see if the e.m.f. induced in the rotor coil is proportional to the rate of flux-cutting, i.e. to the speed of rotation of the rotor, the drill is turned steadily and the number of turns made in, say, 10 seconds is counted. The reading (steady) on the d.c. voltmeter across the rotor is noted. The procedure is repeated for different drill speeds.

The effect on the induced e.m.f. of increasing the field current (and so also B and Φ) can be found.

(ii) *Using a.c., two coils and a CRO.* Alternating current from the low impedance output of a signal generator is passed through a solenoid and the peak-to-peak value of the e.m.f. induced (by mutual induction) in a ten-turn coil wound round the middle of the solenoid is measured on a CRO (used as a voltmeter and set on its most sensitive range, e.g. 0.1 V cm^{-1}) for different frequencies between 1 kHz and 3 kHz, Fig. 4.6. If the frequency of the a.c. in the solenoid is doubled, but the value of the current remains the same (as shown by the a.c. ammeter), the rate of change of the flux ' linking ' the ten-turn coil doubles, as should the induced e.m.f.

Fig. 4.6

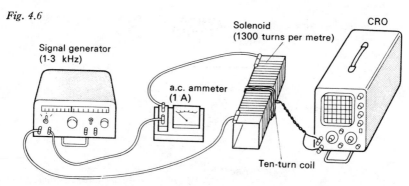

(d) *Further experimental investigations.* These are concerned with properties of the coil itself which affect the e.m.f. induced in it. They help further with the understanding of Faraday's law.

(i) *Number of turns.* The effect of this can be investigated by comparing the vertical heights of the CRO traces due to ten-turn and five-turn coils round the middle of the solenoid, the same frequency of a.c. (say 2 kHz) being used in each case.

(ii) *Area.* In this case a.c. of frequency 2 kHz is passed through two solenoids in series, each having the same number of turns per metre, e.g. about 1300 (so that $B (= \mu_0 n I)$ is the same inside both), but one with twice the cross-sectional area of the other, Fig. 4.7. The flux linking ten-turn coils round the middle of each will be in the ratio of the areas (i.e. 2:1), as should be the ratio of the e.m.f.s induced in the coils and displayed on a CRO, preferably double beam.

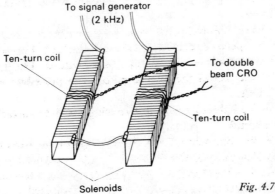

Fig. 4.7

(*iii*) *Orientation.* If the plane of the coil is at an angle (other than 90°) to the magnetic field, the e.m.f. induced in it is less, as can be shown by the apparatus of Fig. 4.8. What angle will the handle of the search coil make with the plane of the coil when the induced e.m.f. has half its maximum value?

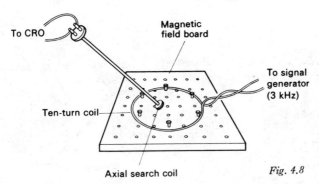

Fig. 4.8

(*e*) *Equivalence of flux cutting and flux linking.* This can be shown using the arrangement of Fig. 4.9. The outer solenoid is in series with a *smooth*, low voltage variable d.c. supply, an ammeter and a rheostat, and it carries a steady current of 1–2 A. The inner solenoid is connected to a light-beam galvanometer on ' direct ' setting and is pulled out slowly so that the galvanometer deflection remains steady at, say, 50 mm. The time taken for the complete removal of the solenoid is noted.

The inner solenoid is reinserted and the current reduced to zero in the outer solenoid by turning down the output control on the low voltage supply. This is

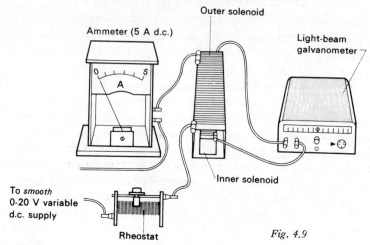

Fig. 4.9

done at a rate which gives the same, as steady as possible, reading on the galvanometer as before. The time required is again measured.

In both cases the flux changes are the same and if the times are similar then the rate of flux-cutting equals the rate of change of flux-linkage.

The actions are apparently different, the first involves relative motion and the second a changing magnetic field. Nevertheless, the same law describes both.

Lenz's law

Whilst the magnitude of the induced e.m.f. is given by Faraday's law, its direction can be predicted by a law due to the Russian scientist Lenz. It may be stated as follows:

The direction of the induced e.m.f. is such that it tends to oppose the flux-change causing it, and does oppose it if induced current flows.

Fig. 4.10

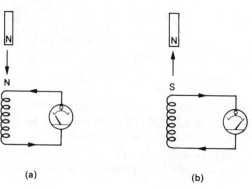

(a)   (b)

Thus in Fig. 4.10a a bar magnet is shown approaching the end of a coil, north pole first. If Lenz's law applies, the induced current should flow in a direction which makes the coil behave like a magnet with a north pole at the top. The downward motion of the coil and the accompanying flux-change will then be opposed. When the magnet is withdrawn, the top of the coil should behave like a south pole, Fig. 4.10b, and attract the north pole of the magnet, so hindering its removal and again opposing the flux-change. The induced current is therefore in the opposite direction to that when the magnet approaches. (Polarities may be checked using the right-hand screw rule if the direction of the windings on the coil are known and the current directions observed on the galvanometer.)

Lenz's law is an example of the principle of conservation of energy; here, energy would be created from nothing if the e.m.f.s and currents acted differently. Thus if a south pole were produced at the top of the coil in Fig. 4.10a, attraction would occur between coil and magnet and the latter would, if it was allowed to, accelerate towards the coil, gaining kinetic energy as well as generating electrical energy. In practice, work has to be done to overcome the forces that arise, i.e. energy transfer occurs which in this case is from the mechanical to the electrical form.

For straight conductors moving at right angles to a magnetic field a more useful version of Lenz's law is *Fleming's right-hand rule* (also called the *dynamo rule*; his left-hand rule is often referred to as the *motor rule*, p. 76). It states that *if the thumb and first two fingers of the right hand are held so that each is at right angles to the other with the First finger pointing in the direction of the Field and the thuMb in the direction of Motion of the conductor, then the seCond finger indicates the direction (conventional) of the induced Current*, Fig. 4.11.

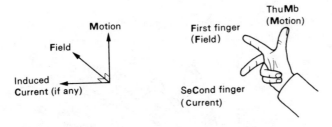

Fig. 4.11

Lenz's law is incorporated in the mathematical expression of Faraday's law by including a negative sign to show that current due to the induced e.m.f. produces an opposing flux-change, thus we write

$$\mathscr{E} = -\frac{d}{dt}(N\Phi).$$

115

Eddy currents

Any piece of metal moving in a magnetic field, or exposed to a changing one, has e.m.f.s induced in it, as we might expect. These can cause currents, called *eddy currents*, to flow inside the metal and they may be quite large because of the low resistance of the paths they follow. Their magnetic and heating effects are both helpful and troublesome.

(*a*) *Magnetic effect.* According to Lenz's law, eddy currents will circulate in directions such that the magnetic fields they create oppose the motion (or flux-change) producing them. This acts as a brake on the moving body and may be shown simply with the arrangement of Fig. 4.12. The solid copper cylinder

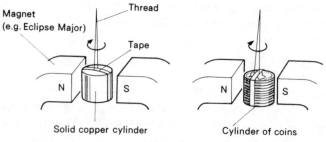

Fig. 4.12

quickly comes to rest if it is spun between the poles of the magnet. With the cylinder of coins there is very little braking because dirt on the coins increases the resistance of the cylinder as a whole, thereby reducing the eddy current flow.

Use is made of the effect for the electromagnetic damping of moving coil

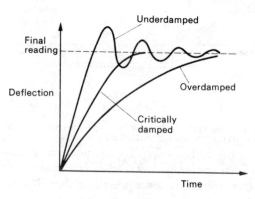

Fig. 4.13

116

meters so that the coil takes up its deflected position quickly without overshooting and oscillating about its final reading. In most *pointer* instruments the coil is wound on a metal frame in which large eddy currents are induced and cause opposition to the motion of the coil as it cuts across the radial magnetic field of the permanent magnet. When oscillation is *just* prevented, the meter is said to be *critically damped* and its movement is ' dead-beat '. The curves in Fig. 4.13 show the effect of different degrees of damping.

In *light-beam* galvanometers the coil is not usually wound on a frame but is glued together and near-critical damping is achieved by having appropriate internal shunts across the coil on all ranges (except ' direct '). Suitable eddy currents then flow round the coil and shunt, whatever the current to be measured. On ' direct ' setting, the electromagnetic damping is made a minimum (no internal shunts are connected) and the coil swings freely. Its first deflection can be shown to be proportional to the charge passing; the galvanometer is then said to be used *ballistically* (see p. 144). On ' short ' the coil is short-circuited internally and the eddy currents induced in it bring it to rest quickly.

(*b*) *Heating effect.* In induction or eddy current heating, a coil carrying high frequency a.c. surrounds the material to be heated and the rapidly changing magnetic flux induces large eddy currents in the conducting parts of it. For example, in the zone refining of metals and semiconductors a narrow crucible containing the material is passed very slowly through the heating coil, Fig. 4.14*a*. The impurities tend to collect in the molten zone which moves to one end. After cooling this end is removed leaving a very pure, single crystal sample. In practice, multiple zone refining is employed as shown in Fig. 4.14*b* (for germanium).

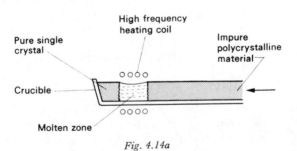

Fig. 4.14a

Electric motors, dynamos and transformers contain iron that experiences flux changes when the device is in use. To minimize energy loss through eddy currents the iron parts consist of sheets, called *laminations*, insulated from each other by thin paper, varnish or some other insulator. The resistance of eddy current paths is thereby increased.

Fig. 4.14b

Calculation of induced e.m.f.s

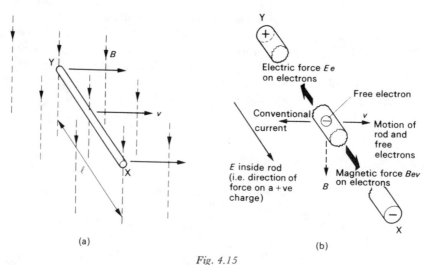

Fig. 4.15

(*a*) *Straight conductor*. In this case it is more helpful to consider flux-cutting. Suppose a conducting rod XY of length *l* moves sideways with steady velocity *v* through and at right angles to a uniform magnetic field of flux density *B* directed into the paper, Fig. 4.15*a*. The area swept out per second by XY is *lv* and therefore the flux cut per second is *Blv*. Assuming Faraday's law, we can say that the e.m.f. $\mathscr{E}$ induced in the rod is given numerically by

$$\mathscr{E} = \text{flux cut per second}$$

$$\therefore \quad \mathscr{E} = Blv.$$

118

ELECTROMAGNETIC INDUCTION

The e.m.f. induced in a rod cutting magnetic flux can be explained in terms of the forces acting on the charged particles in it and the above expression derived without recourse to Faraday's law. In a moving conductor both positive ions and electrons are carried along and both experience a force (magnetic) at right angles to the field and to the direction of motion of the conductor. However, only electrons are free to move inside the conductor and Fleming's left-hand (motor) rule indicates they will be forced to end X, making X negative and Y positive, Fig. 4.15b. (The conventional Current direction (seCond finger) will be opposite to the motion of the conductor since we are dealing with negative charges.) As a result of the charge separation and electron accumulation, an electric field is created inside the conductor which causes a repulsive electric force to be exerted on other electrons being urged towards X by the magnetic force.

These two forces act oppositely and when they become equal there is no further charge accumulation and we can say

$$Ee = Bev$$

where E is the equilibrium electric field strength, e the charge on an electron, B the magnetic flux density and v the velocity of the conductor. Hence

$$E = Bv.$$

If V is the p.d. developed between the ends X and Y of the conductor and l is its length, then E = potential gradient = V/l and so $V = Blv$. The rod is on open circuit, therefore V equals the induced e.m.f. $\mathscr{E}$. We therefore get, as before,

$$\mathscr{E} = Blv.$$

(b) *Spinning disc.* The first dynamo was made by Faraday and consisted of a metal disc rotated in a magnetic field, Fig. 4.16. A modern version of his apparatus is shown in Fig. 4.17. When the disc is driven at a steady speed by the motor a steady deflection is obtained on the light-beam galvanometer connected to two sliding contacts (e.g. 4-mm plugs) held one at the centre and the other at the edge of the disc between the poles of the magnet. The effect of changing the speed of rotation and the position of the contacts may be investigated.

We can consider that an e.m.f. is induced in the circuit because the radius of the disc between the contacts at any instant is cutting the flux there, i.e. we regard the disc as a many-spoked wheel. If the disc makes f revolutions per second and has radius r, the area swept out per second by a radius is $\pi r^2 f$. The flux cut per second = $B\pi r^2 f$ where B is the flux density (assumed uniform) between the contacts. Hence the induced e.m.f. $\mathscr{E}$ is, by Faraday's law,

$$\mathscr{E} = B\pi r^2 f.$$

The ' homopolar generator ' is a recent form of Faraday's disc and can deliver a very large direct current with very small e.m.f. for the production of powerful magnetic fields.

Fig. 4.16

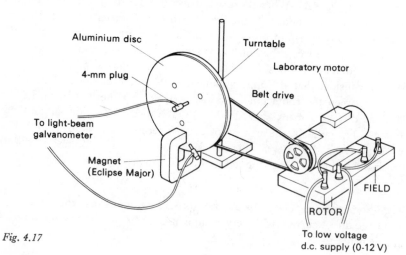

Aluminium disc

Turntable

4-mm plug

Laboratory motor

Belt drive

To light-beam
galvanometer

Magnet
(Eclipse Major)

FIELD

ROTOR

Fig. 4.17

To low voltage
d.c. supply (0-12 V)

(c) *Rotating coil.* The coil in Fig. 4.18 has N turns each of area A and is being rotated about a horizontal axis in its own plane at right angles to a uniform magnetic field of flux density B. If the normal to the coil makes an angle θ with the field at time t (measured from the position where $\theta = 0$) then the flux Φ linking each turn is given by

$$\Phi = BA \cos \theta.$$

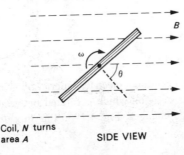

Coil, N turns
area A SIDE VIEW

Fig. 4.18

But $\theta = \omega t$ where ω is the steady angular velocity of the coil, therefore

$$\Phi = BA \cos \omega t.$$

By Faraday's law, the e.m.f. $\mathscr{E}$ induced in N turns is

$$\mathscr{E} = -\frac{d}{dt}(N\Phi) = -N\frac{d}{dt}(BA \cos \omega t)$$

$$= -BAN\frac{d}{dt}(\cos \omega t)$$

$$= -BAN(-\omega \sin \omega t)$$

$$= BAN\omega \sin \omega t.$$

The e.m.f. is thus an alternating one which varies sinusoidally with time and would cause a similar alternating current in an external circuit connected across the coil.

When the plane of the coil is parallel to B, $\theta = \omega t = 90°$, $\sin \omega t = 1$ and $\mathscr{E}$ has its maximum value $\mathscr{E}_0$ given by

$$\mathscr{E}_0 = BAN\omega.$$

Hence we can write $\mathscr{E} = \mathscr{E}_0 \sin \omega t.$

When is $\mathscr{E}$ equal to (i) zero, and (ii) $\mathscr{E}_0/2$?

ELECTROMAGNETIC INDUCTION

If a coil has area 1.0×10^{-2} m² (i.e. 100 cm²), 800 turns and makes 600 revolutions per minute in a magnetic field of flux density 5.0×10^{-2} T, then $\mathscr{E}_0$ is given by

$$\mathscr{E}_0 = BAN\omega$$

$$= (5.0 \times 10^{-2}\,\text{T}) \times (1.0 \times 10^{-2}\,\text{m}^2) \times (800) \times (2\pi \times 600/60\,\text{s}^{-1})$$

$$= 5.0 \times 10^{-2} \times 1.0 \times 10^{-2} \times 800 \times 20\pi\ \text{Wb s}^{-1} \qquad (1\ \text{T} = 1\ \text{Wb m}^{-2})$$

$$= 25\ \text{V}.$$

Generators (a.c. and d.c.)

A generator or dynamo produces electrical energy by electromagnetic induction. In principle it consists of a coil which is rotated between the poles of a magnet so that the flux-linkage changes, Fig. 4.19a. The flux Φ linking each turn of a

Fig. 4.19

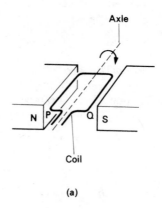

(a)

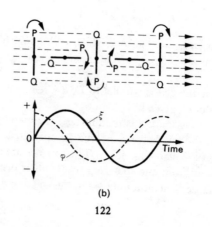

(b)

coil of area A having N turns, rotating with angular velocity ω in a uniform flux density B, is given at time t (measured from the vertical position) by

$$\Phi = BA \cos \omega t.$$

By Faraday's law, the induced e.m.f. $\mathscr{E}$ is, as we found in the previous section,

$$\mathscr{E} = - \frac{d}{dt} (N\Phi) = BAN\omega \sin \omega t.$$

Both Φ and $\mathscr{E}$ alternate sinusoidally and their variation with the position of the coil is shown in Fig. 4.19b.

We see that although Φ is a maximum when the coil is vertical (i.e. perpendicular to the field), $\mathscr{E}$ is zero because the *rate of change* of Φ is zero *at that instant*, i.e. the tangent to the Φ-graph is parallel to the time axis and so has zero gradient—in calculus terms $d\Phi/dt = 0$. Also, when $\Phi = 0$, its rate of change is a maximum, therefore $\mathscr{E}$ is a maximum.

The expression for $\mathscr{E}$ shows that its instantaneous values increase with B, A, N and the angular velocity of the coil. If the coil makes one complete revolution, one cycle of alternating e.m.f. is generated, i.e. for a simple, single coil generator the frequency of the supply equals the number of revolutions per second of the coil.

In an *a.c. generator* (or alternator) the alternating e.m.f. is taken off and applied to the external circuit by two spring-loaded graphite blocks (called ' brushes ') which press against two copper *slip-rings*. These rotate with the axle, are insulated from one another and each is connected to one end of the coil, Fig. 4.20.

Fig. 4.20

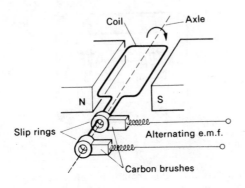

In a *d.c. generator* a *commutator* is used instead of slip-rings. This consists of a split-ring of copper, the two halves of which are insulated from each other and joined to the ends of the coil, Fig. 4.21a. The brushes are arranged so that the change-over of contact from one split ring to the other occurs when the coil is vertical. In this position the e.m.f. induced in the coil reverses and so one brush is always positive and the other negative. The graphs of Fig. 4.21*b* show the e.m.f.s in the coil and at the brushes; the latter, although varying, is unidirectional and produces d.c. in an external circuit.

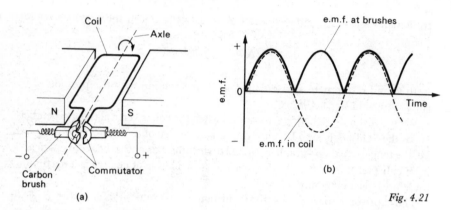

(a)

(b)

Fig. 4.21

In actual a.c. and d.c. generators, several coils are wound in uniformly spaced slots in a soft iron cylinder which is laminated to reduce eddy currents. The whole assembly is known as the *armature*. In the d.c. case the use of many coils and a correspondingly greater number of commutator segments gives a much steadier e.m.f. Also, in practical generators the magnetic field is produced by electromagnets (except in a cycle dynamo where a permanent magnet is used) and the coils which energize them are called *field* coils.

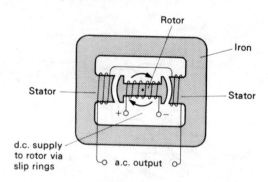

Fig. 4.22

In power station alternators the armature coils and their iron cores are stationary (and are called the *stator*) whilst the field coils and their core (i.e. the electromagnets) rotate (and are called the *rotor*); Fig. 4.22 shows a simplified alternator. The advantage of this is that only the relatively small direct current needed for the field coils is fed through the rotating slip-rings. The large p.d.s and currents induced in the armature coils (25 kV and thousands of amperes in some modern alternators) are then led away through fixed connections. The rotor is driven either by a steam or water turbine which also powers a small d.c. dynamo (called the *exciter*) for supplying current to the field coils. In Fig. 4.23*a* a rotor is being inserted into the stator of an alternator and Fig. 4.23*b* shows a complete power station layout.

Fig. 4.23a and b.

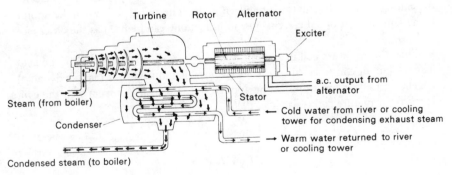

Electric motors (d.c. and a.c.)

Electric motors form the heart of a whole host of devices ranging from domestic appliances such as vacuum cleaners and washing machines to electric locomotives and lifts. In a car the windscreen wipers are usually driven by one and the engine is started by another. There are many types, most of which work off a.c.; the principles involved in a few cases will be outlined.

(a) *d.c. motors*. Basically a d.c. motor consists of a coil on an axle, carrying a current in a magnetic field. The coil experiences a couple as in a moving coil galvanometer (see pp. 95 and 97) which makes it rotate. When its plane is perpendicular to the field, a *split-ring commutator* reverses the current in the coil and, as Fig. 4.24 shows, ensures that the couple continues to act in the same direction thereby maintaining the rotation.

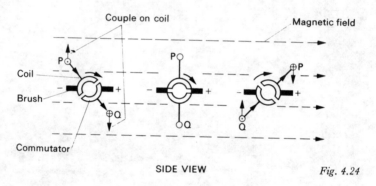

SIDE VIEW

Fig. 4.24

In practice several coils are wound in equally spaced slots in a laminated soft iron cylinder (the *armature* or *rotor*) and are connected to a commutator with many segments. Greater and steadier torque is thus obtained. Electromagnets with concave pole pieces frequently provide the magnetic field but the use of modern permanent magnets (e.g. of Magnadur) for this is increasing, especially in small motors. When electromagnets are used the field coils may be in series (a series-wound motor) or in parallel (a shunt-wound motor) with the armature, depending on what the motor is required to do. The construction of a d.c. motor is the same as that of a d.c. dynamo and in fact one can be used as the other (see p. 112).

When the armature coil in a motor rotates it cuts the magnetic flux of the field magnet and an e.m.f. $\mathscr{E}$, called the *back e.m.f.*, is induced in it (as in a dynamo) which, by Lenz's law, opposes the applied p.d. V causing current I in the coil. If r is the armature coil resistance, then

$$V - \mathscr{E} = Ir.$$

Multiplying by I we get

$$VI = \mathscr{E}I + I^2r.$$

126

Now VI is the power supplied to the motor and I^2r is the power dissipated as heat in the armature coil. The difference, $\mathscr{E}I$, must be the mechanical power output of the motor; it is also the rate of working against the induced e.m.f.

The armature resistance r of a d.c. motor is small (e.g. 1 Ω or less) to make I^2r small and give high efficiency. However when the motor is started, the armature is at rest and the back e.m.f. $\mathscr{E}$ is zero. The armature current I then equals V/r and would be so large as to burn out the armature coils. This is prevented by connecting a 'starting' resistance in series with the motor and gradually reducing it as the motor speeds up. The back e.m.f. then limits the current and will normally only be slightly less than the applied p.d. V.

(b) a.c. motors. A d.c. motor may be used on a.c. if the armature and field coils are in series. The current then reverses simultaneously in each and rotation in the same direction continues. (The torque developed in a shunt-wound motor on a.c. is very small due to inductive effects causing the armature and field currents to reach their maxima at different times, p. 168.)

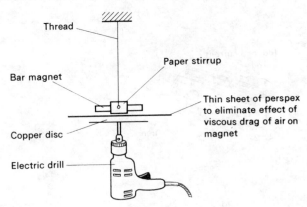

Fig. 4.25

The *induction motor* is widely used in industry and is the commonest type of a.c. motor. Its action depends on the fact that a moving magnetic field can set a neighbouring conductor into motion. The converse is also true, namely that a moving conductor can cause a magnetic field to move and is readily demonstrated with the arrangement in Fig. 4.25 in which the bar magnet starts spinning when the copper disc is rotated rapidly. Whether it be the conductor or the field that moves, eddy currents are induced in the conductor and these try to reduce the effect causing them, i.e. the relative motion between conductor and field. If the conductor is stationary it starts moving in the same direction as the field and tries to catch it up in an (unsuccessful) attempt to eliminate the relative motion between them.

Moving magnetic fields are produced in various ways in actual induction motors. In large rotary machines three pairs of stationary electromagnets (the stator) are arranged at equal angles round a conductor (the rotor) and each pair is

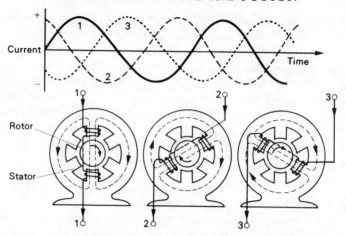

Note Only the stator winding carrying maximum current is shown in each case; the other windings then carry smaller currents

(a)

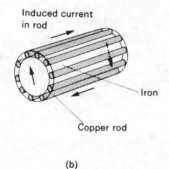

(b)

Fig. 4.26

connected to one phase of a 3-phase a.c. supply, Fig. 4.26*a*; the graph shows how the phases reach their maximum values one after the other. The rotor is generally of the squirrel-cage pattern comprising a number of copper rods in an iron cylinder, Fig. 4.26*b*, and it 'interprets' the alternations of the magnetic field as a field sweeping round it, i.e. a rotating field. The eddy currents induced in the copper rods set it into rotation as explained above. Linear induction motors operate on the same principle except that the field travels in a straight line; such motors are the subject of much research at present.

The *shaded-pole* induction motor, used in record players, produces a 'rotating' magnetic field by covering part of the pole of an electromagnet carrying a.c. with a thick conducting plate. The alternating field of the electromagnet induces eddy

128

currents in the plate and these create another field, adjacent to the main electro-magnet field. There is a phase difference between the two fields and a nearby metal disc regards this as a moving field and responds by rotating. A model shaded-pole motor is shown in Fig. 4.27. (*Note.* The small phase difference between the fields may be shown by holding two search coils over the ' shaded ' and ' unshaded ' parts of the pole and examining the traces on a double beam CRO.)

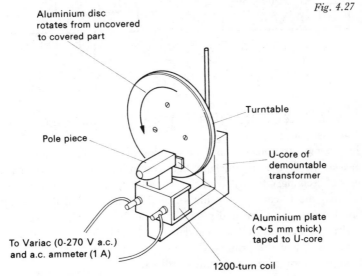

Fig. 4.27

Aluminium disc rotates from uncovered to covered part

Pole piece

Turntable

U-core of demountable transformer

Aluminium plate (∼5 mm thick) taped to U-core

To Variac (0-270 V a.c.) and a.c. ammeter (1 A)

1200-turn coil

Transformers

(*a*) *Action.* A transformer changes, i.e. transforms, an alternating p.d. from one value to another of greater or smaller value using the mutual induction principle (p. 107).

Two coils, called the *primary* and *secondary* windings, which are not connected to one another in any way, are wound on a complete soft iron core, either one on top of the other as in Fig. 4.28*a* or on separate limbs of the core as in Fig. 4.28*b*. When an alternating p.d. is applied to the primary, the resulting current produces a large alternating magnetic flux which links the secondary and induces an e.m.f. in it. The value of this e.m.f. depends on the number of turns on the secondary and we will show shortly that under certain conditions it is approximately true to say

$$\frac{\text{e.m.f. induced in secondary}}{\text{p.d. applied to primary}} = \frac{\text{secondary turns}}{\text{primary turns}}.$$

A ' step-up ' transformer has more turns on the secondary than the primary and the e.m.f. induced in the secondary is greater than the p.d. applied to the primary, e.g. if the turns are stepped-up in the ratio 1:2, the secondary e.m.f.

129

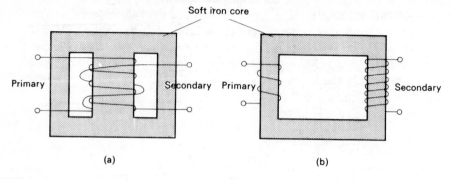

Soft iron core

Primary   Secondary   Primary   Secondary

(a)   (b)

Fig. 4.28

will be about twice the primary p.d. In a ' step-down ' transformer the secondary e.m.f. is less than the primary p.d.

Three demonstrations to show the working of a transformer are illustrated in Fig. 4.29*a*, *b*, *c*. In the first the lamp lights up to full brightness when a sufficient number of secondary turns have been wound on. The effect of placing the iron yoke across the U-core can be investigated. In the second demonstration the relation between the turns and p.d. ratios may be studied, a CRO being used as a voltmeter to measure the secondary e.m.f. The third demonstration shows how the current in the primary depends on that in the secondary, the latter being increased by connecting lamps across it.

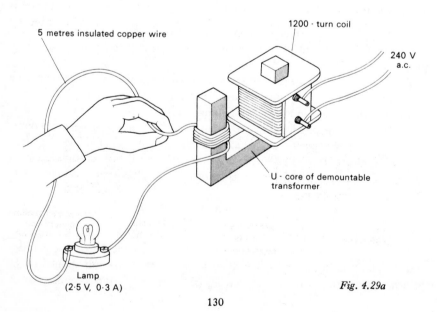

5 metres insulated copper wire

1200 · turn coil

240 V
a.c.

U · core of demountable
transformer

Lamp
(2·5 V, 0·3 A)

Fig. 4.29a

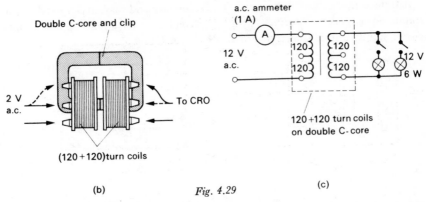

Double C-core and clip

2 V
a.c.

To CRO

(120 + 120)turn coils

a.c. ammeter
(1 A)

12 V
a.c.

120 · 120
120 · 120

12 V
6 W

120 + 120 turn coils
on double C-core

(b)   *Fig. 4.29*   (c)

If the p.d. is stepped up by a transformer, the current is stepped down, roughly in the same ratio. This follows if conservation of energy is assumed, because, taking the transformer to be 100 per cent efficient (many approach this), if all the electrical energy supplied to the primary appears in the secondary, then

$$\text{power in primary} = \text{power in secondary}$$

that is,

$$\text{primary p.d.} \times \text{primary current} = \text{secondary e.m.f.} \times \text{secondary current}$$

or,

$$\frac{\text{secondary current}}{\text{primary current}} = \frac{\text{primary p.d.}}{\text{secondary e.m.f.}}$$

The stepping up of current can be demonstrated effectively using the apparatus of Figs. 4.30a and b. In the first the iron nail melts spectacularly and in the second the water boils very quickly.

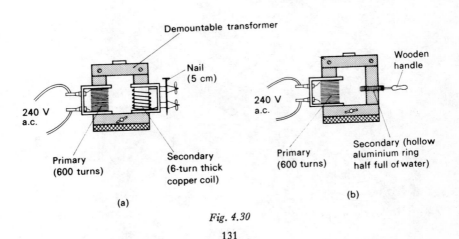

Demountable transformer

Nail
(5 cm)

Wooden
handle

240 V
a.c.

240 V
a.c.

Primary
(600 turns)

Secondary
(6-turn thick
copper coil)

Primary
(600 turns)

Secondary (hollow
aluminium ring
half full of water)

(a)   (b)

Fig. 4.30

A transformer can have more than one secondary and may step up and down simultaneously. A transformer (with two secondaries) for a mains-operated h.t. power supply unit is shown diagrammatically in Fig. 4.31.

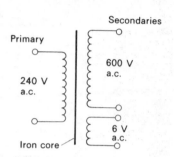

Fig. 4.31

(b) *Energy losses.* Although transformers are very efficient devices, small energy losses do occur in them due to four main causes:

(i) *Resistance of windings.* The copper wire used for the windings has resistance and so ordinary (I^2R) heat losses occur. In high-current, low-p.d. windings these are minimized by using thick wire.

Fig. 4.32a

Fig. 4.32b

(*ii*) *Eddy currents*. The alternating magnetic flux induces eddy currents in the iron core and causes heating. The effect is reduced by having a laminated core (see p. 117).

(*iii*) *Hysteresis*. The magnetization of the core is repeatedly reversed by the alternating magnetic field. The resulting expenditure of energy in the core appears as heat and is kept to a minimum by using a magnetic material (such as mumetal) which has a low hysteresis loss (p. 149).

(*iv*) *Flux leakage*. The flux due to the primary may not all link the secondary if the core is badly designed or has air gaps in it.

Very large transformers like those in Fig. 4.32a and b (400 kV and 11 kV respectively) have to be oil-cooled to prevent overheating.

(*c*) *Theory*. The complete theory is complex and before we tackle even a simple version of it, consideration of the following will be helpful.

The primary winding of a mains transformer is found to have a resistance of 10 Ω, on a 240-V supply; the primary current should therefore be (by Ohm's law) $240/10 = 24$ A. An a.c. ammeter connected in the primary records 0.10 A. The difference is very large and is due to the fact that the alternating flux (arising from the a.c. in the primary), which induces an e.m.f. in the secondary,

133

induces an e.m.f. called a *back e.m.f.* in the primary as well. (The primary is said to have *self-inductance*, see p. 138.) This e.m.f. opposes the applied p.d. and is nearly but not quite equal to it at every instant. The *net* e.m.f. in the primary is therefore quite small. We can now proceed to an approximate theoretical treatment.

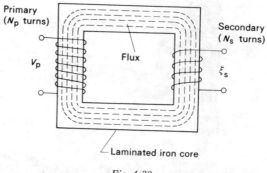

Fig. 4.33

Consider an ideal transformer in which the primary has negligible resistance and all the flux in the core links both primary and secondary windings, Fig. 4.33. If Φ is the flux in the core at time t due to the current in the primary when a p.d. V_p is applied to it, then the back e.m.f. $\mathscr{E}_p$ induced in the primary of N_p turns (due to its self-inductance) is given by

$$\mathscr{E}_p = \frac{d}{dt}(N_p\,\Phi) = N_p \cdot \frac{d\Phi}{dt}.$$

But, $\qquad\qquad \mathscr{E}_p = V_p.$

If this were not so, the primary current would be infinite since the primary has zero resistance. Hence

$$V_p = N_p \cdot \frac{d\Phi}{dt}. \tag{1}$$

The e.m.f. $\mathscr{E}_s$ induced in the secondary (N_s turns) by the same flux in the core is

$$\mathscr{E}_s = \frac{d}{dt}(N_s\Phi) = N_s \cdot \frac{d\Phi}{dt}.$$

If the secondary is on open circuit or the current taken from it is small then, to a good approximation,

$$\mathscr{E}_s = V_s$$

where V_s is the p.d. across the secondary. Thus

$$V_s = N_s \cdot \frac{d\Phi}{dt}. \qquad (2)$$

From (1) and (2),

$$\frac{V_s}{V_p} = \frac{N_s}{N_p}.$$

This expression would be roughly true for an actual transformer if (*i*) the primary resistance and current were small, (*ii*) very little flux escaped from the core, and (*iii*) the secondary current was small.

When a greater load (i.e. a smaller resistance) is connected to the secondary it causes the secondary current to increase and this acts to reduce the flux in the core (since the secondary current opposes the change producing it). The back e.m.f. in the primary therefore falls and so the primary current increases. Eventually the flux is restored to its previous value and as a result the back e.m.f. rises and becomes nearly equal to the applied p.d. The net effect is an increase of primary current, i.e. more energy is drawn from the source connected to the primary.

Transmission of electrical power

(*a*) *Grid system.* The Grid system in Britain is a network of cables, most of it supported on pylons, which connects over 200 power stations, situated at convenient places throughout the country and carrying electrical energy from them to consumers.

In the largest modern power stations electricity is generated at about 25 kV (50 Hz) and stepped up in a transformer to 275 kV or 400 kV for transmission over long distances. The p.d. is subsequently reduced in sub-stations by other transformers for distribution to local users at suitable p.d.s—33 kV for heavy industry, 11 kV for light industry and 240 V for homes, schools, shops, farms etc., Fig. 4.34. In the latest rail electrification schemes working at 25 kV, there are special sub-stations alongside the track taking their supply from the Grid system. This is fed to step-down transformers in the electric locomotive and then rectified (see p. 185) for driving d.c. traction motors operating at about 900 V and 650 A.

To supervise the operation of the power stations, England and Wales are divided into eight areas, each with a Grid Control Centre. At these, engineers assess the demand, direct the flow and reroute it when breakdowns occur. In this way the electricity supply is made more reliable, requires less reserve plant to

Fig. 4.34

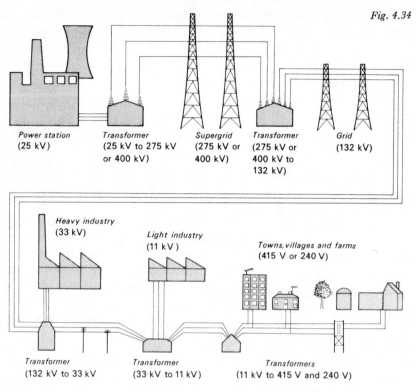

| Power station (25 kV) | Transformer (25 kV to 275 kV or 400 kV) | Supergrid (275 kV or 400 kV) | Transformer (275 kV or 400 kV to 132 kV) | Grid (132 kV) |

Heavy industry (33 kV)

Light industry (11 kV)

Towns, villages and farms (415 V or 240 V)

| Transformer (132 kV to 33 kV | Transformer (33 kV to 11 kV) | Transformers (11 kV to 415 V and 240 V) |

Fig. 4.35

cover maintenance etc. and cuts costs by enabling smaller, less efficient power stations to be shut down at off-peak periods. All eight Grid Control Centres are in direct communication with the National Control Centre in London, Fig. 4.35.

(*b*) *Why high p.d.s are used.* Suppose electrical power P has to be delivered at a p.d. V by supply lines of total resistance R, Fig. 4.36. The current $I = P/V$ (since $P = IV$) and the power loss in the lines $= I^2R = (P/V)^2R$. Clearly, the greater V the smaller is the loss—in fact, doubling V quarters the loss. Electrical power is thus transmitted more economically at high p.d.s but on the other hand they create insulation problems and raise installation costs. In the 400-kV Supergrid, currents of 2500 A are typical and the power loss is about 200 kW per kilometre of cable, i.e. a 0.02 per cent loss per kilometre.

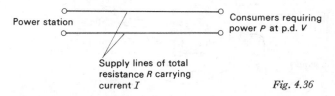

Power station

Consumers requiring power P at p.d. V

Supply lines of total resistance R carrying current I

Fig. 4.36

The ease and efficiency with which alternating p.d.s are stepped up and down in a transformer and the fact that alternators produce much higher p.d.s than d.c. generators (25 kV compared with several thousand volts) are the main considerations influencing the use of high alternating, rather than direct, p.d.s in most situations. An exception to this is the cross-channel link between England and France where the underground cables favour a d.c. supply because of the high dielectric losses with a.c. in such cables.

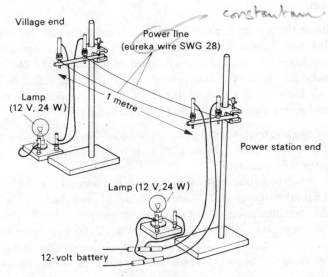

Village end

constantan

Power line (eureka wire SWG 28)

Lamp (12 V, 24 W)

1 metre

Power station end

Lamp (12 V, 24 W)

12-volt battery

Fig. 4.37a

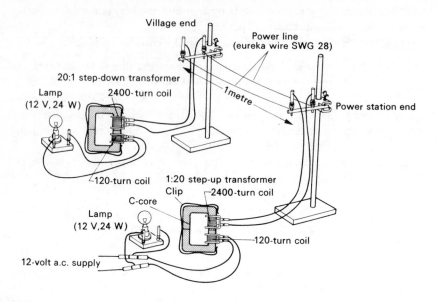

Fig. 4.37b

The advantages of ' high ' alternating p.d. power transmission may be shown with the model power line arrangements of Figs. 4.37a and b.

Self-inductance

The flux due to the current in a coil links that coil and if the current changes the resulting flux change induces an e.m.f. in the coil itself. This changing magnetic field type of electromagnetic induction is called *self-induction,* the coil is said to have *self-inductance* or simply *inductance* (symbol L) and is called an *inductor* (symbol ∿ if air-cored and ∿̄ if it has a core of magnetic material). The induced e.m.f. obeys Faraday's law like other induced e.m.f.s.

(*a*) *Some demonstrations.* From Lenz's law we would expect the induced e.m.f. to oppose the current change causing it. That it does so in both d.c. and a.c. circuits may be demonstrated.

In Fig. 4.38a, L is an iron-cored inductor and R is a variable resistor adjusted to have the same resistance as L. When the current is switched on, the lamp in series with L lights up a second or two *after* that in series with R. This can be attributed to the induced e.m.f. in L opposing the change and trying to drive a current against the increasing current due to the battery. The growth of the current to its steady value (when no self-induction occurs) is thus delayed.

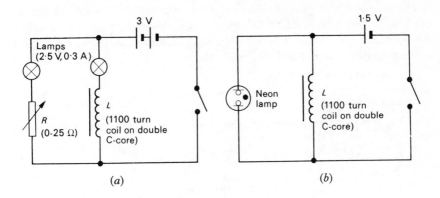

Fig. 4.38

The effect of the induced e.m.f. when the current is switched off is more striking and is shown with the circuit of Fig. 4.38b. Opening the switch causes the current to fall very rapidly to zero and the rate of change of flux is large. The induced e.m.f. is therefore large and in a direction which tries to maintain the current in its original direction. It is sufficiently great in this case (more than 100 V) to produce a brief flash of the neon lamp across L. When circuits carrying large currents in large inductors are switched off, the induced e.m.f. can cause sparking between the switch contacts and may even fuse them together. In the Grid network special circuit breakers are used.

Inductors are important components of a.c. circuits where, as we shall see later, the induced e.m.f. opposes the applied p.d. continuously. In Fig. 4.38a if the 3-V d.c. supply is replaced by a 3-V a.c. supply, the lamp in series with L does not light—unless the iron core is removed.

(b) *Definition and unit.* It would seem reasonable to say that a coil (or circuit) has a large inductance if a small rate of change of current in it induces a large back e.m.f. This is the basis of the following definition of inductance. If the e.m.f. induced in a coil is $\mathscr{E}$ when the rate of change of current in it is dI/dt, the inductance L of the coil is defined by the equation

$$L = - \frac{\mathscr{E}}{dI/dt}.$$

The negative sign is inserted to make L a positive quantity since $\mathscr{E}$ and dI/dt act in opposite directions and are given opposite signs.

The unit of inductance is the *henry* (H), defined as *the inductance of a coil (or circuit) in which an e.m.f. of 1 volt is induced when the current changes at the rate of 1 ampere per second.* That is, $1 \text{ H} = 1 \text{ Vs A}^{-1}$.

139

(c) *Inductance of a solenoid.* Calculation of inductance is possible in certain cases, as it was of capacitance. Consider a long, air-cored solenoid of length l, cross-sectional area A having N turns and carrying current I. The flux density B is almost constant over A and, neglecting the ends, is given by

$$B = \mu_0 \frac{N}{l} I. \qquad \text{(see p. 82)}$$

The flux Φ through each turn of the solenoid is BA and for the flux-linkage we have

$$N\Phi = BAN$$

$$= \left(\mu_0 \frac{N}{l} I \right) AN$$

$$= \frac{\mu_0 A N^2}{l} . I.$$

If the current changes by dI in time dt causing a flux-linkage change $d(N\Phi)$ then by Faraday's law the induced e.m.f. $\mathscr{E}$ is

$$\mathscr{E} = -\frac{d}{dt}(N\Phi)$$

$$= -\frac{\mu_0 A N^2}{l} . \frac{dI}{dt}.$$

If L is the inductance of the solenoid, then from the defining equation we get

$$\mathscr{E} = -L\frac{dI}{dt}.$$

Comparing these two expressions it follows that

$$L = \frac{\mu_0 A N^2}{l}.$$

L depends only on the geometry of the solenoid. If $N = 400$ turns, $l = 25$ cm $= 25 \times 10^{-2}$ m, $A = 50$ cm^2 $= 50 \times 10^{-4}$ m^2 and $\mu_0 = 4\pi \times 10^{-7}$ H m^{-1}, then $L = 4.0 \times 10^{-3}$ H $= 4.0$ mH.

A solenoid having a core of a magnetic material would have a much greater inductance but the value would vary depending on the current in the solenoid.

(d) *Energy stored by an inductor.* A current-carrying inductor stores energy in the magnetic field associated with it and it can be shown that for current I in an inductance L this equals $\frac{1}{2}LI^2$. Compare this with the analogous case of $\frac{1}{2}Q^2/C$ for the energy stored in the electric field of a capacitor.

Since every current produces a field, every circuit must have some self-inductance. On switching on any circuit some time is necessary to provide the energy in the magnetic field and so no current can be brought instantaneously to a non-zero value. Similarly on switching off any circuit, the energy of the magnetic field must be dissipated somehow, hence the spark. A capacitor across the switch can ' suppress ' sparking.

Unit of μ_0

Reference was made to this on p. 81 and it is convenient to consider it now. The permeability of free space μ_0 was defined by the Biot-Savart law

$$\delta B = \frac{\mu_0 I \, \delta l \sin \theta}{4\pi r^2}.$$

The unit of μ_0 from this equation is

$$\frac{(\text{Wb m}^{-2}) \times (\text{m}^2)}{(\text{A}) \times (\text{m})} \text{ or Wb A}^{-1} \text{ m}^{-1}. \tag{1}$$

From Faraday's law, $\mathscr{E} = \dfrac{d}{dt}(N\Phi)$, we can say that

$$1 \text{ Wb} = 1 \text{ V s}. \tag{2}$$

Also, from the inductance defining equation $L = \mathscr{E}/(dI/dt)$ we have

$$1 \text{ H} = 1 \text{ V s A}^{-1}. \tag{3}$$

From (2) and (3)

$$1 \text{ H} = 1 \text{ Wb A}^{-1}.$$

Hence from (1), μ_0 can be expressed in

$$\text{H m}^{-1} \text{ (henry per metre)}.$$

This is the SI unit of μ_0 (and μ); it may be compared with F m^{-1} (farad per metre) for the unit of ϵ_0 (and ϵ), the permittivity of free space.

Mutual inductance

(a) *Definition and unit.* In mutual induction, current changing in one coil or circuit (the primary) can induce an e.m.f. in a neighbouring coil or circuit (the secondary), as we saw earlier. The *mutual inductance, M*, of two coils, Fig. 4.39, is defined by the equation

$$M = -\frac{\mathscr{E}}{dI_\text{p}/dt}$$

where $\mathscr{E}$ is the e.m.f. induced in the secondary when the rate of change of current in the primary is dI_p/dt.

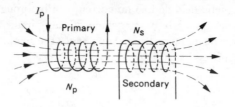

Fig. 4.39

It follows from the definition that M has the same unit as L, i.e. henry (H). *Two coils are said to have a mutual inductance of 1 henry if an e.m.f. of 1 volt is induced in the secondary when the primary current changes at the rate of 1 ampere per second.*

It can be shown that the mutual inductance of two coils is the same if current flows in the secondary and flux links the primary causing an induced e.m.f. when a flux-linkage change occurs.

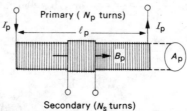

Fig. 4.40

(*b*) *Mutual inductance of a solenoid and a coil.* In Fig. 4.40 the long air-cored solenoid (the primary) with N_p turns, length l_p and cross-sectional area A_p carries current I_p. The flux density B_p in the centre of the solenoid is nearly constant over A_p and is given by

$$B_p = \mu_0 \frac{N_p}{l_p} I_p . \qquad \text{(see p 82)}$$

The flux Φ_s linking each of the N_s turns of the short coil (the secondary) round the middle of the solenoid is $B_p A_p$. The flux-linkage of the short coil is therefore

$$N_s \Phi_s = B_p A_p N_s$$

$$= \left(\mu_0 \frac{N_p}{l_p} I_p \right) A_p N_s.$$

If the current in the primary changes by dI_p in time dt causing a flux-linkage change $d(N_s \Phi_s)$ in the secondary, the e.m.f. $\mathscr{E}_s$ induced in the secondary is

$$\mathscr{E}_s = - \frac{d}{dt} (N_s \Phi_s)$$

$$= - \frac{\mu_0 A_p N_p N_s}{l_p} \cdot \frac{dI_p}{dt} .$$

ELECTROMAGNETIC INDUCTION

If M is the mutual inductance, we can also say

$$\mathscr{E}_s = - M \frac{dI_p}{dt}$$

Hence
$$M = \frac{\mu_0 A_p N_p N_s}{I_p}.$$

When the coils have a ferromagnetic core, the value of M is very much greater especially if the core is complete as in a transformer, but it does vary with the current.

Induced charge and flux change

When the flux linking or cutting a *complete* circuit changes, an e.m.f. is induced in it and current flows. A simple connection exists between the total charge circulation that constitutes the current and the flux change.

Consider a coil of N turns in a circuit of total resistance R in which the flux linking each turn is changing and has value Φ at time t. The induced e.m.f. at t will be (in magnitude)

$$\mathscr{E} = \frac{d}{dt}(N\Phi).$$

Also, the induced current I at time t will be

$$I = \frac{\mathscr{E}}{R} = \frac{1}{R} \cdot \frac{d}{dt}(N\Phi).$$

Now I = rate of flow of charge = dQ/dt

$$\therefore \frac{dQ}{dt} = \frac{1}{R}\frac{d}{dt}(N\Phi) = \frac{N}{R}\frac{d\Phi}{dt}.$$

If the flux changes from say Φ_1 to Φ_2 the total charge Q that passes is

$$Q = \int_0^Q dQ = \frac{N}{R}\int_{\phi_1}^{\phi_2} d\Phi$$

$$\therefore Q = \frac{N(\Phi_2 - \Phi_1)}{R}$$

$$= \frac{\text{flux-linkage change}}{R}.$$

We see that Q does not depend on the time taken by the flux change.

Consider a numerical example. A search coil of average cross-sectional area 3.0 cm² has 400 turns and is in a circuit of total resistance 200 Ω. It is inserted into a magnetic field of flux density 2.5×10^{-3} T so as to produce the maximum

143

flux change. We have, $N = 400$ turns, $A = 3.0 \times 10^{-4}$ m², $R = 200 \ \Omega$, $B = 2.5 \times 10^{-3}$ T. Hence $\Phi_2 = BA = 2.5 \times 10^{-3} \times 3.0 \times 10^{-4}$ Wb and $\Phi_1 = 0$. The induced charge Q is given by

$$Q = \frac{N(\Phi_2 - \Phi_1)}{R}$$

$$= \frac{400 \times 2.5 \times 10^{-3} \times 3.0 \times 10^{-4}}{200} \ C$$

$$= 1.5 \times 10^{-6} \ C = 1.5 \ \mu C.$$

Measuring B by ballistic galvanometer and search coil

(a) *Ballistic galvanometer.* A moving coil galvanometer will measure charge if (i) the period of oscillation of the movement is large (e.g. 2 seconds) so that all the charge passes through the coil before it moves appreciably, and (ii) the damping is very small. It is then called a ballistic galvanometer (since it is set into motion by an impulse, as is a projectile whose motion is under study in ballistics) and theory shows that the first deflection or 'throw' θ is proportional to the total charge Q that has passed. Hence

$$\theta \propto Q$$

or

$$\theta = bQ$$

where b is a constant called the *charge sensitivity* of the galvanometer. It is expressed in *mm per μC* and must either be known (from information supplied by the manufacturer) or found by a calibration experiment (see later).

Generally only light-beam galvanometers are suitable for ballistic use. Damping due to the air and the suspension are negligible and on 'direct' setting there are no internal shunts across the coil which, in modern instruments like that shown in Fig. 3.35c, is not wound on any kind of former. Electromagnetic damping is therefore due only to eddy currents in the coil. These depend solely on the resistance of the external circuit which should consequently be high enough to allow the coil to swing to and fro freely.

(b) *Measuring B.* A search coil in series with a ballistic galvanometer and a suitable high resistance (to reduce damping and adjust the sensitivity of the galvanometer) is placed in the magnetic field to be measured so that the flux links it normally, Fig. 4.41. It is then *quickly* removed (Why?) from the field and the first 'throw' θ produced by the flux change is noted. The charge Q driven through the coil (e.g. 1–2 μC) is proportional to θ.

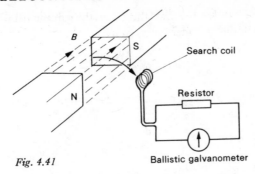

Fig. 4.41

Ballistic galvanometer

If B is the flux density of the field and A is the cross-sectional area of the coil which has N turns, then

$$\text{flux-linkage change} = NAB$$

$$\therefore Q = \frac{NAB}{R}$$

where R is the *total* resistance of the circuit. Hence

$$B = \frac{RQ}{NA}.$$

R, N and A can be measured or are given. Q can be found from the charge sensitivity of the galvanometer—known or determined from (c)—and hence B calculated.

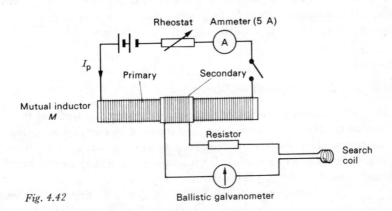

Fig. 4.42

Ballistic galvanometer

(c) *Calibrating a ballistic galvanometer.* A known current I_p is passed through the primary of a mutual inductance M in the circuit of Fig. 4.42. Let flux Φ_p link the secondary. I_p is switched off causing a flux-linkage change of $N_s\Phi_p$ with the

145

secondary (N_s turns). Let the first 'throw' on the galvanometer be θ. The charge Q driven through the secondary is given by

$$Q = \frac{N_s \Phi_p}{R}$$

where R is the *total* resistance of the secondary circuit. If $\mathscr{E}_s$ is the e.m.f. induced in the secondary, then

$$\mathscr{E}_s = -\frac{d}{dt}(N_s \Phi_p) = -N_s \frac{d\Phi_p}{dt}.$$

Also,
$$\mathscr{E}_s = -M \frac{dI_p}{dt}.$$

Therefore,
$$N_s d\Phi_p = M dI_p.$$

But here the change of current is I_p and the flux linkage change is $N_s \Phi_p$, and so

$$N_s \Phi_p = M I_p$$

$$\therefore \quad Q = \frac{M I_p}{R}.$$

M can be calculated from

$$M = \frac{\mu_0 A_p N_p N_s}{l_p}$$

where the symbols have their previous meanings (p. 142). Hence Q and b (the charge sensitivity $= \theta/Q$) follow.

The search coil and the secondary of the mutual inductor are *both* in circuit during the calibration of the ballistic galvanometer; if they are so when the measurement of B is made, the damping is the same in both cases.

Absolute measurement of resistance

In an absolute method an electrical quantity is measured in terms of the basic mechanical quantities, i.e. mass, length and time and no electrical measurements are necessary. The absolute measurement of current was considered earlier (p. 93), the principle of a method for resistance, due to Lorenz and based on Faraday's disc dynamo (p. 119) is shown in Fig. 4.43.

A metal disc is rotated with its plane at right angles to the uniform flux density B at the centre of a long current-carrying solenoid having n turns per metre. The e.m.f. $\mathscr{E}$ induced between the centre and rim of the disc for the radius joining the sliding contacts is balanced against the p.d. across part of a low resistance R (a copper rod) which carries the same current I as flows through the solenoid. R is to be measured.

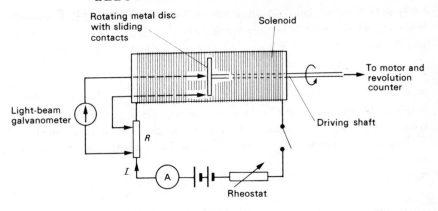

Fig. 4.43

If the disc makes f revolutions per second at balance and has radius r, the area swept out per second by a radius is $\pi r^2 f$. The flux cut per second is $B\pi r^2 f$, hence when there is no deflection on the light-beam galvanometer,

$$\mathscr{E} = IR = B\pi r^2 f$$

$$\therefore R = \frac{B\pi r^2 f}{I}.$$

But at the centre of a long solenoid

$$B = \mu_0 nI \qquad \text{(p. 82)}$$

$$\therefore R = \mu_0 n\pi r^2 f.$$

R can thus be calculated if n, r and f are measured.

This method is used to measure the resistance of coils kept as standards in laboratories such as the NPL, but various modifications and precautions are necessary. Thermoelectric e.m.f.s, comparable with the small induced e.m.f., arise at the sliding contacts due to frictional heating and they must be allowed for. The earth's magnetic field has to be taken into account and allowance also made for the field inside the solenoid not being perfectly uniform over the disc.

Ferromagnetic materials

Iron, cobalt, nickel and substances containing them, are strongly magnetic and are called *ferromagnetic* materials. Many other materials exhibit large magnetic effects at very low temperatures.

147

ELECTROMAGNETIC INDUCTION

(a) *Relative permeability*. The flux density in a coil increases many times when it has a ferromagnetic core because the core becomes magnetized and contributes flux. The relative permeability μ_r of a material is defined by the equation

$$\mu_r = \frac{B}{B_0}$$

where B_0 is the flux density in a current-carrying toroid (an end-less solenoid) containing air (strictly a vacuum) and B is the flux density when the same toroid is filled with the material, Fig. 4.44a. Since B and B_0 are both measured in teslas, μ_r has no units.

Fig. 4.44

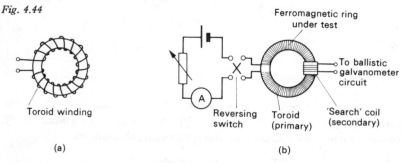

Toroid winding

(a)

Ferromagnetic ring under test

To ballistic galvanometer circuit

Reversing switch

Toroid (primary)

'Search' coil (secondary)

(b)

A toroid rather than a solenoid is specified so that the magnetization of the material is nearly uniform, if the difference between the external and internal radii is relatively small. A rod-shaped specimen in a solenoid would have poles at its ends which tend to demagnetize the rod near the ends (hence the use of keepers to store magnets and reduce self-demagnetization). Uniform magnetization, such as can be achieved all along the material under test when it forms a closed magnetic loop in a toroid, is impossible in a solenoid.

In a measurement of μ_r, B is found using a ' search ' coil, connected to a calibrated ballistic galvanometer and wound round part of the toroid, Fig. 4.44b. B_0 is calculated since it can be shown that it equals $\mu_0 n I$ (the same as for the middle of a long solenoid) where n is the number of turns per metre on the toroid and I is the current in it.

(b) *Magnetization curve*. If B is measured for a ferromagnetic material as the magnetizing current is increased from zero, a magnetization curve of B against B_0 (which is proportional to the current) can be obtained and has the form of Fig. 4.45a.

Along OP the magnetization is small and reversible, i.e. it returns to zero when the magnetizing field B_0 is removed. Between P and Q the magnetization increases rapidly as B_0 increases and is irreversible, i.e. the specimen remains magnetized when the field is reduced to zero. For values of B_0 beyond Q very little increase of B occurs and the specimen is said to be approaching full magnetization or ' saturation ' along QT.

148

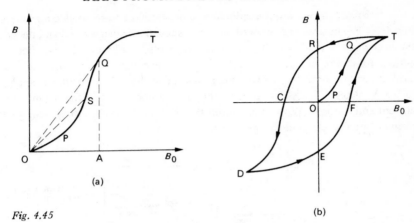

Fig. 4.45

(a)

(b)

The relative permeability μ_r ($= B/B_0$) at a point such as S is the gradient of the line joining O to S. Its value varies along the graph and is a maximum at Q. The value of B_0 at Q (i.e. OA) therefore gives the most efficient flux production and would be achieved by a correct choice of n and I for the magnetizing toroid. When values of μ_r are quoted (and they can be as high as 10^5) they usually refer to point Q on the magnetization curve.

(c) *Hysteresis loop.* When a specimen of a ferromagnetic material has reached saturation and the magnetizing field is reduced to zero, it remains quite strongly magnetized. The flux density B_r it retains is called the *remanence* or *retentivity* of the material, OR in Fig. 4.45b. A reverse magnetizing field is required to de-magnetize it completely and the value of B_0 which makes B zero is called the *coercive force* or *coercivity* of the material, OC in Fig. 4.45b. If the reverse field is increased more, the specimen becomes saturated in the reverse direction (D). Decreasing the field and again reversing to saturation in the first direction gives the rest of the loop, DEFT.

The curve in Fig. 4.45b is called a *hysteresis loop*. It shows that the magnetization of a material (i.e. B) lags behind the magnetizing field (i.e. B_0) when it is taken through a complete magnetization cycle, the effect is called *magnetic hysteresis*; the term *hysteresis* is derived from a Greek word meaning ' lagging behind '.

The shape of a hysteresis loop provides useful information to the designer of electrical equipment. For example, it can be shown that the area of the loop is proportional to the energy required to take unit volume of the material round one cycle of magnetization. This energy increases the internal energy of the specimen. It is called the *hysteresis loss* and is important when materials are subject to alter-nating fields which take them through many cycles of magnetization per second.

A hysteresis loop can be displayed on a CRO with an easily accessible tube from which any magnetic screen must be removed. The specimen (e.g. a strip of soft iron tinplate or a length of steel clockspring) is inserted in a magnetizing coil set at right angles to the oscilloscope tube and close to the deflecting plates, Fig. 4.46. When current flows in the coil the specimen is magnetized and deflects the electron beam in the Y-direction (Fleming's left-hand rule). The Y-deflection is thus a measure of the flux density B of the specimen. The magnetizing coil current also passes through a variable resistor, and the p.d. across this is applied to the X-plates. The X-deflection is therefore proportional to the magnetizing current and so to B_0. With an a.c. input the specimen is taken through complete magnetization cycles and the spot produces a hysteresis loop.

Fig. 4.46

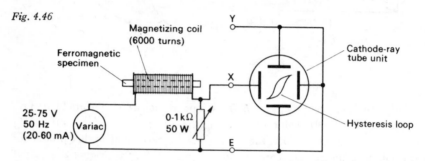

(*d*) *Demagnetization.* A simple but effective way of demagnetizing a magnetic material is to insert it in a multi-turn coil carrying a.c. and then either to reduce the current to zero or to withdraw the specimen from the coil. In both cases the material is taken through a series of ever-diminishing hysteresis loops.

Properties and uses of magnetic materials

Ferromagnetic materials can be classified into two broad groups—'soft' and 'hard'. Soft magnetic materials are easily magnetized and demagnetized, hard materials require large magnetizing fields and retain their magnetization.

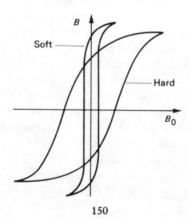

Fig. 4.47

Originally, the characteristics of the two groups were displayed by soft iron and hard steel but modern magnetic materials surpass these in performance. Typical hysteresis loops for soft and hard materials are shown in Fig. 4.47.

(*a*) *Permanent magnets* are made of hard magnetic materials with high remanence to give them 'strength' and high coercivity so that they are not easily demagnetized by stray magnetic fields or mechanical ill-treatment. Such materials are either (*i*) *alloys* containing, for example, iron, aluminium, nickel, copper and cobalt and having trade names like 'Ticonal', 'Alnico' and 'Alcomax', or (*ii*) *ceramics*, made by heat and pressure treatment from powders of iron oxide and barium oxide ($BaFe_{12}O_{19}$): they belong to the group of materials called *ferrites* and have hexagonal crystal structures; one is called 'Magnadur'.

Ceramic magnets are brittle, like china. The powder can be bonded with plastics and rubber to give flexible magnets of any shape. Very fine powder is used to coat tapes for tape recorders. In large fast computers many tiny ceramic ring magnets, each about 1 mm in diameter, Fig. 4.48*a*, are threaded together to act as memory stores. A close-up of a memory is shown in Fig. 4.48*b*.

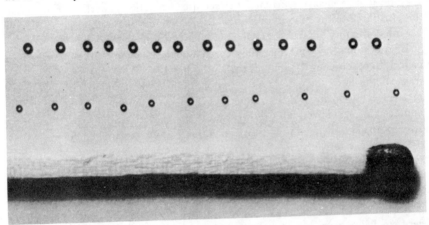

Fig. 4.48a and b

(b) *Electromagnets* require a core of soft magnetic material which will give a strong but temporary magnet. Small coercivity is essential.

(c) *Transformer cores* are subject to many cycles of magnetization and must be 'soft' with a narrow hysteresis loop to prevent heating from hysteresis loss. They should also have high resistivity to reduce eddy current loss (p. 116) and must never be saturated when in normal use or the flux will not follow the changes in primary current. Silicon iron (e.g. 'Stalloy') and mumetal are used at mains frequencies and ferrite materials with cubic crystal structures (e.g. 'Ferroxcube'—general formula MFe_2O_4, where M is a divalent atom of copper, zinc, magnesium, manganese or nickel) and very high resistivities are suitable for high frequency applications in, for example, radio work.

Domain theory of ferromagnetism

(a) *Electrons, atoms and domains.* The magnetic field produced by a magnet can, in general, also be produced by a current in a suitably-shaped conductor. This suggests that possibly all magnetic effects, including permanent magnetism, may be due to electric currents. In fact the magnetic properties of materials are attributed to the motion of electrons inside atoms and each electron may be regarded as a tiny 'current-carrying coil' having a magnetic field.

In the atoms of some materials the magnetic effects of different electrons cancel; in others they do not and each atom has a resultant magnetic field. With most of the latter materials, the vibratory motion of the atoms (due to their internal energy) causes their magnetic axes to have random orientations and no appreciable magnetization is shown by the material as a whole, even when we attempt to align them all in the same direction by a strong applied field.

However, in ferromagnetic materials each atom has a resultant field and there is a force (explicable in terms of quantum mechanics) which causes *neighbouring* atoms to react on one another so that all their magnetic axes are lined up in the same direction even when there is no external magnetizing field. They do this in groups of about 10^{10} atoms to form regions called *domains* which behave like very small but very strong permanent magnets, each roughly 10^{-3} mm wide. The directions of alignment of the magnetic axes vary from one domain to another and in an unmagnetized specimen they form closed magnetic loops, Fig. 4.49a, with the 'closure' domains acting like the keepers on a pair of bar magnets. The magnetic fields of the domains thus neutralize one another and no detectable external magnetic effect is produced.

(b) *Explanation of magnetization and hysteresis.* The domain theory offers the following account. When a small field is applied to an unmagnetized specimen those domains whose magnetic axes are most in line with the field grow at the expense of others and a movement of domain 'walls' results, Fig. 4.49b. This first stage of magnetization, OP in the magnetization curve of Fig.

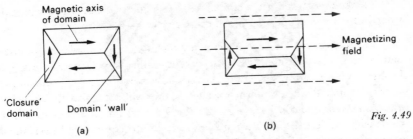

Magnetic axis of domain

Magnetizing field

'Closure' domain   Domain 'wall'

(a)   (b)

Fig. 4.49

4.50, is almost wholly reversible and if the applied field is removed the walls return to their previous positions and the magnetization is again zero.

Larger magnetizing fields cause the magnetic axes of entire domains to ' jump ' round quite suddenly in succession into alignment with the field and the magnetization increases sharply, PQ in Fig. 4.50. This stage is largely irreversible and the specimen retains its magnetization if the field is reduced to zero. When the field is great enough, more or less all domains are in line with the field and saturation occurs; QT on the curve. If a sufficiently strong reverse field is applied the domains can be re-aligned and saturation obtained in the opposite direction.

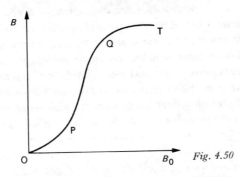

Fig. 4.50

Hysteresis is considered to be due to domain walls being unable to move across grain boundaries and other defects in the polycrystalline specimen until a reverse field of sufficient strength is applied. The magnetization thus lags behind the magnetizing field.

(c) *Evidence for domains.* Various effects support their existence. Two will be considered briefly:

(i) *Bitter patterns.* These were first obtained by Bitter in 1931 when he allowed very fine iron powder in a colloidal suspension to settle on a single ferromagnetic crystal with a smooth surface. At the domain walls there is slight leakage of magnetic flux, the powder collects there and a ' maze ' pattern like that in Fig. 4.51 is seen through a microscope, whether the specimen is magnetized or not.

153

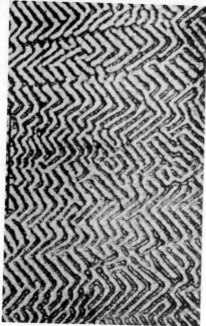

Fig. 4.51

(*ii*) *Barkhausen effect*. This may be readily demonstrated using the apparatus of Fig. 4.52. When, say, the north pole of the magnet is drawn across and a little above the end of the bundle of ferromagnetic wires, a rushing noise is produced in the loudspeaker due to induced e.m.f.s in the coil arising from the succession of 'jumps' made by the magnetic axes of domains during magnetization. There is no repetition of the effect on subsequent transits of the magnet unless the south pole of the magnet is nearest the wires. Neither do copper wires give any effect.

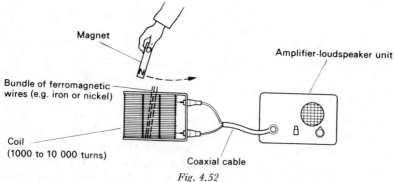

Magnet

Amplifier-loudspeaker unit

Bundle of ferromagnetic wires (e.g. iron or nickel)

Coil
(1000 to 10 000 turns)

Coaxial cable

Fig. 4.52

(*d*) *Curie temperature*. At a certain temperature, called the *Curie point*, a ferromagnetic material loses its ferromagnetic properties. This is attributed to the internal energy and vibration of the atoms becoming so vigorous as to destroy

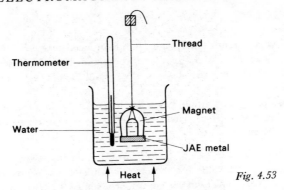

Fig. 4.53

the domain structure. The Curie point of iron is 770 °C, for the ferromagnetic alloy JAE metal (70 per cent Ni, 30 per cent Cu) it is about 70 °C. The latter can be used to show the effect, Fig. 4.53. The JAE metal drops off the magnet at the Curie point but becomes magnetic again below it.

ELECTROMAGNETIC INDUCTION

QUESTIONS

1. Under what circumstances is an e.m.f. induced in a conductor? What factors govern the magnitude and direction of the induced e.m.f.? Describe a quantitative experiment which demonstrates how the magnitude of the induced e.m.f. depends on one of these factors.

By considering any simple case, show that if the induced e.m.f. acted in the opposite direction to that in which it does act the law of conservation of energy would be contravened.

(S.)

2. Write down an expression for the e.m.f. induced between the ends of a rod of length l moving with velocity v so as to cut a flux density B normally.

A straight wire of length 50 cm and resistance 10 Ω moves sideways with a velocity of 15 m s^{-1} at right angles to a uniform magnetic field of flux density 2.0 × 10^{-3} T. What current would flow if its ends were connected by leads of negligible resistance?

3. State Lenz's law of electromagnetic induction and describe an experiment by which it may be demonstrated.

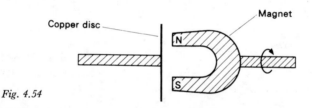

Copper disc

Magnet

N

S

Fig. 4.54

A circular disc of copper and a horseshoe magnet are mounted as shown in Fig. 4.54. The disc is free to rotate and the magnet can be rotated on the axle as shown. Describe and explain what happens when the magnet is set in rotation.

An aircraft is flying horizontally at 800 km h^{-1} at a point where the horizontal component of the earth's magnetic flux density is 2.0 × 10^{-5} T and the angle of dip is 60°. If the wing-span of the aircraft is 50 m calculate the potential difference in volts which is produced between the wing-tips of the aircraft.

(A.E.B. part qn.)

4. A copper disc of radius 10 cm is situated in a uniform field of magnetic flux density 1.0 × 10^{-2} T with its plane perpendicular to the field.

The disc is rotated about an axis through its centre parallel to the field at 3.0 × 10^{3} rev min^{-1}. Calculate the e.m.f. between the rim and centre of the disc.

Draw a circle to illustrate the disc. Show the direction of rotation as clockwise and consider the field directed into the plane of the diagram.

Explaining how you obtain your result, state the direction of the current flowing in a stationary wire whose ends touch the rim and centre of the disc. (J.M.B.)

5. State an expression for the e.m.f. induced in a conductor moving in a magnetic field and show in a diagram the directional relations involved.

ELECTROMAGNETIC INDUCTION

A rectangular coil 30.0 cm long and 20.0 cm wide has 25 turns. It rotates at the uniform rate of 3000 rev min⁻¹ about an axis parallel to its long side and at right angles to a uniform magnetic field of flux density 5.00×10^{-2} T. Find (a) the frequency, and (b) the peak value of the induced e.m.f. in the coil.

Describe with the aid of diagrams how you would arrange for the rotating coil to supply to an external circuit (i) direct current, and (ii) alternating current. (L.)

6. State the laws of electromagnetic induction and describe briefly experiments (one in each case) by which they may be demonstrated.

An electromagnet is in series with a 12-V battery and a switch across which is connected a 230-V neon lamp. Explain why, when the switch is closed, the neon remains unlit but, when the switch is opened, it flashes momentarily. Explain the importance of this observation in connection with large power switching.

A flat circular coil of 100 turns of mean radius 5.0 cm is lying on a horizontal surface and is turned over in 0.20 s. Calculate the mean e.m.f. induced if the vertical component of the earth's magnetic flux density is 4.0×10^{-5} T. (A.E.B.)

7. The e.m.f. generated by a simple single-coil a.c. dynamo may be represented by the equation $E = E' \sin \omega t$.

(i) State the meanings of, and give the units for, the symbols employed.

(ii) Draw diagrams showing the relative position of the coil and the magnetic field (a) when $t = 0$, and (b) when $E = E'$.

(iii) Discuss the factors which, in practice, determine the maximum current which may be generated by such a dynamo.

(iv) Deduce a formula for the torque on the coil at the moment when the maximum current is flowing. Assume that the coil is rectangular, and state the units of any new symbols you employ. (C.)

8. A designer has suggested the arrangement shown in the diagram for a heavy vehicle braking system to reduce fatigue by eliminating the need for large forces to be applied to the brake pedal.

Fig. 4.55

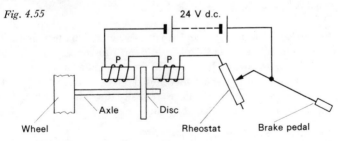

A disc is mounted near each road wheel, as shown in Fig. 4.55. When the driver depresses the brake pedal, the effect is gradually to increase the current in the circuit and hence also the magnetic flux in the region of the disc.

Explain how braking is achieved and discuss its effectiveness: (a) at motorway speeds; (b) in city centre traffic conditions; and (c) when the vehicle is stationary.

157

ELECTROMAGNETIC INDUCTION

Select from the materials listed below the one you would use for each of the following, giving a reason for your choice in each case: (i) the disc, and (ii) the formers P on which the coils are wound.

Materials: copper, steel, cast iron, wood, rigid polythene.　　(*J.M.B. Eng. Sc.*)

9. Draw a simple diagram to illustrate the essential electrical connections of a shunt-wound d.c. motor.

The armature resistance of such a motor is 0.75 Ω and it runs from a 240–V d.c. supply. When the motor is running freely, i.e. under no applied load, the current in the armature is 4.0 A and the motor makes 400 revolutions per minute. What is the value of the back e.m.f. produced in the motor and what is the rate of working?

When a load is placed on the motor the armature current increases to 60 A. What is now the back e.m.f., the rate of working, and the speed of rotation? It may be assumed that the field current remains constant. (*Hint.* For a shunt-wound motor, torque ∝ armature current since field current and so flux-leakage is constant.)

(*W.*)

10. Describe the construction of a simple form of alternating current transformer.

If the secondary coil is on open circuit explain, without calculation, the effect on the current flowing in the primary of (a) a fall in the supply frequency, and (b) a reduction in the number of primary turns.

Calculate the current which flows in a resistance of 3 Ω connected to a secondary coil of 60 turns if the primary has 1200 turns and is connected to a 240-V a.c. supply, assuming that all the magnetic flux in the primary passes through the secondary and that there are no other losses. (*O. and C.*)

11. By describing a suitable experiment using direct current explain what is meant by *self inductance*.

The current in a coil of inductance 0.10 H rises from zero to its maximum value at a mean rate of 2.0 A s^{-1}. Estimate the mean magnitude of the self-induced e.m.f. and state, giving your reason, its direction. (*J.M.B.*)

12. A choke of large self inductance and small resistance, a battery and a switch are connected in series. Sketch and explain a graph illustrating how the current varies with time after the switch is closed. If the self inductance and resistance of the coil are 10 H and 5.0 Ω respectively and the battery has an e.m.f. of 20 V and negligible resistance, what are the greatest values after the switch is closed of (a) the current, and (b) the rate of change of current? (*J.M.B.*)

13. The terminals of a moving coil ballistic galvanometer are short-circuited after a charge has been passed through it. Explain why the oscillations of the coil are damped out and the coil returns slowly to its zero position.

A ballistic galvanometer is connected to a flat coil having 40 turns of mean area 3.0 cm^2 to form a circuit of total resistance 80 Ω. The coil, held between the poles of an electromagnet with its plane perpendicular to the field, is suddenly withdrawn from the field, producing a throw of 30 scale divisions. Find the magnetic induction (flux density) of the field at the place where the coil was held, assuming that the sensitivity of the galvanometer under the conditions of the experiment is 0.40 division per microcoulomb. *(L. part qn.)*

14. Explain the special features that are necessary in a moving coil galvanometer intended for ballistic use.

A ballistic galvanometer is connected in series with a search coil and the secondary winding of a mutual inductance. When a current is reversed in the primary winding of the inductance a charge of 90 μC flows through the galvanometer. After switching off the primary current the search coil (which has 200 turns of mean diameter 1.00 cm) is placed in and perpendicular to a magnetic field. The deflection of the galvanometer caused by the rapid removal of the search coil from the magnetic field is the same as was observed when the primary current was reversed. The total resistance of the galvanometer circuit is 250 Ω. Calculate the magnetic induction (flux density) of the magnetic field. Draw a complete circuit diagram of the arrangement. *(J.M.B.)*

15. Explain what is meant by *cycle of magnetization*, and *hysteresis*.

Sketch on the same diagram the hysteresis loops for soft iron and hardened steel, indicating on the axes the physical quantities that have been plotted.

What information of practical importance can be obtained from a hysteresis loop? State, with reasons, which of the above metals would be suitable for (a) a permanent magnet, and (b) the core of a transformer.

Describe, with circuit diagram, an effective electrical method of demagnetizing a steel bar magnet and explain what is happening during the process of demagnetization. *(L.)*

16. Give a general account of the magnetization of iron. Show how the processes of magnetization and demagnetization, the phenomenon of hysteresis, and the existence of a Curie temperature can be explained in terms of elementary magnets and a domain structure.

What magnetic properties are desirable for the material of
(a) the core of a transformer,
(b) the core of a relay electromagnet,
(c) the tape of a magnetic tape recorder? *(O.)*

17. A student who was particularly interested in physics wrote at the end of his school course that he had found it interesting because ' you do a *few experiments* in the laboratory, you get out *a few rules and a few ideas about the sizes of some physical quantities*, and then presto—you find you can *understand and explain or predict* a whole range of phenomena and applications '.

Discuss this opinion illustrating your agreement or disagreement with it by choosing any *one* of the topics given in the list below or one stated topic of your own choice. You should explain, giving as many examples as possible, how each of the phrases given in italics applies to your study of the topic.

Topics:
The inverse square law for electric charges.
Electromagnetic induction.
The random behaviour of atoms and the energy they share.
(O. and C. Nuffield)

159

5 Alternating current

Introduction

(*a*) *a.c. and d.c.* In a direct current (d.c.) the drift velocity superimposed on the random motion of the charge carriers (e.g. electrons) is in one direction only. In an alternating current (a.c.) the direction of the drift velocity reverses, usually many times a second.

The effects of a.c. are essentially the same as those of d.c. Both are satisfactory for heating and lighting purposes. The magnetic field due to a.c. fluctuates with time and although, for example, a couple is exerted on a current-carrying coil, the inertia of the coil may prevent it responding, unless the frequency of the a.c. is very low. Thus a moving coil meter gives a reading with d.c. but normally not with a.c. Chemical effects are observed in some cases. The electrolysis of acidulated water by mains frequency a.c. produces a hydrogen-oxygen mixture at both platinum electrodes. There is no resultant effect when a.c. passes through copper sulphate solution using copper electrodes.

As we have seen (p. 137) a.c. is more easily generated and distributed than d.c. and for this reason the mains supply is a.c. However, processes such as electro-plating and battery charging require d.c., as does electronic equipment like radios and television receivers. When necessary a.c. can be rectified to give d.c.

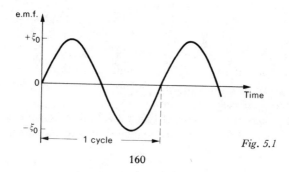

Fig. 5.1

(b) *Terms.* An alternating current or e.m.f. varies periodically with time in magnitude and direction. One complete alternation is called a *cycle* and the number of cycles occurring in one second is termed the *frequency* (*f*) of the alternating quantity. The unit of frequency is the *hertz* (Hz) and was previously the cycle per second. The frequency of the electricity supply in Britain is 50 Hz which means that the duration of one cycle, known as the *period* (*T*), is $1/50 = 0.02$ s. In general $f = 1/T$.

The simplest and most important alternating e.m.f. can be represented by a sine curve and is said to have a *sinusoidal waveform*, Fig. 5.1. It can be expressed by the equation

$$\mathscr{E} = \mathscr{E}_0 \sin \omega t$$

where $\mathscr{E}$ is the e.m.f. at time t, $\mathscr{E}_0$ is the peak or maximum e.m.f. and ω is a constant which equals $2\pi f$ where f is the frequency of the e.m.f. Similarly, for a sinusoidal alternating current we may write

$$I = I_0 \sin \omega t.$$

In the previous chapter we saw that a sinusoidal e.m.f. is induced in a coil rotating with *constant* speed in a *uniform* magnetic field. In that case ω was the angular velocity of the coil (in rad s^{-1}) and f equalled the number of complete revolutions of the coil per second. The mains supply is very nearly sinusoidal.

Alternating e.m.f.s and currents of many different waveforms can be produced and have their uses, Fig. 5.2. All, however irregular, can be shown to be combinations of sinusoidal e.m.f.s or currents.

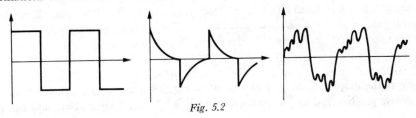

Fig. 5.2

Root mean square (r.m.s.) values

The value of an alternating current (and e.m.f.) varies from one instant to the next and the problem arises of what value we should take to measure it. The average value over a complete cycle is zero; the peak value is a possibility. However, the *root mean square* (r.m.s.) value is chosen because by using it many calculations can be done as they would be for direct currents.

The r.m.s. value of an alternating current (also called the *effective* value) is *the steady direct current which converts electrical energy to other forms of energy in a given resistance at the same rate as the a.c.*

Thus if the lamp in the circuit of Fig. 5.3 is lit first from a.c. and the brightness noted, then if 0.3 A d.c. produces the same brightness, the r.m.s. value of the

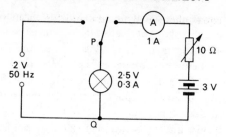

Fig. 5.3

a.c. is 0.3 A. A lamp designed to be fully lit by a current of 0.3 A d.c. will there-fore be fully lit by an a.c. of r.m.s. value 0.3 A. Although the value (I) of the a.c. is varying, the *average* rate at which it supplies electrical energy to the lamp equals the *steady* rate of supply by the d.c. ($I_{d.c.}$) and in practice it is this aspect which is often important.

In general, considering energy supplied to a resistance R we can say

$$I_{d.c.}^2 R = \text{(mean value of } I^2) \times R$$

$$\therefore I_{d.c.} = \sqrt{\text{mean value of } I^2}$$

$$= \text{square } root \text{ of the } mean \text{ value of the } squares \text{ of the current}$$

$$= I_{r.m.s.}$$

If the a.c. is sinusoidal then

$$I = I_0 \sin \omega t$$

$$\therefore I_{r.m.s.} = \sqrt{\text{mean value of } I_0^2 \sin^2 \omega t}$$

$$= I_0 \sqrt{\text{mean value of } \sin^2 \omega t.}$$

From the graphs of $\sin \omega t$ and $\sin^2 \omega t$ in Fig. 5.4 it can be seen that $\sin^2 \omega t$ is always positive and varies between 0 and 1. The shaded areas above and below

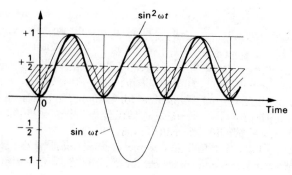

Fig. 5.4

the dotted line are equal, there is symmetry and the mean value for $\sin^2 \omega t$ is therefore 1/2. Hence

$$I_{\text{r.m.s.}} = I_0 \sqrt{1/2} = \frac{I_0}{\sqrt{2}} = 0.707\, I_0.$$

The r.m.s. current is 0.707 times the peak current. Similar relationships hold for e.m.f.s and p.d.s, and the r.m.s. value is the one usually quoted. Thus 240 V is the r.m.s. value of the electricity supply and the peak value $\mathscr{E}_0 = \sqrt{2}\mathscr{E}_{\text{r.m.s.}} = \sqrt{2} \times 240 = 339$ V.

The circuit of Fig. 5.3 may be used to check roughly that the peak value of an alternating p.d. is 1.4 times its r.m.s. value. The lamp is adjusted to the same brightness on d.c. as on a.c. The CRO is then connected across PQ and is used as a voltmeter. It measures the r.m.s. value of the p.d. across the lamp when it is lit by d.c. and *twice* the peak value using the a.c. supply.

Most voltmeters and ammeters for a.c. use are calibrated to read r.m.s. values and give correct readings only if the waveform is sinusoidal.

Meters for a.c.

The deflection of an a.c. meter must not depend on the direction of the current.

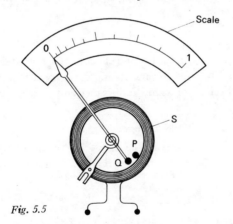

Fig. 5.5

(a) *Moving-iron meter*. The repulsion type consists of two soft iron rods P and Q mounted inside a solenoid S and parallel to its axis, Fig. 5.5. P is fixed and Q is carried by the pointer. Current passing either way through S magnetizes P and Q in the same direction and they repel each other. Q moves away from P until stopped by the restoring couple due, for example, to hair-springs. In many cases air damping is provided by attaching to the movement an aluminium pointer or vane which moves inside a curved cylinder.

The deflecting force is a function of the average value of the square of the current. Hence a moving-iron meter can be used to measure either d.c. or a.c.

and in the latter case r.m.s. values are recorded. The scale is not divided uniformly being closed up for smaller currents.

A moving-iron voltmeter is a moving-iron milliammeter with a suitable (non-inductive) multiplier connected in series.

(*b*) *Thermocouple meter.* One junction of a thermocouple (i.e. two wires of dissimilar metals) is joined to the centre of the wire XY carrying the current to be measured and is heated by it; the other junction is at room temperature, Fig. 5.6. When a.c. flows in XY, a thermoelectric e.m.f. is generated and produces a direct current that can be measured by a moving-coil microammeter (previously calibrated by passing known values of d.c. through XY). The hot junction is enclosed in an evacuated bulb B to shield it from draughts.

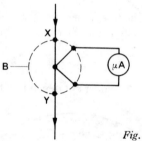

Fig. 5.6

This type of meter relies on the heating effect of a current. It therefore measures r.m.s. values and can be used for alternating currents of high frequency (up to several megahertz) because of its low inductance and capacitance compared with other meters.

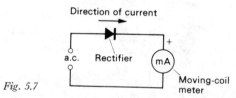

Fig. 5.7

(*c*) *Rectifier meter.* A rectifier is a device with a low resistance to current flow in one direction and a high resistance for the reverse direction. When connected to an a.c. supply it allows pulses of varying but direct current to pass. In a rectifier-type meter the average value of these is measured by a moving-coil meter, Figs.

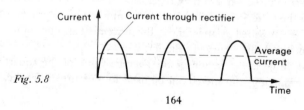

Fig. 5.8

164

5.7 and 5.8. Rectification, i.e. the conversion of a.c. to d.c., thus occurs. There are various kinds of rectifier (p. 185); those in many instruments are semi-conducting (germanium) diodes.

Rectifier instruments, being based on the moving-coil meter, are much more sensitive than other a.c. meters and are used in multimeters that have a.c. as well as d.c. ranges. The scale of a rectifier meter is calibrated to read r.m.s. values of currents and p.d.s with sinusoidal waveforms.

Capacitance in a.c. circuits

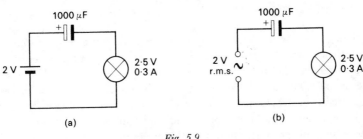

Fig. 5.9

(*a*) *Flow of a.c. ' through ' a capacitor.* If a 1000-μF capacitor is connected in series with a 2.5-V, 0.3-A lamp, and a 2-V d.c. supply, Fig. 5.9*a*, the lamp, as expected, does not light. Is there *any* current flow? With a 2-V r.m.s. 50-Hz supply, Fig. 5.9*b*, it is nearly fully lit.

The a.c. is apparently flowing through the capacitor. In fact, the capacitor is being charged, discharged, charged in the opposite direction and discharged again, fifty times per second (the frequency of the a.c.), and the charging and discharging currents flowing through the lamp light it. No current actually passes through the capacitor (since its plates are separated by an insulator) but it appears to do so and we talk as if it did. A current would certainly be recorded by an a.c. milliammeter.

When the 1000-μF capacitor is replaced by one of 100 μF, the charging and discharging currents are too small to light the lamp. Larger capacitances thus offer less 'opposition' to a.c. Increasing the frequency of the a.c. (at constant p.d. and capacitance) increases the current 'through' a capacitor since the same charge has to flow on and off the plates in a shorter time.

(*b*) *Phase relationships.* When a.c. flows through a resistor (having no capacitance or inductance) the current and p.d. reach their peak values at the same instant, i.e. they are in phase. This is not so for a capacitor.

The circuit of Fig. 5.10*a* enables phase relationships to be studied using 'slow a.c.' of frequency less than 1 Hz. With a 2000-Ω resistor between X and Y the current (shown on the milliammeter) and the p.d. (shown on the voltmeter) rise

165

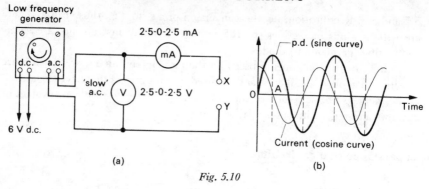

(a)

(b)

Fig. 5.10

and fall together. With a 250-μF capacitor replacing the resistor, the current through the capacitor is seen to lead the p.d. across it by one-quarter of a cycle, i.e. the current reaches its maximum value one-quarter of a cycle before the p.d. reaches its peak value, as shown in Fig. 5.10*b* by the cosine and sine curves.

The circuit for an alternative demonstration at 50 Hz using a double-beam CRO is given in Fig. 5.11. (A method of using an electronic beam splitter to convert a single beam CRO into a double beam one is given in Appendix 13, p. 543.) The Y_2 trace is the p.d. across R and this gives the waveform of the current ' through ' C since C and R are in series and the current in a resistor is in phase with the p.d. across it. The Y_1 trace is the p.d. across C and R in series and not just C, which accounts for the phase difference being less than a quarter of a cycle.

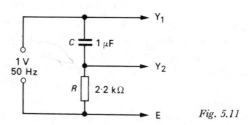

Fig. 5.11

Current and applied p.d. are out of step because current flow is a maximum immediately an uncharged capacitor is connected to a supply be it d.c. (p. 35) or a.c. There is as yet no charge on the capacitor to oppose the arrival of charge. Thus at O the applied p.d. though momentarily zero, is increasing at its maximum rate (the slope of the tangent at O to the p.d. graph is a maximum) and so the rate of flow of charge—the current—is also a maximum. Between O and A the p.d. is increasing but at a decreasing rate, the charge on the capacitor is increasing but less quickly, which means that the charging current is less. At A the applied p.d. is a maximum and for a brief moment is constant. The charge on the capacitor will also be a maximum and constant. The rate of flow of charge is therefore zero, i.e. the current is zero. The phase difference between V and I can thus be explained.

(c) *Mathematical treatment.* Let a p.d. V be applied across a capacitance C and let its value at time t be given by

$$V = V_0 \sin \omega t$$

where V_0 is its peak value and $\omega = 2\pi f$ where f is the frequency of the supply. The charge Q on the capacitance at time t is

$$Q = VC.$$

For the current I flowing 'through' the capacitor we can write

$$I = \text{rate of change of charge} = \frac{dQ}{dt}$$

$$= \frac{d}{dt}(VC) = C\frac{dV}{dt} = C\frac{d}{dt}(V_0 \sin \omega t)$$

$$= CV_0 \frac{d}{dt}(\sin \omega t)$$

$$\therefore I = \omega C V_0 \cos \omega t.$$

The current 'through' C (a cosine function) thus leads the applied p.d. (a sine function) by one quarter of a cycle or, as is often stated, by $\pi/2$ radians or $90°$ (1 cycle being regarded as 2π radians or $360°$). This confirms the results of the demonstrations and the 'physical' explanation outlined above. We can also write

$$I = I_0 \cos \omega t$$

where I_0 is the peak current and is given by $I_0 = \omega C V_0$

$$\therefore \frac{V_0}{I_0} = \frac{1}{\omega C}$$

But

$$\frac{V_{\text{r.m.s.}}}{I_{\text{r.m.s.}}} = \frac{V_0}{I_0}$$

$$\therefore \frac{V_{\text{r.m.s.}}}{I_{\text{r.m.s.}}} = \frac{1}{\omega C} = \frac{1}{2\pi fC}.$$

This expression resembles $V/I = R$ which defines resistance, $1/(2\pi fC)$ replacing R. The quantity $1/(2\pi fC)$ is taken as a measure of the opposition of a capacitor to a.c. and is called the *capacitive reactance* X_c. Hence

$$X_c = \frac{V_{\text{r.m.s.}}}{I_{\text{r.m.s.}}} = \frac{1}{2\pi fC}.$$

The ohm is the unit of X_C since the unit of f is s^{-1} and that of C is C V^{-1}. The term $1/(fC)$ therefore has units V/(C s^{-1}) = V A^{-1} = Ω. If f is in hertz and C in farads then X_C is in ohms. It is clear that X_C decreases as f and C increase. A 10-μF capacitor has a reactance of 320 Ω at 50 Hz. Check this. What will it be at 1 kHz?

Reactance is not to be confused with resistance, in the latter, electrical power is dissipated, it is not in a reactance as we shall see later.

Inductance in a.c. circuits

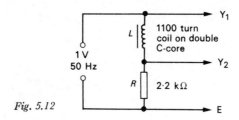

Fig. 5.12

(*a*) *Phase relationships.* An inductor in an a.c. circuit behaves like a capacitor in that it causes a phase difference between the applied p.d. and the current. In this case, however, the current lags on the p.d. by one-quarter of a cycle (i.e. 90°).

The effect may be observed using ' slow a.c. ' and the circuit of Fig. 5.10*a* (p. 166) with a 12 000–turn coil on a complete iron core (from a demountable transformer) as the inductor (500 H) connected to XY. Alternatively a double-beam CRO may be used with 50 Hz a.c. as in Fig. 5.12. In both cases the phase difference is less than 90° due to the resistance of the inductor and in the CRO demonstration because the Y_1 trace gives the p.d. across L and R in series; the graphs in Fig. 5.13 are for a pure inductor.

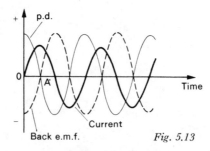

Fig. 5.13

We can explain the effects as follows. At O the current is zero but its rate of increase is a maximum (as given by the slope of the tangent to the current graph at O) which means, for an inductor of constant inductance (e.g. an air-cored

168

coil), that the rate of change of flux is also a maximum. Therefore by Faraday's law the back e.m.f. is a maximum but, by Lenz's law, of negative sign since it acts to oppose the current change. At A the current and flux are momentarily a maximum and constant. Their rate of change is zero (slope of tangent to current graph is zero at A) and so the back e.m.f. is zero. If the inductor has negligible resistance, then at every instant the applied p.d. must be nearly equal and opposite to the back e.m.f. The applied p.d. curve is therefore as shown. The p.d. acts on the coil whilst the e.m.f. acts back upon the source, just like two forces acting on different bodies.

(b) *Mathematical treatment.* In this case it is simpler to start with the current. Consider an inductance L through which current I flows at time t where

$$I = I_0 \sin \omega t.$$

I_0 is the peak current and $\omega = 2\pi f$ where f is the frequency of the a.c. The back e.m.f. $\mathscr{E}$ in the inductor due to the changing current is

$$\mathscr{E} = - L \frac{dI}{dt} \qquad \text{(p. 139)}$$

$$= - L \frac{d}{dt}(I_0 \sin \omega t)$$

$$= - \omega L I_0 \cos \omega t.$$

Assuming the inductor has zero resistance, then for current to flow the applied p.d. V must be equal and opposite to the back e.m.f., hence

$$V = - \mathscr{E} = \omega L I_0 \cos \omega t.$$

The applied p.d. thus leads the current by 90°. We can also write

$$V = V_0 \cos \omega t$$

where V_0 is the peak value of the applied p.d. and is given by

$$V_0 = \omega L I_0$$

$$\therefore \quad \frac{V_0}{I_0} = \frac{V_{\text{r.m.s.}}}{I_{\text{r.m.s.}}} = \omega L = 2\pi f L.$$

The quantity $2\pi f L$ is called the *inductive reactance* X_L of the inductor and like X_c it is measured in ohms. Thus

$$X_L = \frac{V_{\text{r.m.s.}}}{I_{\text{r.m.s.}}} = 2\pi f L.$$

X_L increases with f and L and is a measure of the opposition of the inductor to a.c. If $L = 10$ H for an inductor, X_L at 50 Hz is 3.1 kΩ and 63 MΩ at 1 MHz.

ALTERNATING CURRENT

Vector diagrams

A sinusoidal alternating quantity can be represented by a *rotating vector* (often called a *phasor*). Suppose the graph in Fig. 5.14 represents an alternating quantity $y = Y_0 \sin \omega t$ where Y_0 is the peak value of the quantity and its frequency $f = \omega/2\pi$. If the line OP has length Y_0 and rotates in an anticlockwise direction about O with uniform angular velocity ω, the projection O'P' of OP on O'y at time t (measured from the time when OP passes through OO') is $Y_0 \sin \omega t$. If OP is directed as shown by the arrow on it, then we can say that the projection on O'y of the rotating vector OP, gives the value at any instant of the sinusoidal quantity y. (Simple harmonic motion, being sinusoidal, can be derived similarly from uniform motion in a circle.)

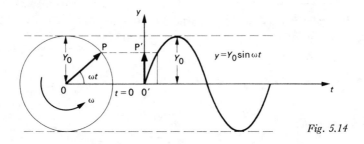

Fig. 5.14

The method is very useful for representing two sinusoidal quantities which have the same frequency but are not in phase. In Fig. 5.15 the waveforms of two such quantities y_1 and y_2 and the corresponding vector diagram are shown for time t. The phase difference between them is ϕ, with y_2 lagging and this phase angle is maintained between them as the vectors rotate. Being vectors they can be added by the parallelogram law if they represent similar quantities, e.g. p.d.s.

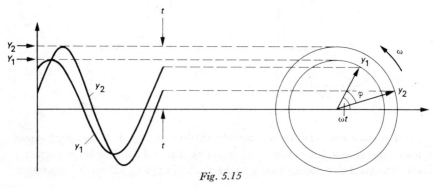

Fig. 5.15

170

Algebraically y_1 and y_2 are expressed by the equations

$$y_1 = Y_1 \sin \omega t$$

and
$$y_2 = Y_2 \sin (\omega t - \phi).$$

The vector diagram for a pure capacitance (i.e. infinite dielectric resistance) in an a.c. circuit is shown in Fig. 5.16a; the current I leads the applied p.d. V by 90°. That for a pure inductance (i.e. zero resistance) is given in Fig. 5.16b; in this case the current I lags on the applied p.d. V by 90°.

We shall see presently how vector diagrams are used to solve problems involving capacitance, inductance and resistance in a.c. circuits.

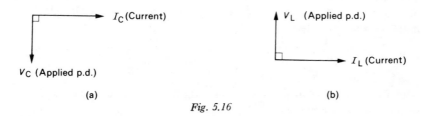

Fig. 5.16

Series circuits

When drawing vector diagrams a vector representing a quantity which is the same for all the circuit components should be drawn first. For a series circuit it would be the current vector. What would it be for a parallel circuit? This reference vector is drawn horizontal, directed to the right and the other vectors are then drawn so that their phases with respect to it are correct. The r.m.s. values of currents and p.d.s are used.

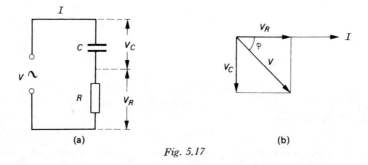

Fig. 5.17

(a) *Resistance and capacitance.* Suppose an alternating p.d. V is applied across a resistance R and a capacitance C in series, Fig. 5.17a. The same current I flows through each component and so the reference vector will be that representing I. The p.d. V_R across R is in phase with I and V_C, that across C, lags on I by $90°$. The vector diagram is as shown in Fig. 5.17b, V_C and V_R being drawn to scale.

The vector sum of V_R and V_C equals the applied p.d. V, hence

$$V^2 = V_R^2 + V_C^2.$$

But $V_R = IR$ and $V_C = IX_C$ where X_C is the reactance of C and equals $1/\omega C$, hence

$$V^2 = I^2(R^2 + X_C^2)$$

$$\therefore \ V = I\sqrt{R^2 + X_C^2}.$$

The quantity $\sqrt{R^2 + X_C^2}$ is called the *impedance* Z of the circuit and measures its opposition to a.c. It has resistive and reactive components and like both is measured in ohms. Hence

$$Z = \frac{V}{I} = \sqrt{R^2 + X_C^2}.$$

Also, from the vector diagram we see that the current I leads V by a phase angle ϕ which is less than $90°$ and is given by

$$\tan \phi = \frac{V_C}{V_R} = \frac{IX_C}{IR} = \frac{X_C}{R}.$$

(*b*) *Resistance and inductance.* The analysis is similar but in this case the p.d. V_L across L leads on the current I and the p.d. V_R across R is again in phase

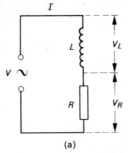

(a)

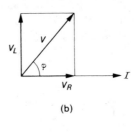

(b)

Fig. 5.18

with I, Fig. 5.18a. As before the applied p.d. V equals the vector sum of V_L and V_R, Fig. 5.18b, and so

$$V^2 = V_R^2 + V_L^2.$$

But $V_R = IR$ and $V_L = IX_L$ where X_L is the reactance of L and equals ωL, hence

$$V^2 = I^2(R^2 + X_L^2)$$

$$\therefore \ V = I\sqrt{R^2 + X_L^2}.$$

172

Here the *impedance* Z is given by

$$Z = \frac{V}{I} = \sqrt{R^2 + X_L^2}.$$

The phase angle ϕ by which I lags on V is given by

$$\tan \phi = \frac{V_L}{V_R} = \frac{IX_L}{R} = \frac{X_L}{R}.$$

(c) *Resistance, capacitance and inductance.* An R, C, L series circuit is shown in Fig. 5.19a; in practice R may be the resistance of the inductor. V_L leads the

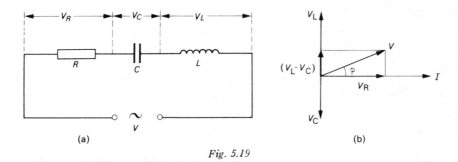

(a) (b)

Fig. 5.19

current (reference) vector I by $90°$, V_C lags on it by $90°$ and V_R is in phase with it. V_L and V_C are therefore $180°$ (half a cycle) out of phase, i.e. in antiphase. If V_L is greater than V_C, their resultant $(V_L - V_C)$ is in the direction of V_L, Fig. 5.19b. The vector sum of $(V_L - V_C)$ and V_R equals the applied p.d. V, therefore

$$V^2 = V_R^2 + (V_L - V_C)^2.$$

But $V_R = IR$, $V_L = IX_L$ and $V_C = IX_C$, hence

$$V^2 = I^2 \left[R^2 + (X_L - X_C)^2 \right]$$

$$\therefore V = I\sqrt{R^2 + (X_L - X_C)^2}.$$

The impedance Z is given by

$$Z = \frac{V}{I} = \sqrt{R^2 + (X_L - X_C)^2}.$$

The phase angle ϕ by which I lags on V is given by

$$\tan \phi = \frac{V_L - V_C}{V_R} = \frac{X_L - X_C}{R}.$$

Electrical resonance

(*a*) *Series resonance.* The expression just derived for the impedance Z of an RCL series circuit shows that Z varies with the frequency f of the applied p.d. since X_L and X_C both depend on f. $X_L = 2\pi fL$ and increases with f, $X_C = 1/(2\pi fC)$ and decreases with f, R is assumed to be independent of f (but it can vary). Fig. 5.20a shows how X_L, X_C, R and Z vary with f.

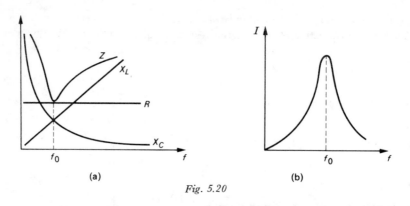

(a)   (b)

Fig. 5.20

At a certain frequency f_0, called the *resonant frequency*, $X_L = X_C$ and Z has its minimum value, being equal to R. The circuit behaves as a pure resistance and the current I has a maximum value (given by $I = V/R$), Fig. 5.20b. The phase angle ϕ (given by $\tan \phi = (X_L - X_C)/R$) is zero, the applied p.d. V and the current I are in phase and there is said to be *resonance*. A series resonant circuit is called an *acceptor* circuit.

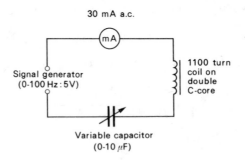

Fig. 5.21

Series or current resonance may be shown using the arrangement of Fig. 5.21. As the frequency increases the milliammeter reading rises to a maximum and then falls.

An expression for f_0 is obtained from $X_L = X_C$, that is

$$2\pi f_0 L = \frac{1}{2\pi f_0 C}$$

or $$4\pi^2 f_0^2 LC = 1$$

$$\therefore f_0 = \frac{1}{2\pi\sqrt{LC}}.$$

If L is in henrys and C in farads, f_0 will be in hertz.

At resonance V_L and V_C can both be very much greater than the total p.d. V applied across the whole circuit. Thus, if I is the resonance current, we have

$$I = \frac{V}{R}$$

$$\therefore V_L = IX_L = \frac{V}{R} X_L$$

$$\therefore \frac{V_L}{V} = \frac{X_L}{R}.$$

Since R (which is mostly due in practice to the resistance of the inductor L) is usually very small compared with X_L, V_L (and V_C) can be large compared with V. The magnification or *Q-factor* (Q for quality) of the circuit at resonance is defined by

$$Q = \frac{V_L}{V} = \frac{X_L}{R}.$$

In actual circuits Q-factors of over 200 are realized.

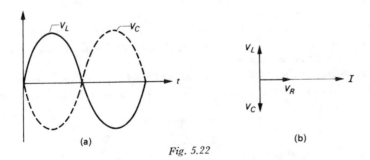

Fig. 5.22

It may seem strange for a small applied p.d., V, to give rise to two large p.d.s, V_L and V_C. The explanation is that V_L and V_C are in antiphase (i.e. 180° out of phase) and so their vector sum can be quite small—zero, in fact, if they are equal in magnitude as they are at resonance, Fig. 5.22. The circuit of Fig. 5.23*a*

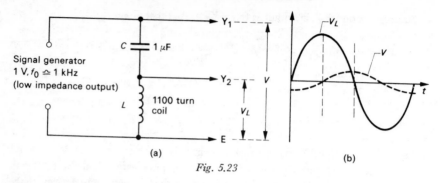

(a)   (b)

Fig. 5.23

may be used to demonstrate on a double-beam CRO that at resonance V_L is much greater than V and leads it by nearly 90°, Fig. 5.23b. If C and L are interchanged, V_C is seen to be equal to V_L and to lag on V by almost 90°. V_L and V_C are thus approximately in antiphase.

(b) *Parallel resonance.* In a parallel *RCL* circuit resonance can also occur but in this case the impedance of the circuit becomes a maximum. The resonant frequency f_0 is also given to a good approximation by $f_0 = 1/(2\pi\sqrt{LC})$. Large

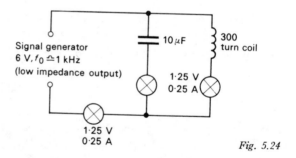

Fig. 5.24

currents I_C and I_L circulate to and fro *within* the circuit which are equal in magnitude but 180° out of phase, the supply current I is thus small. The p.d. across the circuit at resonance is large.

In the circuit of Fig. 5.24 the brightness of the lamps indicates the currents flowing in different parts of the circuit. Their behaviour as the frequency of the

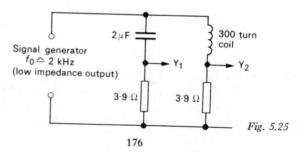

Fig. 5.25

176

p.d. applied from the signal generator increases (from about 0.5 kHz to 2 kHz) is worthy of study. In Fig. 5.25 the p.d.s across the 3.9-Ω resistors give the waveforms of the currents in C and L and show that they have equal amplitude but are in antiphase at resonance.

(c) *Tuned circuits.* The ability of a resonant circuit to select and amplify a p.d. of one particular frequency (strictly, a very narrow band of frequencies) is used in radio and television. For example in the aerial circuit of a radio receiver, Fig. 5.26, radio signals from different transmitting stations induce e.m.f.s of various frequencies in the aerial which cause currents to flow in the aerial coil. These induce currents of the same frequencies in coil L by mutual induction. If the capacitance C is adjusted (tuned) so that the resonant frequency of circuit LC equals the frequency of the wanted station, a large p.d. at that frequency (and no other) is developed across C. This p.d. is then applied to the next stage of the receiver.

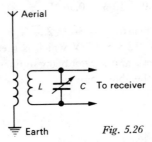

Fig. 5.26

(d) *Mechanical analogy.* The behaviour of a resonant circuit is similar to that of a vibrating system in mechanics. At one particular frequency it responds and stores a large amount of energy which passes to and fro between the electric field of the capacitor and the magnetic field of the inductor. Kinetic and potential energy behave similarly in an oscillating mass-spring system. Energy has then to be supplied only to compensate for that dissipated as heat in doing work against resistance—electrical or air.

Worked examples

1. *A 1000-μF capacitor is joined in series with a 2.5-V, 0.30-A lamp and a 50-Hz supply. Calculate (a) the p.d. (r.m.s.) of the supply to light the lamp to its normal brightness, and (b) the p.d.s across the capacitor and the resistor respectively.*

(a) Reactance X_C of capacitor $= 1/(2\pi fC)$

$$= 1/(2\pi \times 50 \times 10^3 \times 10^{-6})$$

$$= 10/\pi \; \Omega.$$

Resistance R of lamp $= 2.5/0.30 = 8.3 \; \Omega.$

Impedance Z of circuit $= \sqrt{R^2 + X_C^2}$

$$= \sqrt{(8.3)^2 + (10/\pi)^2}$$

$$= \sqrt{69 + 10}$$

$$= 8.9 \ \Omega.$$

The applied r.m.s. p.d. V to cause an r.m.s. current of 0.30 A to flow round the circuit is given by

$$V = IZ$$

$$= 0.30 \times 8.9$$

$$= 2.7 \ \text{V}.$$

(b) p.d. V_C across the capacitor $= IX_C$

$$= 0.30 \times 10/\pi = 0.96 \ \text{V}.$$

p.d. V_R across the resistor $= IR = 2.5 \ \text{V}.$

Note. $V_C + V_R = 0.96 + 2.5 \simeq 3.5$ V which is greater than the applied p.d. V of 2.7 V. This is due to V_C and V_R not being in phase. In fact $V^2 = V_C^2 + V_R^2$ —as you can check from the above figures.

2. A 2.0-H inductor of resistance 80 Ω is connected in series with a 420-Ω resistor and a 240-V, 50-Hz supply. Find (a) the current in the circuit, and (b) the phase angle between the applied p.d. and the current.

(a) Reactance X_L of inductor $= 2\pi f L$

$$= 2\pi \times 50 \times 2$$

$$= 200\pi \ \Omega.$$

Total resistance R of circuit $= 80 + 420$

$$= 500 \ \Omega.$$

Impedance Z of circuit $= \sqrt{R^2 + X_L^2}$

$$= \sqrt{(500)^2 + (200\pi)^2}$$

$$= 800 \ \Omega.$$

Current I in circuit $= V/Z$

$$= 240/800$$

$$= 0.30 \ \text{A}.$$

ALTERNATING CURRENT

(b) Phase angle ϕ between V and I is given by

$$\tan \phi = \frac{X_L}{R}$$

$$= \frac{200\pi}{500}$$

$$\therefore \phi = 52°.$$

V leads I by $52°$.

3. *A circuit consists of an inductor of 200 μH and resistance 10 Ω in series with a variable capacitor and a 0.10-V (r.m.s.), 1.0-MHz supply. Calculate (a) the capacitance to give resonance, (b) the p.d.s across the inductor and the capacitor at resonance, and (c) the Q-factor of the circuit at resonance, Fig. 5.27.*

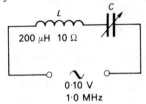

Fig. 5.27

0·10 V
1·0 MHz

(a) We have $L = 200 \times 10^{-6}$ H, $R = 10\ \Omega$, $f_0 = 10^6$ Hz. Also,

$$f_0 = \frac{1}{2\pi\sqrt{LC}} \quad \text{or} \quad 4\pi^2 f_0^2 LC = 1.$$

$$\therefore C = \frac{1}{4\pi^2 f_0^2 L}$$

$$= \frac{1}{4\pi^2 \times (10^6)^2 \times (200 \times 10^{-6})}$$

$$= 0.000\ 13\ \mu\text{F}.$$

(b) At resonance the impedance $Z = R$ and if V is the applied p.d. (0.10 V), the current I is given by

$$I = \frac{V}{R} = \frac{0.10}{10} = 1.0 \times 10^{-2}\ \text{A}.$$

If the inductor has reactance X_L, the p.d. V_L across it is

$$V_L = IX_L = I \times 2\pi f_0 L$$

$$= 1.0 \times 10^{-2} \times 2\pi \times 10^6 \times 200 \times 10^{-6}$$

$$= 4\pi$$

$$\approx 13\ \text{V}.$$

179

Since $V_C = V_L$, we have

$$V_C \simeq 13 \text{ V.}$$

(c) The *Q-factor* at resonance is given by

$$Q = \frac{V_L}{V} = \frac{13}{0.10} = 130.$$

Power in a.c. circuits

(a) *Resistance.* The general expression for the power absorbed by a device at any instant is IV where I and V are the instantaneous values of the current through it and the p.d. across it respectively.

In a resistor I and V are in phase and we can write $I = I_0 \sin \omega t$ and $V = V_0 \sin \omega t$, therefore the *instantaneous* power absorbed at time t is $I_0 V_0 \sin^2 \omega t$. The *mean* power P will be given by

$$P = \text{mean value of } I_0 V_0 \sin^2 \omega t.$$

We saw previously (p. 162, Fig. 5.4) that the mean value of $\sin^2 \omega t$ is $1/2$, thus

$$P = \frac{I_0 V_0}{2}$$

$$= \frac{I_0}{\sqrt{2}} \cdot \frac{V_0}{\sqrt{2}}.$$

For sinusoidal quantities

$$I_{\text{r.m.s.}} = \frac{I_0}{\sqrt{2}} \text{ and } V_{\text{r.m.s.}} = \frac{V_0}{\sqrt{2}}$$

$$\therefore P = I_{\text{r.m.s.}} \times V_{\text{r.m.s.}}$$

For a resistance

$$V_{\text{r.m.s.}} = I_{\text{r.m.s.}} \times R$$

$$\therefore P = I_{\text{r.m.s.}}^2 \times R = \frac{V_{\text{r.m.s.}}^2}{R}.$$

The power varies sinusoidally at twice the frequency of either V or I, as shown in Fig. 5.28.

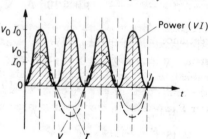

Fig. 5.28

(b) *Inductance.* In a pure inductor V leads I by $90°$ (or $\pi/2$ rad) and if $I = I_0 \sin \omega t$ then $V = V_0 \sin (\omega t + \pi/2) = V_0 \cos \omega t$. Hence the *instantaneous* power absorbed at time t is $I_0 V_0 \sin \omega t \cos \omega t$. But $\sin 2\omega t = 2 \sin \omega t \cos \omega t$, therefore

$$\text{instantaneous power absorbed} = \tfrac{1}{2} I_0 V_0 \sin 2\omega t$$

$$= I_{\text{r.m.s.}} V_{\text{r.m.s.}} \sin 2\omega t.$$

This represents a sinusoidal variation with mean value zero (and frequency twice that of I and V), Fig. 5.29. It follows that the *power absorbed by an inductor in a cycle is zero.*

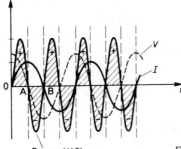

Power (*VI*)   *Fig. 5.29*

To explain this we consider that during the first quarter-cycle OA of current, power is drawn from the source and energy is stored in the *magnetic field* of the inductor. In the second quarter-cycle AB, the current and magnetic field decrease and the e.m.f. induced in the inductor causes it to act as a generator returning the energy stored in its magnetic field to the source. The power so restored is represented by the shaded area in Fig. 5.29. Thus, although the mean power taken over a cycle is zero, large amounts of energy flow in and out of the inductor every quarter-cycle. In practice an inductor has resistance and some energy is drawn from the source on this account and not returned.

(c) *Capacitance.* Similar reasoning shows that zero power is also taken by a pure capacitor over a cycle—since V and I are $90°$ out of phase. In this case energy taken from the source is stored in the *electric field* due to the p.d. between the plates of the charged capacitor. During the next quarter-cycle the capacitor discharges and the energy is returned to the source.

(d) *Formula for a.c. power.* Power is only absorbed by the *resistive* part of a circuit, i.e. in the expression IV for power, V is that part of the applied p.d. across the total resistance in the circuit; it is in phase with the current. Thus if $V_{\text{r.m.s.}}$ is the p.d. applied to an a.c. circuit and it leads the current $I_{\text{r.m.s.}}$ by angle ϕ, Fig. 5.30, the component of p.d. in phase with $I_{\text{r.m.s.}}$ is $V_{\text{r.m.s.}} \cos \phi$. Hence the total power P expended in the circuit is given by

$$P = I_{\text{r.m.s.}} V_{\text{r.m.s.}} \cos \phi.$$

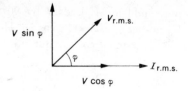

Fig. 5.30

The component $V_{r.m.s.}$ sin ϕ is that part of the applied p.d. across the total reactance of the circuit and is often called the ' wattless ' component of the p.d.

The *apparent power* absorbed in an a.c. circuit is, by comparison with the d.c. case, $I_{r.m.s.}V_{r.m.s.}$ The *real power* is $I_{r.m.s.}V_{r.m.s.}$ cos ϕ. The *power factor* of an a.c. circuit is defined by the equation

$$\text{power factor} = \frac{\text{real power}}{\text{apparent power}}$$

$$= \frac{I_{r.m.s.}V_{r.m.s.} \cos \phi}{I_{r.m.s.}V_{r.m.s.}} = \cos \phi.$$

In a purely resistive circuit cos ϕ has its maximum value of 1.

Electrical oscillations

Electrical oscillators are used in radio and television transmitters and receivers, in signal generators, oscilloscopes and computers, to produce a.c. with waveforms which may be sinusoidal, square, sawtooth etc. and with frequencies from a few hertz up to millions of hertz.

(*a*) *Oscillatory circuit*. When a capacitor discharges through an inductor in a circuit of low resistance an a.c. flows. The circuit is said to oscillate at its *natural frequency* which, as we will show shortly, equals $1/(2\pi\sqrt{LC})$, i.e. its resonant frequency f_0. Electrical resonance thus occurs when the applied frequency equals the natural frequency—as it does in a mechanical system.

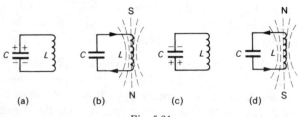

Fig. 5.31

In Fig. 5.31*a* a charged capacitor C is shown connected across a coil L. C immediately starts to discharge, current flows and a magnetic field is created which induces an e.m.f. in L. This e.m.f. opposes the current. C cannot therefore discharge instantaneously and the greater the inductance of L the longer does the discharge take. When C is completely discharged the electrical energy originally stored in the electric field between its plates has been transferred to the magnetic field around L, Fig. 5.31b.

At this instant the magnetic field begins to collapse and a p.d. is induced in L which tries to maintain the field. Current therefore flows in the same direction as before and charges C so that the lower plate is positive. By the time the magnetic field has collapsed, the energy is again stored in C, Fig. 5.31c. Once more C starts to discharge but current now flows in the opposite direction, creating a magnetic field of opposite polarity, Fig. 5.31d. When this field has decayed, C is again charged with its upper plate positive and the same cycle is repeated.

In the absence of resistance in any part of the circuit, an undamped, sinusoidal a.c. would be obtained. In practice, energy is gradually dissipated by the resistance as heat and a damped oscillation is produced, Fig. 5.32.

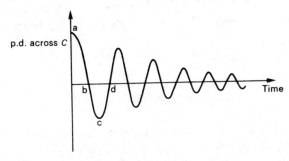

Fig. 5.32

(b) *Demonstrations.* Very slow damped electrical oscillations may be shown using the circuit of Fig. 5.33. Oscillations are started by charging C from a 20-V d.c. supply. Several cycles of ' slow ' a.c. are indicated on the moving coil milliammeter. If either L or C are reduced, the frequency of oscillation increases.

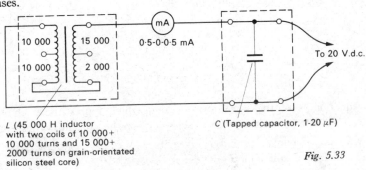

L (45 000 H inductor
with two coils of 10 000+
10 000 turns and 15 000+
2000 turns on grain-orientated
silicon steel core)

C (Tapped capacitor, 1-20 µF)

Fig. 5.33

Slightly higher frequency oscillations (about 2 Hz) are obtained using the arrangement of Fig. 5.34 and a CRO on its slowest time base speed.

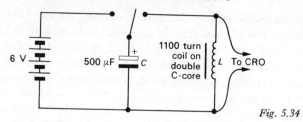

Fig. 5.34

(c) *Frequency of the oscillations.* In Fig. 5.35 the capacitor has capacitance C and is in series with an inductor of inductance L and negligible resistance. If Q

Fig. 5.35

is the charge on the capacitor at time t and I is the current flowing, then

$$\text{p.d. across inductor} = V_L = -L\frac{dI}{dt}$$

$$\text{p.d. across capacitor} = V_C = \frac{Q}{C}.$$

The net p.d. in the circuit is zero (since $R = 0$ and so there is no IR term), therefore $V_C = V_L$, that is,

$$\frac{Q}{C} = -L\frac{dI}{dt}.$$

Also, current is rate of flow of charge, i.e. $I = dQ/dt$ and so

$$\frac{dI}{dt} = \frac{d}{dt}\left(\frac{dQ}{dt}\right) = \frac{d^2Q}{dt^2}$$

$$\therefore \frac{Q}{C} = -L\frac{d^2Q}{dt^2}$$

$$\therefore \frac{d^2Q}{dt^2} = -\frac{1}{LC}\cdot Q.$$

This equation is of the same form as that which represents a s.h.m. (i.e. $d^2x/dt^2 = -\omega^2x$) and indicates that the charge Q on the capacitor varies sinusoidally with time, having a period T given by $T = 2\pi/\omega = 2\pi\sqrt{LC}$. If f is the frequency of the oscillations, $f = 1/T$ and

$$f = \frac{1}{2\pi\sqrt{LC}}.$$

As the resistance of an *LC* circuit increases, the oscillations decay more quickly, and when it is too large the capacitor discharge is unidirectional and no oscillations occur. To obtain undamped oscillations, energy has to be fed into the *LC* circuit in phase with its natural oscillations to compensate for the energy dissipated in the resistance of the circuit. This is done with the help of a valve or a transistor in actual oscillators (pp. 450 and 471).

Rectification of a.c.

Rectification is the conversion of a.c. to d.c. by a rectifier.

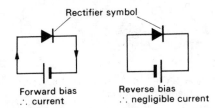

Forward bias
∴ current

Reverse bias
∴ negligible current

Fig. 5.36

(*a*) *Rectifiers*. Rectifiers have a low resistance to current flow in one direction, known as the *forward* direction, and a high resistance in the opposite or *reverse* direction. They are conductors which are largely unidirectional. When connection is made to a supply so that a rectifier conducts it is said to be *forward biased*; in the non-conducting state it is *reverse biased*, Fig. 5.36. The arrowhead on the symbol for a rectifier indicates the forward direction of conventional current flow. Three types of rectifier are

(*i*) the *thermionic diode valve* (p. 466)

(*ii*) the *selenium rectifier*, which is now being replaced by

(*iii*) *the semiconductor diode*, developed as a result of research aimed at producing rectifiers which could supply larger currents, withstand higher p.d.s and be made smaller than metal rectifiers, but yet be reliable and robust. The germanium diode rectifier was introduced in the early 1950s and shortly afterwards the silicon rectifier. The latter has superseded germanium for power rectification: it can operate at a current density of 100 A cm^{-2} of rectifier surface (compared with 100 mA cm^{-2} for metal rectifiers), can withstand reverse p.d.s of several hundred volts and has a working temperature of about 150 °C. Fig. 5.37 is a full-size illustration of a typical silicon diode rectifier. The construction and action of semiconductor diodes will be considered later.

Fig. 5.37

(*b*) *Half-wave rectification*. The rectifying circuit of Fig. 5.38 consists of a rectifier in series with the a.c. input to be rectified and the ' load ' requiring the

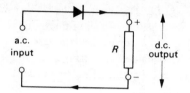

Fig. 5.38

d.c. output. For simplicity the ' load ' is represented by a resistor R but it might be some piece of electronic equipment.

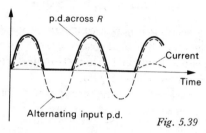

Fig. 5.39

In Fig. 5.39 the alternating input p.d. applied to the rectifier and load is shown. If the first half-cycle acts in the forward direction of the rectifier, a pulse of current flows round the circuit, creating a p.d. across R which will have almost the same value as the applied p.d. if the forward resistance of the rectifier is small compared with R. The second half-cycle reverse biases the rectifier, little or no current flows and the p.d. across R is zero. This is repeated for each cycle of a.c. input. The current pulses are unidirectional and so the p.d. across R is direct, for although it fluctuates it never changes direction.

(*c*) *Full-wave rectification.* In this process both halves of every cycle of input p.d. produce current pulses and the p.d. developed across a load is like that shown in Fig. 5.40. There are two types of circuits.

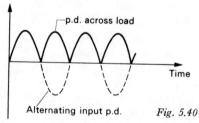

Fig. 5.40

(*i*) *Centre-tap full-wave rectifier.* Two rectifiers and a transformer with a centre-tapped secondary are used, Fig. 5.41. The centre tap O has a potential half-way between that of A and F and it is convenient to take it as a reference point having zero potential. If the first half-cycle of input makes A positive, rectifier B conducts, giving a current pulse in the circuit ABC, R, OA. During

186

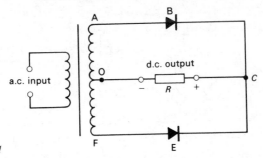

Fig. 5.41

this half-cycle the other rectifier E is non-conducting since the p.d. across FO reverse biases it. On the other half of the same cycle F becomes positive with respect to O and A negative. Rectifier E conducts to give current in the circuit FEC, R, OF; rectifier B is now reverse biased.

In effect the circuit consists of two half-wave rectifiers working into the same load on alternate half-cycles of the applied p.d. The current through R is in the same direction during *both* half-cycles and a fluctuating direct p.d. is created across R like that in Fig. 5.40.

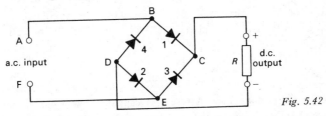

Fig. 5.42

(*ii*) *Bridge full-wave rectifier.* Four rectifiers are arranged in a bridge network as in Fig. 5.42. If A is positive during the first half-cycle, rectifiers 1 and 2 conduct and current takes the path ABC, R, DEF. On the next half-cycle when F is positive, rectifiers 3 and 4 are forward biased and current follows the path FEC, R, DBA. Once again current flow through R is unidirectional during both half-cycles of input p.d. and a d.c. output is obtained.

Smoothing circuits

To produce steady d.c. from the varying but unidirectional output from a half- or full-wave rectifier, smoothing is necessary.

(*a*) *Reservoir capacitor.* The simplest smoothing circuit consists of a large capacitor, 16 μF or more, called a *reservoir* capacitor, placed in parallel with the load R. In Fig. 5.43*a*, C_1 is the reservoir capacitor and its action can be followed from Fig. 5.43*b* where V_1 represents the p.d. developed across C_1 and I is the rectifier (full-wave) current.

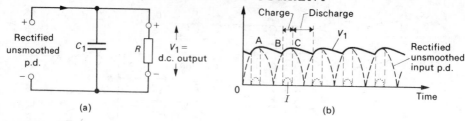

(a) (b)

Fig. 5.43

Initially the rectifier input p.d. causes current to flow through R and at the same time C_1 becomes charged almost to the peak value of the input as shown by OA. At A, the input p.d. falls below V_1 and C_1 starts to discharge. It cannot do so through the rectifier since the polarity is wrong, but it does through the load and thus maintains current flow by its charge storing or reservoir action. Along AB, V_1 falls. At B when the input p.d. equals the value to which V_1 has fallen, rectifier current I again flows to quickly recharge C_1 to the peak p.d., as shown by BC. The cycle of operations is then repeated. The d.c. output developed across R is V_1 and although it fluctuates at twice the frequency of the supply, the amplitude of the fluctuations is much less than when C_1 is absent.

The smoothing action of C_1 arises from its large capacitance making the time constant $C_1 R$ large so that the p.d. across it cannot follow the variations of input p.d. A very large value of C_1 would give better smoothing but initially the uncharged reservoir capacitor would act almost as a short-circuit and the resulting surge of current might damage the rectifier.

(*b*) *Capacitor-input filter.* A reservoir capacitor has a useful smoothing effect but it is usually supplemented by a filter circuit consisting of a choke L (i.e. an iron-cored inductor) having an inductance of about 15 H and a large capacitor C_2 arranged as in Fig. 5.44. The reservoir capacitor C_1 and the filter capacitor C_2 may be electrolytics enclosed in the same can.

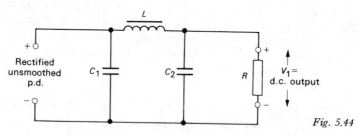

Fig. 5.44

C_1 behaves as explained previously and the p.d. across it is similar to V_1 in Fig. 5.43*b*. The action of the filter circuit $L - C_2$ can be understood if V_1 is resolved into a steady direct p.d. (the d.c. component) and an alternating p.d. (the a.c. component). This procedure is often used when dealing with a varying d.c. and is illustrated in Fig. 5.45.

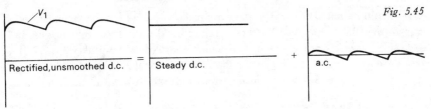

Fig. 5.45

By redrawing the smoothing circuit as in Fig. 5.46 we see that V_1 is applied across L and C_2 in series. L offers a much greater impedance to the a.c. component than C_2 and most of the unwanted ripple p.d. appears across L. For the d.c. component, C_2 has infinite resistance and the whole of this component is developed across C_2 except for the small drop due to the resistance of the choke. The filter thus acts as a potential divider, separating d.c. from a.c. and giving a steady d.c. output across C_2.

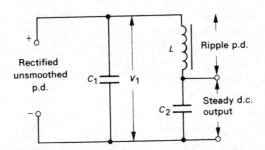

Fig. 5.46

(c) *Laboratory h.t. power supply unit* (a power pack). A typical circuit for a laboratory power pack is shown in Fig. 5.47; it employs a silicon diode or selenium bridge rectifier fed by a step-up mains transformer and has a capacitor-input smoothing filter. A steady d.c. output of 0–400 V at 100 mA is produced as well as a 6.3-V a.c. heater supply for thermionic devices such as valves and cathode-ray tubes. The 25-kΩ variable resistor acts as a potential divider giving a continuously variable output and also allowing the capacitors to discharge when the pack is switched off—this eliminates risk of shock should the output terminals be touched subsequently.

Fig. 5.47

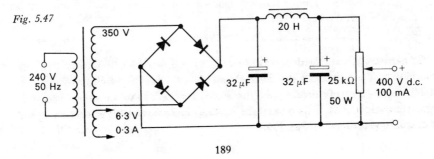

ALTERNATING CURRENT

It should be noted that r.m.s. p.d.s are usually quoted for the secondary of a transformer but, depending on the current taken by the load, rectification and smoothing can result in a d.c. output which approaches the peak value, i.e. 1.4 times the r.m.s. value. In the circuit shown the nominal 350-V secondary gives a d.c. output of about 500 V at no load current and 400 V when 100 mA is supplied.

(*d*) *Demonstrations of rectification and smoothing using slow a.c.* The circuits are shown in Fig. 5.48. An a.c. of frequency about $\frac{1}{2}$ Hz is obtained by applying 6 V d.c. to the input of a low frequency generator which is rotated by hand. The action of the various stages of a rectifying (half-wave) and smoothing circuit can be followed from the voltmeter reading using each arrangement in turn. The voltmeter acts as the ' load '. A resistor is used instead of a choke but in practice the latter is preferred in power units where currents of more than a few milliamperes have to be supplied.

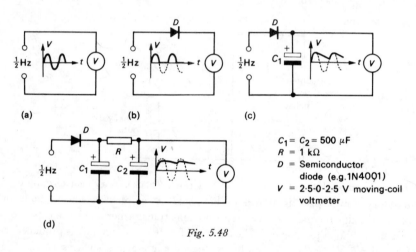

$C_1 = C_2 = 500 \ \mu F$
$R = 1 \ k\Omega$
D = Semiconductor
    diode (e.g.1N4001)
V = 2·5-0-2·5 V moving-coil
    voltmeter

Fig. 5.48

An alternative demonstration using a CRO is given in Appendix 4 (p. 532).

QUESTIONS

1. When a certain a.c. supply is connected to a lamp it lights with the same brightness as it does with a 12-V battery.
   (*a*) What is the r.m.s. value of the a.c. supply?
   (*b*) What is the peak p.d. of the a.c. supply?
   (*c*) The 12-V battery is connected to the Y-plates of a CRO and the gain adjusted so that it deflects the spot by 1.0 cm. What will be the total length of the trace on the CRO screen when the a.c. supply replaces the battery?

2. Electrical energy is supplied to a distant consumer through a power line which has a total resistance of 5 Ω. Explain why an input potential difference of 100 kV

190

ALTERNATING CURRENT

r.m.s. would be more satisfactory than one of 10 kV r.m.s., and calculate in each case (a) the output voltage, and (b) the output power when the power input is 20 MW.

If the figure of 100 kV refers to the r.m.s. value of a sinusoidal a.c. what will be the maximum voltage for which the line must be insulated? (O. and C. part qn.)

3. A sinusoidal p.d. of r.m.s. value 10 V is applied across a 50–μF capacitor.

(a) What is the peak charge on the capacitor?

(b) Draw a graph of charge Q on the capacitor against time.

(c) When is the current flowing into the capacitor (i) a maximum, and (ii) a minimum?

(d) Draw a graph of current against time.

(e) If the a.c. supply has frequency of 50 Hz, calculate the r.m.s. current through the capacitor.

4. In Fig. 5.49 V_1 and V_2 are identical high resistance a.c. voltmeters. Explain why the sum of the r.m.s. p.d.s measured across XY and YZ by V_1 and V_2 exceeds the applied r.m.s. p.d. of 15.0 V. Calculate the value of C.

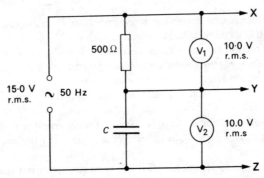

Fig. 5.49

5. As soon as the switch is closed in Fig. 5.50 what is the initial value of

(a) the p.d. across the capacitor

(b) the current through the resistor

(c) the rate at which the p.d. across the capacitor rises?

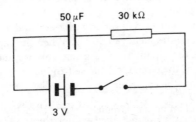

Fig. 5.50

6. What is meant by the *root mean square value* of a sinusoidal alternating current? Why is this a useful measure of alternating current?

A 3.0-Ω resistor is joined in series with a 10-mH inductor of negligible resistance,

ALTERNATING CURRENT

and a potential difference V (= 5.0 V r.m.s.) alternating at $200/\pi$ Hz is applied across the combination.

(a) Calculate the p.d. V_R across the resistor and V_L across the inductor.

(b) Showing clearly your procedure draw a vector diagram representing the relation between V, V_R and V_L.

(c) Determine the phase difference between V and V_L.

(d) How would you use a cathode-ray oscilloscope to show that there is a phase difference between V and V_L? (*J.M.B.*)

7. What is the purpose of a *choke* in a circuit? Describe an experiment to demonstrate the action of a choke, explaining what happens.

When an impedance, consisting of an inductance L and a resistance R in series, is connected across a 12-V, 50-Hz supply a current of 0.050 A flows which differs in phase from that of the applied potential difference by 60°. Find the values of R and L.

Find the capacitance of the capacitor which, connected in series in the above circuit, has the effect of bringing the current into phase with the applied potential difference. (*L.*)

8. A sinusoidal alternating current is represented by $I = I_0 \cos \omega t$, where I_0 is its peak value and ω its angular frequency in rad s^{-1}.

Derive an expression for the voltage necessary to send such a current through (a) a pure resistance R, (b) a pure inductance L, and (c) a pure capacitance C. Sketch the voltage waveforms in the three cases on the same axis as the current waveforms, and explain the terms *inductive* and *capacitative reactance, lag,* and *lead*.

What is the expression for the supply voltage if this current is to flow through a circuit consisting of R, L, and C, in series? What is meant by the impedance of the circuit?

If $R = 500$ Ω, $L = 0.50$ H, and $C = 1.0$ μF, draw the vector impedance diagram, for (i) $\omega = 1000$ rad s^{-1}, and (ii) $\omega = 2000$ rad s^{-1}. For what frequency is the impedance of the circuit smallest? If $I_0 = 10$ A, what would than be the peak voltage across (a) L, (b) C, and (c) the L-C combination? (*W.*)

9. Explain what is meant by *resonance* in an alternating current circuit containing inductance, resistance and capacitance in series. Give *one* practical application of this effect.

A variable capacitor is connected in series with a coil and a sinusoidal alternating supply of 20 V (r.m.s.) at a frequency of 50 Hz. When the capacitor has a value of 1.0 μF, the current in the circuit reaches a maximum value of 0.50 A (r.m.s.). Find (a) the resistance of the circuit, (b) the self-inductance of the coil, and (c) the potential difference across the capacitor. (*L.*)

10. Define the *impedance* of an a.c. circuit.

A 2.5-μF capacitor is connected in series with a non-inductive resistor of 300 Ω across a source of p.d. of r.m.s. value 50 V alternating at $1000/2\pi$ Hz. Calculate (a) the r.m.s. values of the current in the circuit and the p.d. across the capacitor, and (b) the mean rate at which energy is supplied by the source. (*J.M.B.*)

192

11. A simple alternator when rotating at 50 revolutions per second gives a 50-Hz alternating voltage of r.m.s. value 24 V. A 4.0-Ω resistance R and a 0.010-H inductance L are connected in series across its terminals.

(*i*) Assuming that the internal impedance of the generator can be neglected, find the r.m.s. current flowing, the power converted into heating, and the r.m.s. potential difference across each component.

(*ii*) Draw a vector diagram showing the relative phases of the applied voltage and the potential differences across R and L. (*O. part qn.*)

12. An alarm bell is driven off a 20:1 step-down mains transformer from a primary voltage of 240 V at 50 Hz. If the contact points fail to open, the coil presents an inductance of 20 mH and resistance of 10 Ω. For this fault condition, determine the current flowing through and the power dissipated in the coil. (*J.M.B. Eng. Sc.*)

13. Describe what will happen in the first few seconds after the switch is moved from position 1 to position 2 in the apparatus represented in Fig. 5.51. (*S.*)

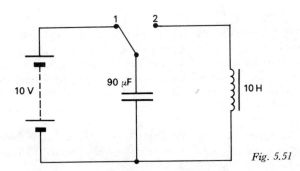

Fig. 5.51

14. Draw a circuit and explain the action of components to provide full-wave rectification of an alternating supply. Explain the action of a capacitor which could be used to smooth the output. (*A.E.B. part qn.*)

Part 2 | WAVES

6 Wave motion

Progressive waves

The idea of a wave is useful for dealing with a wide range of phenomena and is one of the basic concepts of physics. A knowledge of wave behaviour is also important to engineers.

A *progressive* or travelling wave consists of a disturbance moving from a source to surrounding places as a result of which energy is transferred from one point to another.

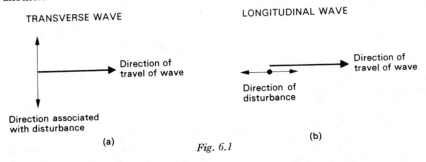

Fig. 6.1

There are two types of progressive wave. In the *transverse* type the direction associated with the disturbance is at right angles to the direction of travel of the wave, Fig. 6.1a. In the *longitudinal* type the disturbance is in the same direction as that of the wave, Fig. 6.1b. Transverse and longitudinal pulses (i.e. waves of short duration) can be sent along a Slinky spring, Fig. 6.2a and b. In both cases the disturbance generated by the hand is passed on from one coil of the spring to the next which performs the same motion but at a slightly later time. The pulse travels along the spring, the coils propagating the disturbance merely by vibrating to and fro (transversely or longitudinally) about their undisturbed positions. A succession of disturbances creates a continuous train of waves, i.e. a wave-train. The wave machine of Fig. 6.3a and b is useful for showing transverse and longitudinal waves.

197

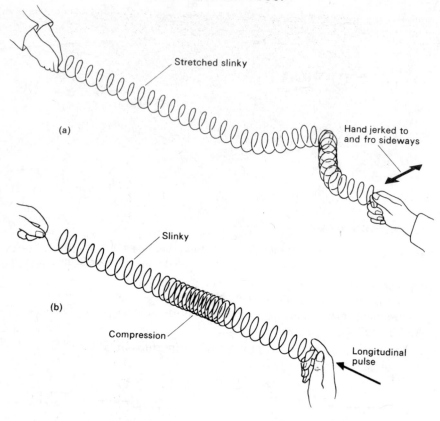

Stretched slinky

(a)

Hand jerked to
and fro sideways

Slinky

(b)

Compression

Longitudinal
pulse

Fig. 6.2

Waves may be classified as *mechanical* or *electromagnetic*. Mechanical waves are produced by a disturbance (e.g. a vibrating body) in a material medium and are transmitted by the particles of the medium oscillating to and fro. Such waves

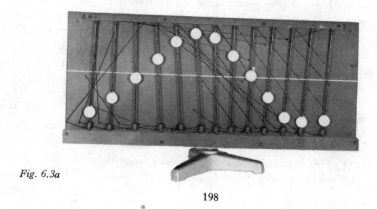

Fig. 6.3a

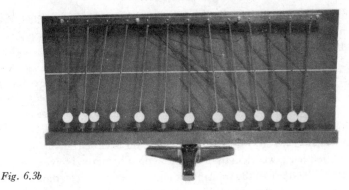

Fig. 6.3b

can be seen or felt and include waves on a spring, water waves, waves on stretched strings (e.g. in musical instruments), sound waves in air and in other materials. Many of the properties of mechanical waves can be shown using water waves in a ripple tank. Such waves are less complex than sea waves and are transmitted by the surface layer, often being called *surface water waves*. However, the displacement of the water is not a simple up-and-down, transverse motion (although the effects we shall observe will be due to this motion); the water particles move in elliptical or circular paths in the direction of the wave at a crest and in the opposite direction in a trough.

Fig. 6.4

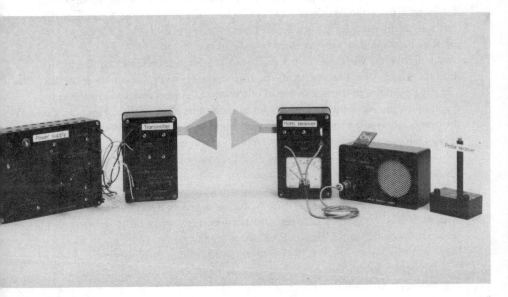

Electromagnetic waves, as we shall see later (p. 308), consist of a disturbance in the form of varying electric and magnetic fields. No medium is necessary and they travel more easily in a vacuum than in matter. Radio signals, light and X-rays are examples of this type. In this chapter microwaves will be used to investigate those properties of electromagnetic radiation which are explained in terms of waves. Microwaves are radar-type waves and are generated by an oscillator of very high frequency a.c. (10 000 MHz). In the transmitter of Fig. 6.4 the oscillator feeds a small aerial in a rectangular metal tube called a *wave guide* which opens into a horn at one end and is closed at the other. The radiation emitted has a wavelength of about 3 cm. The horn receiver has a wave guide and horn like those of the transmitter and contains a silicon diode mounted to ' detect ' (see p. 463) the signal. The probe receiver is less sensitive and non-directional, and simply consists of a diode. The signal from the transmitter can be modulated so that when picked up by either receiver, a note is heard in an amplifier-loudspeaker unit. Otherwise it can produce current in a micro-ammeter connected to the receiver.

Mechanical and electromagnetic waves give effects which are explicable by the same general principles as we shall see presently.

Describing waves

(a) *Graphical representation.* Two kinds of graph may be drawn. A *displacement-distance* graph for a transverse mechanical wave shows the displacements y of the vibrating particles of the transmitting medium at different distances x from the source *at a certain instant*. The dots in Fig. 6.5a show the positions of the particles at a particular time and the corresponding displacement-distance graph is given in Fig. 6.5b.

A longitudinal wave can also be represented by a transverse displacement-distance graph, the displacements in this case, however, being those of the vibrating particles in the line of travel of the wave. Thus y and x are in the same direction in the wave but at right angles to each other in the graph, Fig. 6.5c, i.e. the graph represents a longitudinal displacement as if it were transverse. Regions of high particle density are called *compressions* and regions of low particle density are called *rarefactions*. Both move in the direction of travel of the wave whilst the particles of the medium vibrate to and fro about their undisturbed positions.

The maximum displacement of each particle from its undisturbed position is the *amplitude* of the wave; in Fig. 6.5b this is OA or OB. The *wavelength* λ of a wave is the distance between two consecutive points on it which are in step, i.e. have the same phase. Thus for a transverse wave it is the distance between two successive crests or two successive troughs; for a longitudinal wave it is the separation of consecutive compressions or consecutive rarefactions. In Fig. 6.5b, $\lambda = OS = PT = RV$.

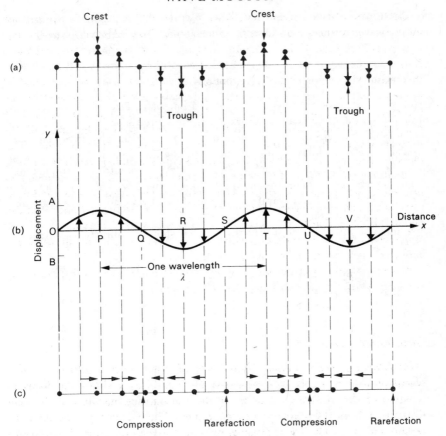

Fig. 6.5

Electromagnetic waves behave as transverse waves (p. 304) and may be represented by the same kind of displacement-distance graph, y being the instantaneous value of the electric field strength E or the magnetic flux density B at different distances from the source.

A *displacement-time* graph may also be drawn for a wave motion showing how the displacement of one particle (or the value of E or B) at a particular distance from the source varies with time. If this is a simple harmonic variation the graph is a sine curve.

(*b*) *Wavelength, frequency, speed.* If the source of a wave makes f vibrations per second, so too will the particles of the transmitting medium in the case of a mechanical wave and as will E or B in an electromagnetic wave. That is, the frequency of the wave equals the frequency of the source.

When the source makes one complete vibration, one wave is generated and the disturbance spreads out a distance λ from the source. If the source continues

201

to vibrate with constant frequency f, then f waves will be produced per second and the wave advances a distance $f\lambda$ in one second. If v is the wave speed then

$$v = f\lambda.$$

This relationship holds for all wave motions.

(*c*) *Wavefronts and rays*. A wavefront is a line or surface on which the disturbance has the same phase at all points. If the source is periodic it produces a succession of wavefronts, all of the same shape. A point source S generates circular wavefronts in two dimensions, Fig. 6.6*a* and spherical wavefronts in three dimensions. A line source S_1S_2 (e.g. the straight vibrator in a ripple tank) creates wavefronts that are straight in two dimensions, Fig. 6.6*b*, and cylindrical in three dimensions. Plane wavefronts are produced by a plane source or by *any* source at a distant point.

A line at right angles to a wavefront which shows its direction of travel is called a *ray*.

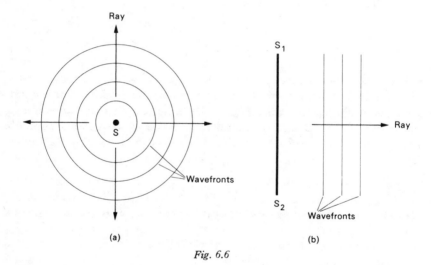

Fig. 6.6

Huygens' construction

The construction proposed by Huygens (a contemporary of Newton) enables the new position of a wavefront to be found, knowing its position at some previous instant. It can be used to explain the reflection, refraction and dispersion of waves as we shall see presently.

According to Huygens, *every point on a wavefront may be regarded as a source of secondary spherical wavelets which spread out with the wave velocity. The new wavefront is the envelope of these secondary wavelets*, i.e. the surface which touches all the wavelets.

As a simple example, suppose AB is a straight wavefront travelling from left to right, Fig. 6.7. The wavefront at time t later is the common tangent CD of the spherical wavelets of radius vt, drawn with centres every point on AB, where v is the speed of the waves in the medium.

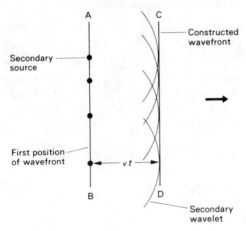

Fig. 6.7

The construction suggests that waves should also travel backwards to the left. This does not occur in practice and the assumption (which can be justified) has to be made that the amplitude of the secondary wavelets varies from a maximum in the direction of travel of the wave to zero in the opposite direction. This is indicated in Fig. 6.7 by drawing only arcs of the wavelets in the forward direction.

(*a*) *Reflection.* In Fig. 6.8*a* the wavefront AB is incident obliquely on the reflecting surface and A has just reached it. To find the new position of the wavefront when B is about to be reflected at B', we draw a secondary wavelet with centre A

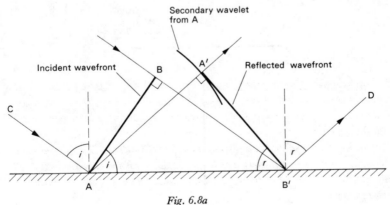

Fig. 6.8a

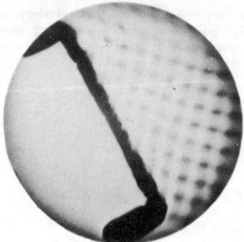

Fig. 6.8b

and radius BB'. The tangent B'A' from B' to this wavelet is the required reflected wavefront.

In triangles AA'B' and ABB'

$$\text{angle } AA'B' = \text{angle } ABB' = 90°$$

$$AB' \text{ is common}$$

$$AA' = BB' \text{ (by construction).}$$

Therefore the triangles are congruent (rt : h : s), hence

$$\text{angle } BAB' = \text{angle } A'B'A.$$

These are the angles made by the incident and reflected wavefronts AB and A'B' respectively, with the reflecting surface. The incident and reflected rays, e.g. CA and B'D are at right angles to the wavefronts as are the normals to the

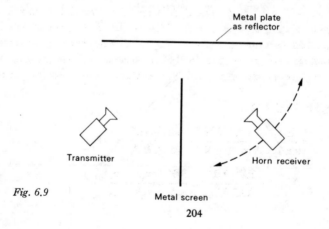

Fig. 6.9

204

surface (the dotted lines at A and B') and so the angle of incidence i equals the angle of reflection r. This is the law of reflection and Fig. 6.8b shows that it holds for the reflection of straight water waves in a ripple tank at a straight barrier. It may be tested for the reflection of microwaves from a metal plate with the apparatus of Fig. 6.9.

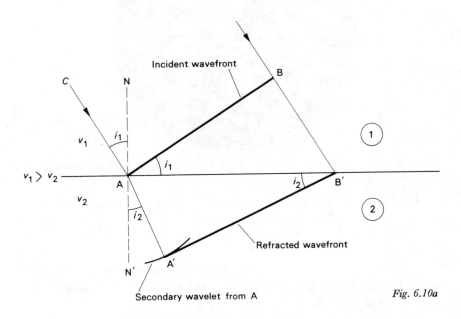

Fig. 6.10a

(*b*) *Refraction.* Fig. 6.10a shows the end A of a plane wavefront AB about to cross the boundary between media ① and ② in which its speeds are v_1 and v_2 respectively. The new position of the wavefront at time t later when B has travelled a distance $BB' = v_1t$ and reached B' is found by drawing a secondary wavelet with centre A and radius $AA' = v_2t$. The tangent B'A' from B' to this wavelet is the required wavefront and the ray AA', which is normal to the wavefront, is a refracted ray. If v_2 is less than v_1 the refracted ray is bent towards the normal in medium ②, as shown. When v_2 is greater than v_1 refraction away from the normal occurs.

In triangles BAB' and A'AB'

$$\frac{\sin i_1}{\sin i_2} = \frac{\sin \text{CAN}}{\sin \text{A'AN'}} = \frac{\sin \text{BAB'}}{\sin \text{A'B'A}}$$

$$= \frac{BB'/AB'}{AA'/AB'} = \frac{BB'}{AA'} = \frac{v_1t}{v_2t} = \frac{v_1}{v_2}.$$

205

Fig. 6.10b

But v_1 and v_2 are constants for given media and a particular wavelength

$$\therefore \quad \frac{\sin i_1}{\sin i_2} = \text{a constant.}$$

This is Snell's law of refraction and it holds for electromagnetic, sound and water waves. The constant is called the *refractive index*, $_1n_2$ for waves passing from medium ① to medium ②. Hence we can write

$$_1n_2 = \frac{v_1}{v_2} = \frac{\sin i_1}{\sin i_2} = \frac{n_2}{n_1}$$

where n_1 and n_2 are the *absolute* refractive indices of media ① and ②.

Huygens' construction thus explains the refraction of, for example, light, on the assumption that the speed of light decreases when it passes into a medium in which the refracted ray is bent towards the normal, i.e. when it enters an optically denser medium. This assumption is verified by experiment.

Refraction occurs when a wave passes from one medium into another in which it has a different speed. The speed of surface waves depends mostly on the depth in shallow liquids, i.e. the depth is not large compared with the wavelength (in deep liquids it depends on wavelength, not depth). Water ripples travel more slowly in shallow than in deeper water and are refracted on passing from one to the other. In Fig. 6.10b a glass plate on the bottom lower half of a ripple tank reduces the depth. Note the shorter wavelength (and hence smaller speed since the frequency is unchanged—how would you check that it is so?) as well as the change of direction of the refracted wavefront in the shallower water. The refraction of microwaves may be demonstrated as in Fig. 6.11.

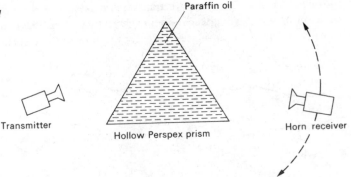

Fig. 6.11

Paraffin oil

Transmitter

Hollow Perspex prism

Horn receiver

(c) *Dispersion.* Huygens' construction predicts that if the speed of waves in a given medium depends on the frequency (and therefore wavelength) of the waves, then when refraction occurs its extent will also depend on the frequency. In Fig. 6.12a the waves of frequencies f_1 and f_2 travel with the same speed in medium ① but in medium ②, v_1 (the speed of wave with frequency f_1) is greater than v_2 and so f_2 suffers greater deviation than f_1. Dispersion occurs; medium ① is called a *non-dispersive* medium and medium ② a *dispersive* medium. Thus, red light travels faster than blue light in glass and if white light travels from air to glass, the blue wavefront is refracted more than the red one, i.e. dispersion is observed, Fig. 6.12b.

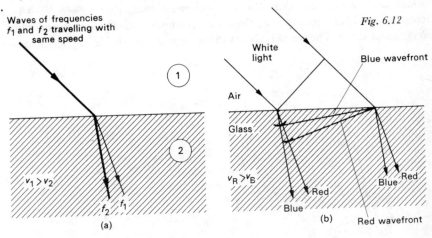

Waves of frequencies f_1 and f_2 travelling with same speed

Fig. 6.12

White light

Blue wavefront

Air

Glass

$v_1 > v_2$

f_2 f_1

(a)

$v_R > v_B$

Red

Blue

Red

Blue Red

Red wavefront

(b)

Principle of superposition

What happens when waves meet? Is their motion changed as it is when solid objects collide? Answers to these questions may be obtained by producing pulses on a long narrow spring.[1]

[1] A PSSC-type spring is better than a Slinky for this.

The drawings in Fig. 6.13a show two transverse pulses (of slightly different shapes) on a spring approaching each other. Where they cross (diagram 3) the resultant displacement of the spring is evidently the sum of the displacements

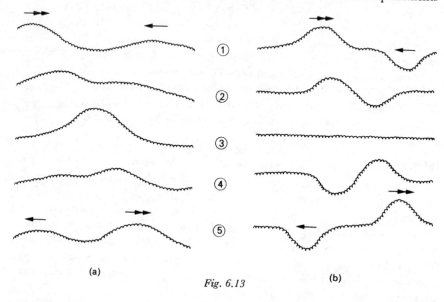

(a)

(b)

Fig. 6.13

which each would have caused at that point, i.e. the pulses superpose. After crossing, each pulse travels along the spring as if nothing had happened and it has its original shape. The result is always the same whatever shapes the pulses have.

Fig. 6.14

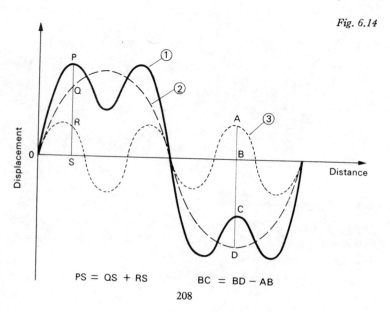

PS = QS + RS   BC = BD − AB

Fig. 6.13*b* shows the superposition of two equal and opposite pulses. In this case, when the pulses meet (diagram 3) they cancel out and the net displacement of the spring is zero.

In general we can conclude that pulses (and waves), unlike particles, pass through each other unaffected and *where they cross the total displacement is the vector sum of the individual displacements due to each pulse at that point.* This statement is called the *principle of superposition* and can be used to explain many wave effects. It is applied in Fig. 6.14 where the resultant, ①, of two waves, ② and ③, of different amplitudes and frequencies is obtained by adding the displacements; ③ has half the amplitude of ② and three times its frequency. The shape of the resultant is often quite different from those of its components, as here.

Interference

In a region where wave–trains from *coherent* sources cross, superposition occurs giving reinforcement of the waves at some points and cancellation at others. The resulting effect is called an *interference pattern* or a *system of fringes.*

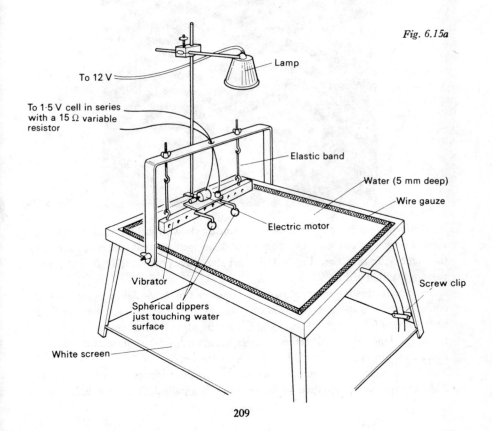

Fig. 6.15a

Lamp

To 12 V

To 1·5 V cell in series with a 15 Ω variable resistor

Elastic band

Water (5 mm deep)

Wire gauze

Electric motor

Vibrator

Screw clip

Spherical dippers just touching water surface

White screen

Coherent sources have a constant phase difference which means they must have the same frequency and for complete cancellation to occur the amplitudes of the superposing waves they produce must be about equal. If their phase difference is not constant, at a certain point there may be reinforcement at one instant and cancellation at the next. If these variations follow one another rapidly, the effect at the point is to produce uniformity, i.e. the interference pattern changes so quickly with the continuously changing phase difference between the sources that the detector—for example, the eye—cannot follow the alterations and records an average effect.

In practice coherent sources are derived from a single source.

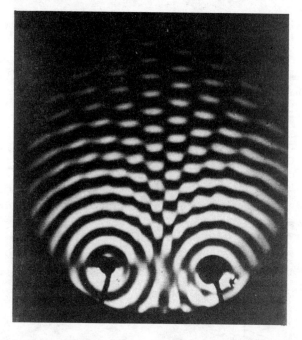

Fig. 6.15b

(*a*) *Mechanical waves.* Interference is readily shown by the transverse component of surface water waves in a ripple tank using two small spheres attached to the same vibrating bar, Fig. 6.15*a*. Circular waves are produced and give a pattern like that of Fig. 6.15*b*.

To explain this interference pattern, consider Fig. 6.16. All points on AB are equidistant from the sources S_1 and S_2 and since the vibrations of these are in phase, crests (or troughs) from S_1 arrive at the same time as crests (or troughs) from S_2. Hence along AB reinforcement occurs by superposition and a wave of double amplitude is formed, Fig. 6.17*a*. Points on CD are half a wavelength (a wavelength being the distance between the centres of successive bright bands in Fig. 6.15*b*) nearer to S_1 than to S_2, i.e. there is a path difference of half a wavelength. Therefore crests (or troughs) from S_1 arrive simultaneously with troughs (or crests) from S_2 and the waves cancel, Fig. 6.17b.

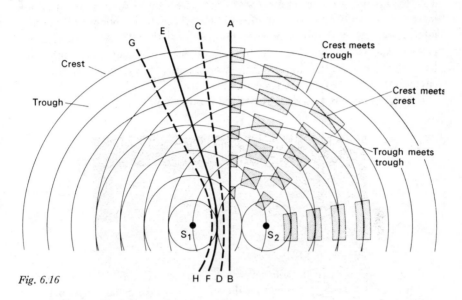

Fig. 6.16

Along EF the difference of the distances from S_1 and S_2 to any point is one wavelength. EF is a line of reinforcement, i.e. constructive interference occurs. GH is a line of cancellation where destructive interference occurs since at every point on GH the distance to S_1 is one and a half wavelengths less than to S_2. Lines such as CD and GH are called *nodal* lines: AB and EF are *antinodal* lines, both sets are hyperbolic curves.

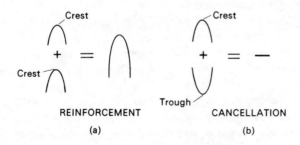

Fig. 6.17

It can be shown that the separation of nodal (and antinodal) lines increases
(*i*) as the distance from S_1 and S_2 increases
(*ii*) the smaller the separation of S_1 and S_2
(*iii*) as the wavelength increases (i.e. as the frequency decreases).

(*b*) *Electromagnetic waves.* Two arrangements for showing the interference of microwaves are given in Fig. 6.18*a* and *b*. In (*a*) interference occurs between the two wave-trains emerging from the 3-cm-wide slits which act as two coherent sources. The receiver detects the maxima and minima of the fringe pattern as it is moved round; if it is a minimum and the waves from one slit are cut off by a metal plate, the signal *increases*—clear evidence of destructive interference.

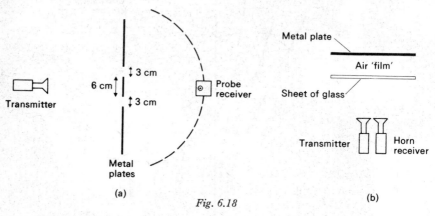

Fig. 6.18

In (*b*), the glass plate partially reflects and partially transmits the microwaves. The transmitted wave-train traverses the air ' film ' to the metal plate where it is reflected back across the film and again transmitted (partially) by the glass. Two wave-trains (derived from the same source) thus reach the receiver and the thickness of the air ' film ' decides the path difference between them (i.e. how much farther one has travelled than the other). If ' crests ' of both arrive simultaneously at the receiver, there is reinforcement but if a ' crest ' arrives with a ' trough ' there is cancellation. Changing the path difference by moving the metal plate towards or away from the glass, produces maxima and minima at the receiver. What is the minimum change in the thickness of the air ' film ' that will cause a maximum to be replaced by another maximum?

The Decca system of navigation is based on the interference of radio waves, the position of the aircraft or ship being plotted automatically. Fig. 6.19 shows a typical layout on a small boat with the track plotter on the extreme left.

Summing up, effects in which disturbances when added produce no disturbance can be regarded as being due to some kind of wave motion. Thus radio signals and light are treated as wave motions because radio signals sometimes ' fade ' and light plus light can result in darkness. However, no energy disappears: it is redistributed (p. 236). In the case of invisible waves where the nature of the disturbance is not revealed directly to our senses we are really using a wave model, based on our experience of visible waves, to enable us to explain their behaviour.

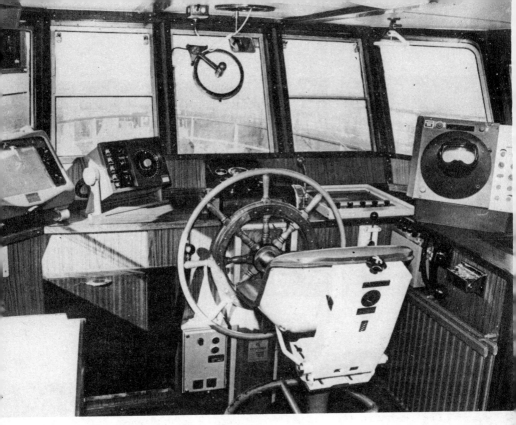

Fig. 6.19

(c) *Rectilinear propagation.* Interference shows the mechanism by which Huygens' secondary wavelets superpose to form the new wavefront. Rectilinear propagation for an infinite plane wave is then explained by the destructive interference of out-of-phase waves in all directions except forwards.

Diffraction

The spreading of waves when they pass through an opening or round an obstacle into regions where we would not expect them is called *diffraction*.

(a) *Mechanical waves.* The diffraction of water ripples at an opening in a barrier may be studied in a ripple tank, a vibrating bar being used to produce straight ripples.

In Fig. 6.20a, the opening is wide and the incident waves emerge almost unchanged (i.e. straight). A fairly sharp 'shadow' of the opening is obtained and diffraction is not marked. Most of the energy propagated through the opening is in the same direction as the incident waves. In Fig. 6.20c, the opening is narrow and the emerging waves are circular, i.e. the opening behaves like a point source. In this case diffraction is appreciable and the energy of the diffracted waves is more or less equally distributed over 180°. In Fig. 6.20b some diffraction

213

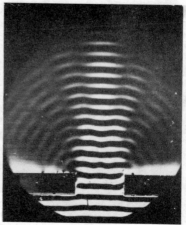

Fig. 6.20a, b and c

Fig. 6.21a and b

is evident at the ' medium-sized ' gap. Diffraction effects are thus greatest when the width of the opening is comparable with the wavelength of the waves.

Huygens' procedure for predicting the future position of a wavefront by replacing it by point sources is able to account for diffraction, as well as for reflection and refraction. Thus the patterns obtained can be considered to arise from the interference (superposition) of the secondary wavelets produced by the point sources imagined to exist at the unrestricted part of the wavefront which falls on the opening. The two ripple tank photographs of Fig. 6.21a and b support this view. The diffraction pattern of straight waves passing through a slit is shown in (a); (b) is the interference pattern of a line of equally-spaced point sources occupying the same position and width as the slit and vibrating together with the same frequency as that of the waves in (a). The two patterns are very similar except near the sources in (b).

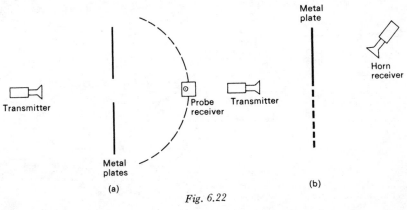

Fig. 6.22

(b) *Electromagnetic waves.* Diffraction of microwaves at a slit may be shown with the apparatus of Fig. 6.22a. Spreading into the regions of geometrical shadow behind the slit is greatest when the slit width is comparable with the wavelength of the microwaves, i.e. 3 cm. (It should now be evident why it is possible to demonstrate the interference of microwaves using the double slit arrangement of Fig. 6.18a.)

It is also instructive to position a single metal plate as in Fig. 6.22b so that it just cuts off the signal to the receiver, placed in the shadow of the plate. When a second plate is slid up (along the dotted line) to make a slit 3–6 cm wide with the first plate, the received signal increases showing that a *smaller* wavefront (less energy) can produce a *stronger* signal. This is due to destructive interference between different parts of the larger wavefront and by removing some of them (with the second plate) the resultant signal increases in this particular direction.

If a circular metal disc (about 15 cm diameter) is used as an ' obstacle ', a signal is received along the axis in the centre of the shadow, indicating appreciable diffraction round the edges of the disc.

WAVE MOTION

Stationary waves

(*a*) *Mechanical waves.* If one end of a long, narrow, stretched spring (or a Slinky) is fixed and the other end is moved continuously from side to side, a progressive transverse wave is generated. At the fixed end it is reflected, travels back to the vibrating end and repeated reflection occurs. Two progressive trains of waves travel along the spring in opposite directions. If the shaking frequency is slowly increased, at certain frequencies one or more vibrating loops of large amplitude are formed in the spring. A *stationary* or *standing wave* is said to have been produced since the waveform does not seem to be travelling along the spring in either direction.

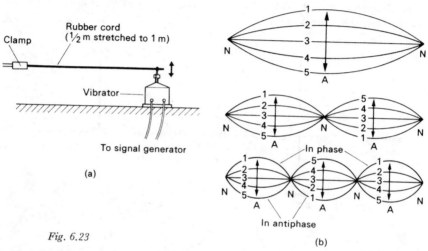

Fig. 6.23

(b)

A more effective demonstration, called Melde's experiment, is shown in Fig. 6.23*a* using a ½-m length of rubber cord (3 mm square section) stretched to 1 m. As the frequency of the a.c. from the low impedance output of the signal generator is increased from 10 Hz to about 100 Hz, standing waves with one, two, three or four loops are obtained. If the cord is illuminated in the dark by a lamp stroboscope and the flashing rate adjusted to be nearly equal to that of the vibrator, the cord can be seen moving up and down slowly, the crests of the standing wave becoming troughs, then crests again and so on, Fig. 6.23*b*. Note that

(*i*) there are points such as N, called *nodes* of the stationary wave, where the displacement is always zero.

(*ii*) within one loop all particles oscillate in phase but with different amplitudes (how does this compare with the behaviour of a progressive wave?) and so all points (except nodes) have their maximum displacement simultaneously; points such as A with the greatest amplitude are called *antinodes*.

(*iii*) the oscillations in one loop are in antiphase with those in an adjacent loop.

216

(*iv*) the frequency of the particle vibration in both the standing wave and the progressive waves is the same, and the wavelength of the stationary wave is *twice* the distance between successive nodes or successive antinodes and equals the wavelength of either of the progressive waves. This offers a solution to the otherwise difficult problem of measuring the wavelength of a progressive wave. We simply produce standing waves from the progressive wave and measure the wavelength of the former directly.

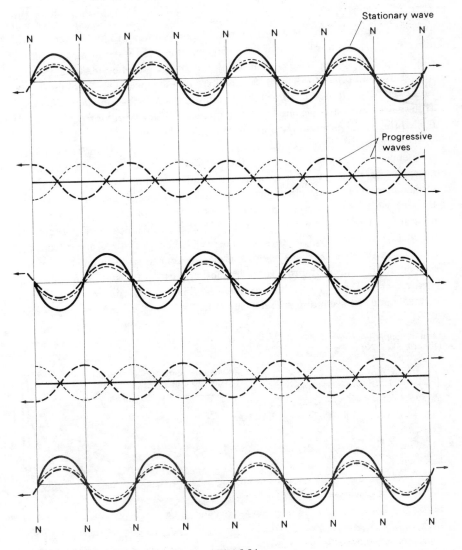

Fig. 6.24

(b) *Explanation of stationary waves.* Stationary waves result from the super-position of two trains of progressive waves of equal amplitude and frequency travelling with the same speed in opposite directions. In Fig. 6.24 the dotted and broken curves are the displacement-distance graphs of two such waves at successive equal time intervals. The continuous curve in each case is their resultant at these instants, formed by superposition. The formation of the stationary wave loops of Fig. 6.23b can now be understood.

In the two previous demonstrations one wave-train is obtained by reflection of the other and if, in the time for a wave to travel along the spring or cord and back again, the vibrator is about to send off the second wave, the latter will reinforce the first. This will also be true if the vibrator produces exactly two, three or any integral number of waves in the time for a wave to make a return journey on the spring or cord. The amplitude of the waves builds up since they are returned in phase to the source, i.e. there is resonance. The coupling is thus improved between the source and the medium, producing ' stimulated emission '. If the stationary wave has one loop at frequency f then a vibrator frequency of $2f$ (called the *second harmonic*) will give two loops and so on.

We can regard the stretched spring or cord as having a number of natural frequencies of vibration and when the applied frequency from the vibrator is near one of these, there is a large-amplitude standing wave vibration, i.e. resonance occurs. A stretched spring thus has a number of natural frequencies. By contrast, a mass hanging from the end of a spring, for example, has only one natural frequency. (We will see later, p. 221, that a system can only vibrate if it has mass and elasticity. The difference between the two systems arises from the fact that in the mass-spring system, the mass is concentrated in one part of the system and the elasticity property is in another part; in the stretched spring or cord system all parts have both properties. The first is said to have *lumped* elements and the second *distributed* elements.)

Fig. 6.25

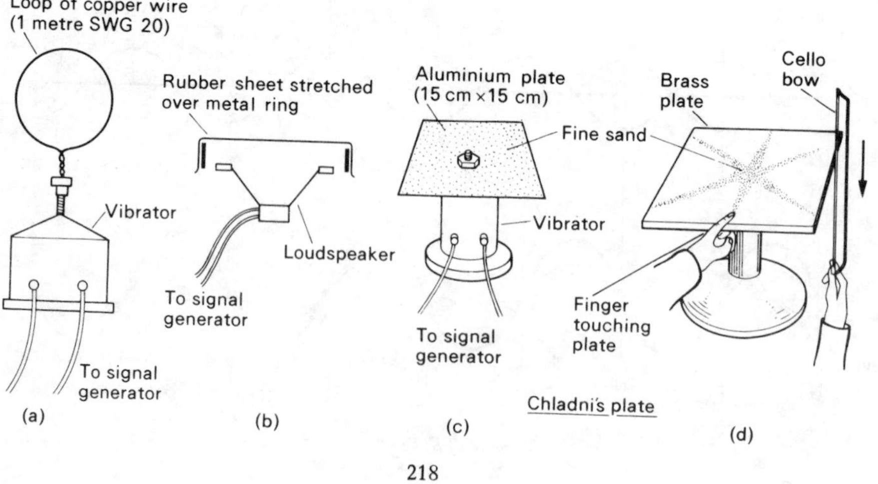

Loop of copper wire
(1 metre SWG 20)

Rubber sheet stretched over metal ring

Aluminium plate
(15 cm × 15 cm)

Brass plate

Cello bow

Fine sand

Vibrator

Vibrator

Loudspeaker

Finger touching plate

To signal generator

To signal generator

To signal generator

Chladni's plate

(a)   (b)   (c)   (d)

218

Standing wave resonance can occur only in systems with 'boundaries' (e.g. the ends of a spring or cord) that restrict wave travel. Progressive waves reflected from the 'boundaries' interfere with progressive waves travelling to the 'boundaries' and the resulting standing waves have to 'fit' into the system, e.g. have nodes at the fixed ends of the cord in Fig. 6.23a, otherwise the pattern is not fixed in form. Large amounts of energy are stored locally in standing waves and become trapped with the waves; there is no energy transmission as with progressive waves. This is an important difference between standing and progressive waves. In 'unbounded' systems the waves are not confined and travel on unrestricted.

Some other demonstrations of standing waves in mechanical systems are illustrated in Figs. 6.25a, b, c and d. The first two should be viewed in the dark in stroboscopic light.

(c) *Electromagnetic waves.* Standing microwave patterns can be obtained with the arrangement of Fig. 6.26. Waves from the transmitter are superposed on those reflected from the metal plate. If the latter is moved slowly towards or away from the transmitter, the signal at the receiver varies and the distance moved by the reflector between two consecutive minima (nodes) equals half the wavelength of the microwaves. Why is the signal not quite zero at a node?

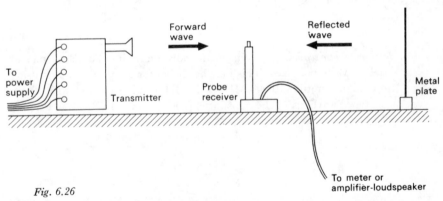

Fig. 6.26

(d) *Importance of standing waves.* Standing waves are a feature of many physical phenomena. The mechanical type are produced on the strings and air columns of musical instruments when sounding a note (pp. 243 and 247). Electromagnetic examples occur in radio and television aerial systems and, as we shall see later (p. 427), the idea that an electron in an atom behaves like a standing wave is used to account for the fact that atoms have definite energy levels (p. 417).

Engineers are often confronted with standing wave problems when dealing with systems such as turbines, propellers, aircraft wings, car bodies etc., which can vibrate and have edges.

Polarization

This effect occurs only with transverse waves.

(a) *Mechanical waves.* In Fig. 6.27 the rope PQRS is fixed at end S and passes through two vertical slits at Q and R. If end P is moved to and fro in all directions (as shown by the short arrowed lines), vibrations of the rope occur in every plane and transverse waves travel towards Q. At Q only waves due to vibrations in a vertical plane can emerge from the slit and the wave between Q and R is said to be *plane polarized* (in the vertical plane containing the slit at Q) in contrast to that between P and Q which is unpolarized. If the slit at R is vertical, as shown, the wave travels on, but if it is horizontal the wave is stopped and the slits are ' crossed '.

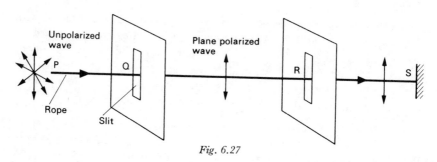

Fig. 6.27

(b) *Electromagnetic waves.* Electromagnetic radiation from radio, television and radar aerials is plane polarized. This may be shown for 3-cm microwaves using the microwave transmitter (10 000 MHz = 10 GHz) and receiver or with 30-cm ' television-type ' waves using the transmitter (1000 MHz = 1 GHz) and receiver

Fig. 6.28

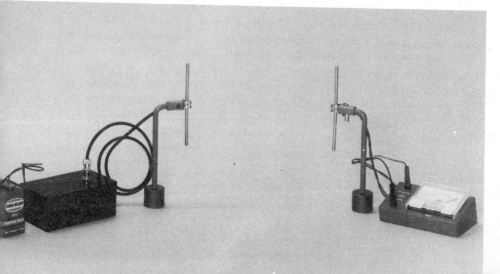

of Fig. 6.28. When the receivers are rotated through 90° in a vertical plane from the maximum signal position, the signal decreases to a minimum. Polarization provides the evidence for the transverse nature of electromagnetic waves.

Consideration of the way in which a transmitting aerial produces radio and other types of waves, shows that the electric field comes off parallel to the aerial and the magnetic field at right angles to it. If the transmitting aerial is vertical, the signal in the receiving aerial is a maximum when it too is vertical. In this case the radio waves are said to be *vertically polarized*, the direction of polarization being given by the direction of the electric field. This follows the practice adopted with light waves where experiments show that the coupling of light with matter is more often through the electric field. The polarization of light will be considered later.

Speed of mechanical waves

(*a*) *Factors affecting.* A mechanical wave is transmitted by the vibration of particles of the propagating medium. A system can vibrate only if it has *mass* and *elasticity*.

Consider a load hung from the lower end of a spring. When the load is pulled down and released, the spring—due to its elasticity—pulls the load up again. However, owing to the load having mass it acquires kinetic energy and overshoots its original position. The opposition of the spring to compression gradually brings it to rest and it starts to fall. It again overshoots its original position and stretches the spring. The cycle of events is then repeated. Proper vibrations of the spring would not occur if it was very light (i.e. had no mass on the end) or was inelastic.

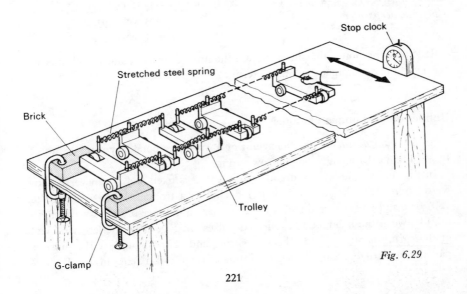

Fig. 6.29

Waves can therefore only be transmitted by a medium if it has mass and elasticity and it is these two factors which determine the speed of a wave. This last point may be demonstrated using a wave model in which trolleys represent particles of the transmitting medium and springs (slightly stretched) represent the bonds between the particles. The time is found for a transverse pulse to travel along to the fixed end and back again of a line of twelve trolleys linked as in Fig. 6.29.

Doubling the mass of each trolley (by adding a load, usually 1 kg) *decreases* the wave speed since it takes longer for the larger mass to acquire a certain speed. Doubling the tension (by connecting another spring in parallel with each one in the model) *increases* the wave speed because the larger force causes the next trolley to respond more quickly. If mass and tension are both doubled, the wave speed has its original value since both quantities change the speed by the same factor.

Note how the energy of the vibrating ' particles ' (trolleys) changes from k.e. to p.e. to k.e. and so on as the wave progresses.

(*b*) *Expressions for wave speeds.* These may be derived for different types of waves using basic mechanical ideas. Some are quoted below.

Transverse waves on a taut string or spring	$v = \sqrt{T/\mu}$	T = tension
		μ = mass per unit length
Longitudinal waves along masses (e.g. trolleys) linked by springs	$v = x\sqrt{k/m}$	x = spacing between mass centres
		k = spring constant
		m = one mass
Short wavelength ripples on deep water	$v = \sqrt{2\pi\gamma/(\lambda\rho)}$	γ = surface tension
		λ = wavelength
		ρ = density

It should be noted that in each case the numerator contains an ' elasticity ' term and the denominator a ' mass ' term.

Reflection and phase changes

(*a*) *Mechanical waves.* The behaviour of a wave at a boundary can be studied by sending pulses along a long narrow spring.

In Fig. 6.30*a*, the left-hand end of the spring is fixed and a transverse ' upward ' pulse travelling towards it is reflected as a trough. A phase change of 180° or π rad has occurred and there is a phase difference of half a wavelength ($\lambda/2$) between the incident and reflected pulses. In Fig. 6.30*b* the left-hand end of the spring is attached to a heavier spring and at the boundary the pulse is partly transmitted and partly reflected, the reflected pulse being inverted.

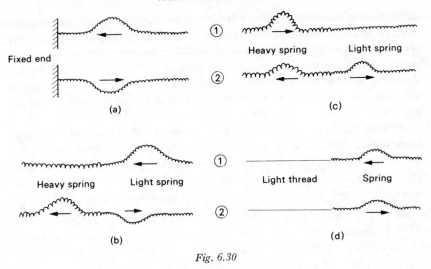

Fig. 6.30

In Fig. 6.30c a pulse passes from a heavy spring on the left to a light spring. Partial reflection and transmission again occur but the reflected pulse is not turned upside down. In Fig. 6.30d the left-hand end of the long narrow spring is fastened to a length of thin string and is in effect 'free'. Here almost the whole of the incident pulse is reflected the right way up, i.e. a crest is reflected as a crest and no phase change occurs.

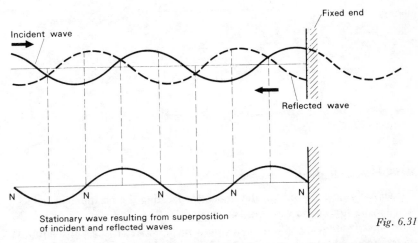

Stationary wave resulting from superposition of incident and reflected waves

Fig. 6.31

These results may be summarized by saying that when a transverse wave on a spring is reflected at a 'denser' medium (e.g. a fixed end or a heavier spring) there is a phase change of 180° (or π rad or $\lambda/2$). The phase change occurs in the case of the spring with one end fixed for example, because there can be no displacement of the fixed end, it must be a node. The incident and reflected waves therefore cause equal and opposite displacements at the fixed end so that they superpose to give resultant zero displacement as shown in Fig. 6.31.

Phase changes also occur when longitudinal waves are reflected, as can be shown by sending pulses along a Slinky spring to ' denser ' and ' less dense ' boundaries, i.e. to fixed and free ends. At a fixed end a compression is reflected as a compression, at a free end it is reflected as a rarefaction. Similar effects are obtained when sound waves are reflected in pipes with closed and open ends; a compression is reflected as a compression at a closed end and as a rarefaction at an open end. In the latter case the air in the compression is able to expand outwards suddenly at the end of the pipe and a rarefaction travels back along the pipe.

(b) *Electromagnetic waves.* The half-wavelength ($\lambda/2$) phase change suffered by microwaves when they are reflected at a metal plate can be shown as in Fig. 6.32. Interference occurs between the waves reaching the probe receiver directly and those arriving at it by reflection from the metal plate. Maxima and minima are detected when the receiver is moved along PQ. At Q the geometrical path difference between the direct and reflected waves is zero and the two sets of waves should reinforce to give a maximum. In fact a *minimum* is obtained due to the $\lambda/2$ phase change of the reflected wave.

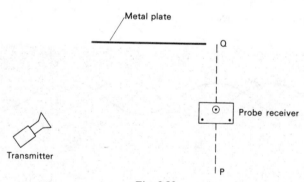

Fig. 6.32

Using reflected waves

Various techniques have been developed for locating the positions of ' objects ' by reflecting from them waves of known speed.

In *radar* (*r*adio *d*etection *a*nd *r*anging), radio waves (e.g. 3-cm microwaves) are emitted in short pulses by a transmitter and picked up after reflection from the ' object ', Fig. 6.33. The received pulse is displayed on a CRO with a calibrated time base which is triggered to start by the transmitted pulse. The time for the waves to travel twice the distance from the transmitting station to the ' object ' is thus found (the interval between transmitted pulses must be greater

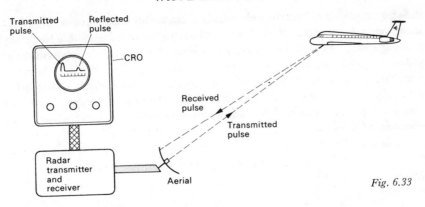

Fig. 6.33

than this to avoid confusion). Military ' objects ' include aircraft and missiles. In ships equipped with radar to assist navigation in fog and at night, a narrow microwave beam is swept continuously through 360° by a rotating aerial. The pulses reflected from land, other ships and buoys are shown on a CRO, called a *plan position indicator* (PPI), which has the time base origin in the centre of the screen and represents the ship; Fig. 6.34 shows Southampton Water on a yacht radar screen. Radar also helps to control aircraft waiting to land.

Fig. 6.34

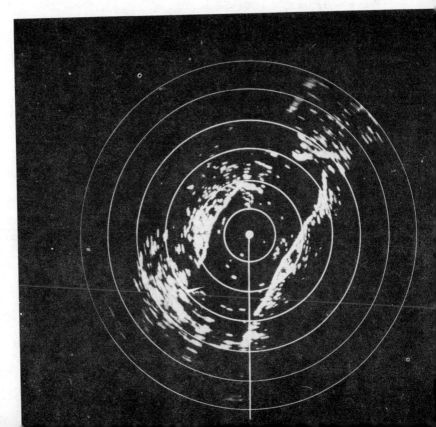

Sonar or sound navigation and ranging is similar to radar but employs ultra-sonic waves, i.e. waves with frequencies above the maximum audible frequency of about 20 kHz. It is used to measure the depth of the sea (i.e. in echo sounding) and to detect shoals of fish.

In the non-destructive testing of materials ultrasonic waves can detect flaws. If three pulses are obtained on the display CRO they are due to the transmitted pulse A, the pulse reflected B from the flaw and pulse C reflected from the boundary of the specimen, Fig. 6.35.

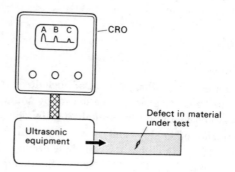

Fig. 6.35

Wave equations

(*a*) *Progressive wave.* Suppose the oscillation of the particle at O in Fig. 6.36 is simple harmonic of frequency f. Its displacement with time t will be given by $a \sin \omega t$ where a is the amplitude of the oscillation and $\omega = 2\pi f$. If the wave generated travels from left to right a particle at P, distance x from O, will lag

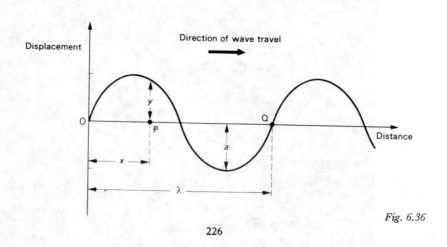

Fig. 6.36

behind the particle at O, say by phase angle ϕ. For the displacement y of the particle at P we can write

$$y = a \sin (\omega t - \phi). \qquad \text{(p. 171)}$$

But $\phi/2\pi = x/\lambda$ since at Q, distance λ from O, the phase difference between the motion of particles at O and Q is 2π (rad). Hence substituting for ϕ in the above equation we get

$$y = a \sin (\omega t - 2\pi x/\lambda)$$

$$= a \sin (\omega t - kx).$$

where $k = 2\pi/\lambda$.

If the wave travelled in the opposite direction, the vibration at P would lead on that at O by ϕ and the displacement y at P would be written

$$y = a \sin (\omega t + kx).$$

This is the equation for a wave travelling from *right to left*.

(b) *Stationary wave.* Let the two progressive waves of equal amplitude a and frequency f ($\omega = 2\pi f$) travelling in opposite directions be represented by

$$y_1 = a \sin (\omega t + kx) \quad \text{(to the left)}$$

and $$y_2 = a \sin (\omega t - kx) \quad \text{(to the right)}.$$

By the principle of superposition, the resultant displacement y is given by

$$y = y_1 + y_2$$

$$= a \sin (\omega t + kx) + a \sin (\omega t - kx).$$

Using the trigonometrical transformation for converting the sum of two sines to a product $\left(\text{i.e. } \sin \alpha + \sin \beta = 2 \sin \dfrac{\alpha + \beta}{2} \cos \dfrac{\alpha - \beta}{2} \right)$ we get

$$y = 2 a \sin \omega t \cos kx.$$

This is the equation of the stationary wave. It can be written

$$y = A \sin \omega t.$$

where $A = 2a \cos kx = 2a \cos (2\pi x/\lambda)$ is the amplitude of oscillation of the various particles. We see that when $x = 0$, $\lambda/2$ etc. A is a maximum and equal to $2a$. These are the antinodes. Nodes occur midway between antinodes since $A = 0$ when $x = \lambda/4$, $3\lambda/4$, $5\lambda/4$ etc.

WAVE MOTION

QUESTIONS

1. A plane wavefront of monochromatic light is incident normally on one face of a glass prism of refracting angle 30°, and is transmitted. Using Huygens' construction trace the course of the wavefront. Explain your diagram and find the angle through which the wavefront is deviated. (Refractive index of glass = 1.5.)

(*J.M.B.*)

2. State Snell's law of refraction and define refractive index.

Show how refraction of light at a plane interface can be explained on the basis of the wave theory of light.

Light travelling through a pool of water in a parallel beam is incident on the horizontal surface. Its speed in water is 2.2×10^{10} cm s^{-1}. Calculate the maximum angle which the beam can make with the vertical if light is to escape into the air where its speed is 3.0×10^{10} cm s^{-1}.

At this angle in water, how will the path of the beam be affected if a thick layer of oil, of refractive index 1.5, is floated on to the surface of the water? (*O. and C.*)

3. S$_1$ and S$_2$ in Fig. 6.37 are two sources of circular water waves of the same frequency, phase and amplitude. There is a maximum disturbance at P, a minimum at Q and a maximum at R.

Fig. 6.37

(*a*) Write *two* expressions for the wavelength λ of the ripples.

(*b*) How does the pattern change if (*i*) S$_1$ and S$_2$ are moved farther apart, and (*ii*) the frequency of the waves decreases?

4. (*a*) What is meant by (*i*) diffraction, and (*ii*) superposition of waves? Describe **one** phenomenon to illustrate each in the case of *sound waves*.

(*b*) The floats of two men fishing in a lake from boats are 21 m apart. A disturbance at a point in line with the floats sends out a train of waves along the surface of the water, so that the floats bob up and down 20 times per minute. A man in a third boat observes that when the float of one of his colleagues is on the crest of a wave that of the other is in a trough, and that there is then one crest between them. What is the velocity of the waves? (*O. and C.*)

5. The range of sound frequencies which can be detected by a certain person is 17.0 Hz to 20.0 kHz. What is the corresponding range of wavelengths if the speed of sound in air is 340 m s^{-1}?

6. (a) BBC Radio 2 broadcasts on a wavelength of 1.50 km. What is its frequency?

(b) What are the wavelengths of a television station which transmits vision on 500 MHz and sound on 505 MHz?

(c) What is the frequency of yellow light of wavelength 0.6 μm?

(Speed of electromagnetic waves in free space = 3×10^8 m s^{-1}; 1 μm (micrometre) = 10^{-6} m.)

7. The map in Fig. 6.38 shows part of a coastline, with two land-based radio navigation stations A and B. Both stations transmit continuous sinusoidal radio waves with the same amplitude and same wavelength (200 m).

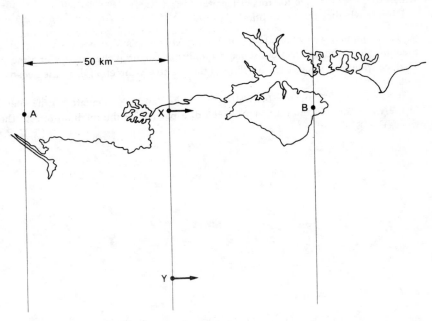

Fig. 6.38

A ship X, exactly midway between A and B, detects a signal whose amplitude is twice that of either station alone.

(a) What can be said about the signals from the two stations?

(b) The ship X travels to a new position by sailing 100 m in the direction shown by the arrow. What signal will it now detect? Explain this.

(c) A ship Y also starts at a position equidistant from A and B and then travels in the direction shown by the arrow. Exactly the same changes to the signal received were observed as in the case of ship X in (b).

Explain whether Y has sailed 100 m, more than 100 m, or less than 100 m.

(O. and C. Nuffield)

8. The velocity v of waves of wavelength λ on the surface of a pool of liquid, whose surface tension and density are σ and ρ respectively, is given by

$$v^2 = \frac{\lambda g}{2\pi} + \frac{2\pi\sigma}{\lambda\rho}$$

where g is the acceleration due to gravity.

Show that the equation is dimensionally correct.

A vibrator of frequency 480 ± 1 Hz produces, on the surface of water, waves whose wavelength is 0.125 ± 0.001 cm. Assuming that for this wavelength the first term on the right-hand side of the equation is negligible, calculate the value which these results give for the surface tension of water.

Discuss whether the assumption is justified.

(O. and C.)

9. The equation $y = a \sin (\omega t - kx)$ represents a plane wave travelling in a medium along the x-direction, y being the displacement at the point x at time t. Deduce whether the wave is travelling in the positive x-direction or in the negative x-direction.

If $a = 1.0 \times 10^{-7}$ m, $\omega = 6.6 \times 10^3$ s^{-1} and $k = 20$ m^{-1}, calculate (a) the speed of the wave, and (b) the maximum speed of a particle of the medium due to the wave.

(J.M.B.)

7 Sound

Sound waves

Sound exhibits all the properties of waves discussed in the previous chapter, except polarization. This suggests it is a longitudinal wave motion and since it cannot travel in a vacuum but requires a material medium it must be of the mechanical type.

A sound wave is produced by a vibrating object which superimposes, on any movement the particles of the transmitting medium have, an oscillatory to-and-fro motion along the direction of travel of the wave. The frequency of these oscillations—i.e. of the sound wave—equals that of the vibrating source and to be audible to human beings must be roughly in the range 20 Hz to 20 000 Hz. The speed of sound in air is about 330 m s^{-1} and the corresponding wavelength limits are therefore (using $v = f\lambda$) 17 m and 17 mm respectively.

A progressive sound wave carries energy from the source to the receiver where, in the case of the ear, the succession of compressions and rarefactions causes the ear-drum to vibrate. The sensation of what we *also* call ' sound ' is then produced by impulses sent to the brain.

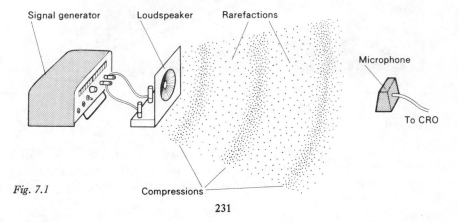

Fig. 7.1

231

It is worthwhile connecting a loudspeaker to a signal generator, Fig. 7.1, varying the frequency from a very low to a very high value (e.g. 10 Hz to 30 kHz) and observing the effect on (*i*) the cone of the loudspeaker, (*ii*) the ear and (*iii*) a CRO with a microphone feeding its Y input. At the lowest frequencies the vibrations of the cone can be seen and felt with the finger, but not heard (except for a weak thudding). As the frequency increases, the pitch of the note rises from that of a hum to a whistle and then to a hiss before it becomes inaudible. Above a few hundred hertz the movement of the speaker cone cannot be felt. The CRO responds throughout.

Sound waves with frequencies greater than 20 kHz or so are called *ultrasonic* waves. They are used by bats as a kind of navigational ' radar ' for night-flying. A correctly-cut quartz crystal generates ultrasonic waves when an alternating p.d. (of ultrasonic frequency) is applied across its faces. Their use in the non-destructive testing of materials has already been mentioned (p. 226) and Fig. 7.2 shows the shades of street lamps being cleaned by immersion in a tank of water which has an ultrasonic vibrator in the base.

Fig. 7.2a and b

Graphical representation

In Fig. 7.3a the circled figures ①, ②, ③ etc. represent parts of the air in which a progressive longitudinal wave is generated when the blade P is plucked. It can be seen that as P and the air oscillate to and fro about their undisturbed positions,

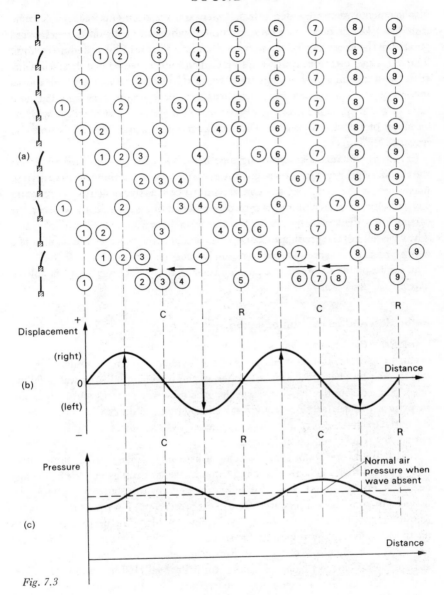

Fig. 7.3

compressions (C) and rarefactions (R) travel to the right. The following conclusions may be drawn.

(*i*) *Displacement*. Air at the centres of compressions and rarefactions has zero displacement. A displacement-*distance* graph for the instant of time corresponding to the last line of Fig. 7.3*a* can be drawn as in Fig. 7.3*b* where the longitudinal

233

displacements are represented as if they were transverse and displacements to the right are taken as positive. The amplitude of vibration of the air is much exaggerated in the graph; even for a very loud sound it is only of the order of 10^{-2} mm. The velocities of the central parts are a maximum but they are directed away from the source in a compression and towards it in a rarefaction, i.e. the air moves forward in a compression and backwards in a rarefaction. The variation with time of the displacement of the air is thus $90°$ ($\pi/2$ rad) out of phase with its variation of velocity with time. Draw two graphs to show this for a sinusoidal wave.

(*ii*) *Pressure excess*. The crowding together of the air in a compression causes the density and pressure of the air to be greater there than normal; both are a maximum at the centre of the compression. Conversely, density and pressure have minimum values at the centre of a rarefaction. Fig. 7.3c shows how the pressure (or density) varies with *distance* in a progressive longitudinal wave. The variations of the graph are exaggerated, actual variations are very small, 30 Pa being typical for a very loud sound (normal atmospheric pressure is about 10^5 Pa). The ear responds to pressure changes rather than displacements and is able to detect variations of as little as 2×10^{-5} Pa.

Reflection and refraction of sound

(*a*) *Reflection*. Sound waves obey the laws of reflection. Regular reflection occurs at a surface if it treats all parts of the incident wavefront similarly. To do this it must be ' flat ' to within a fraction of the wavelength of the waves falling on it. Speech sound waves have wavelengths of several metres and so it is not surprising that surfaces as ' rough ' as cliffs, large buildings etc. reflect sound regularly to give *echoes*, i.e. sound images. (Light requires a highly polished surface for regular reflection. Why?)

In a hall or large room sound reaches a listener by many paths, some much longer than others, as a result of reflection at walls, ceilings etc. *Reverberation* occurs when the sound produced at one instant by the source persists at the listener for some time afterwards. Too short a reverberation time makes a room sound ' dead ', but if it is too long, ' confusion ' results. The best value depends on the function of the building; for speech about 0.5 s is acceptable, whereas for music between one and two seconds is required.

In a modern concert hall such as the Royal Festival Hall in London, Fig. 7.4, the reverberation time is made the same irrespective of the size of the audience by upholstering the seats with material having similar absorption to clothing. Also, to distribute the sound as equally as possible, walls etc., tend to be built up from many small, flat surfaces at various inclinations. Curved surfaces, which might focus the sound strongly in one place, are avoided.

(*b*) *Refraction*. When the speed of sound waves changes at a boundary between two media, refraction occurs, i.e. the direction of travel changes as does the wave-

Fig. 7.4

length (since the frequency is fixed by the source). Snell's law, (p. 206) is obeyed and the refractive index $_1n_2$ for waves of a given frequency, having speed v_1 in medium ① and v_2 in medium ② is defined by

$$_1n_2 = \frac{v_1}{v_2}.$$

Sounds are more audible at night than during the day because the speed of sound in warm air exceeds that in cold air (p. 251) and refraction occurs. At night the air is usually colder near the ground than it is higher up and refraction towards the earth occurs, Fig. 7.5a. During the day the reverse is usually true. Fig. 7.5b.

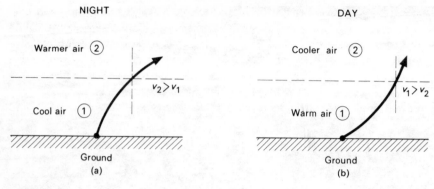

NIGHT   DAY

Warmer air ②   Cooler air ②

$v_2 > v_1$ $v_1 > v_2$

Cool air ①   Warm air ①

Ground   Ground
(a)   (b)

Fig. 7.5

Interference and diffraction of sound

(*a*) *Interference.* The interference of sound waves may be demonstrated with two loudspeakers connected in parallel to the low impedance output of a signal generator, set at 4 kHz, Fig. 7.6. The fringes are detected by moving a microphone feeding a CRO, preferably via a pre-amplifier, along XY. (The effect of unwanted reflections is minimized by working between two benches or, better still, out of doors.)

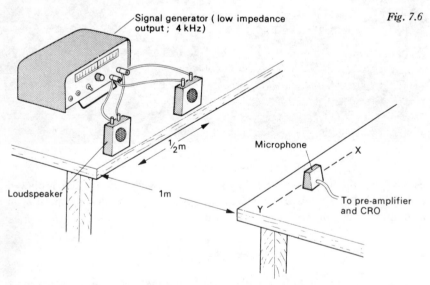

Signal generator (low impedance output ; 4 kHz)

Fig. 7.6

$\frac{1}{2}$m

Microphone

X

Loudspeaker

1m

To pre-amplifier and CRO

Y

When the microphone is at the central maximum of the interference pattern (i.e. along AB in Fig. 6.16, p. 211) and one loudspeaker is covered up, the output p.d. from the microphone (as shown on the CRO) falls to *half* of the value it had with both speakers uncovered. If the CRO were replaced by a resistor, the current through it would also be halved on blocking off the sound from one speaker and so the rate of energy conversion in it would be a *quarter* of its value with two speakers (since power = p.d. × current). Conversely, the energy at a central maximum will be *four* times that emitted in a given time by one of the speakers. We must conclude that since there is zero energy at minima, the energy from them goes to the maxima when two waves superpose to produce interference fringes. That is, energy is redistributed.

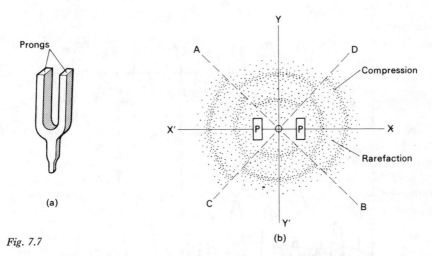

Fig. 7.7
(a)
(b)

A tuning fork, Fig. 7.7a, may also be used to show interference of sound. If it is struck and held close to the ear with the prongs vertical whilst being slowly rotated about a vertical axis, no sound is heard four times in each revolution. When the prongs PP are moving outwards, a rarefaction is produced in the air between them and travels along OY and OY'. Simultaneously, compressions are generated outside the prongs and move along OX and OX', Fig. 7.7b. The compressions and rarefactions spread out as spherical wavefronts and destructive interference occurs along AOB and COD where compressions and rarefactions superpose.

(b) *Diffraction*. The spreading of sound waves round the corners of an open doorway is an everyday event and suggests that the width of such an aperture (say, one metre) is of the same order as the wavelength of sound. A note of frequency 300 Hz has a wavelength in air of about one metre.

Beats

When two notes of slightly different frequencies but similar amplitudes are sounded together, the loudness increases and decreases periodically and *beats* are said to be heard. The effect may be demonstrated with two signal generators, each feeding a loudspeaker. Alternatively, two tuning forks of the same frequency can be used, one having a little Plasticine stuck on to one prong to lower its frequency slightly. They are struck simultaneously and the stems pressed against the bench top. If the difference between the frequencies of the two sources is increased, the beat frequency also rises until it is too high to be counted.

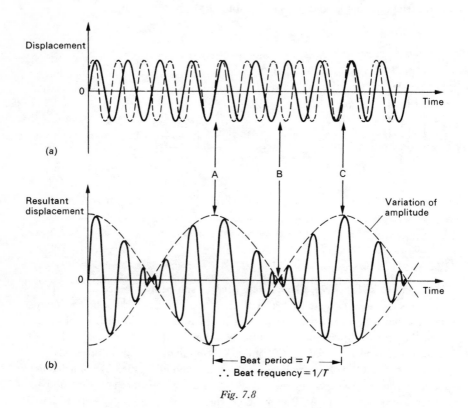

Fig. 7.8

The production of beats is a wave effect explained by the principle of super-position. The displacement-time graphs for the wave-trains from two sources of nearly equal frequency are shown in Fig. 7.8a for a certain observer. At an instant such as A the waves from the sources arrive in phase and reinforce to produce a loud sound. The phase difference then increases until a compression (or rarefaction) from one source arrives at the same time as a rarefaction (or compression) from the other. The observer hears little or nothing, point B. Later the waves are in phase again (point C) and a loud note is heard. Fig. 7.8b gives the resultant variation of amplitude. Beats are thus due to interference but because the sources are not coherent (frequencies different) there is sometimes reinforcement at a given point and at other times cancellation.

We will now show that the *beat frequency equals the difference of the two almost equal frequencies*. Suppose the beat period (i.e. the time between two successive maxima) is T and that one wave-train of frequency f_1 makes one cycle more than the other of frequency f_2 then

$$\text{number of cycles of frequency } f_1 \text{ in time } T = f_1 T$$

and, $$\text{number of cycles of frequency } f_2 \text{ in time } T = f_2 T$$

$$\therefore f_1 T - f_2 T = 1$$

$$\therefore \quad f_1 - f_2 = 1/T.$$

But one beat has occurred in time T and so $1/T$ is the beat frequency, hence

$$\text{beat frequency} = f_1 - f_2.$$

Stationary sound waves

(a) *Production.* The production of stationary or standing waves by the super-position of two trains of progressive waves of equal amplitude and frequency, travelling with the same speed in opposite directions, was considered previously for transverse waves. We saw that nodes and antinodes were formed. At a node the displacement of the vibrating particle was always zero but at an antinode it varied periodically from zero to maxima in opposite directions (see Fig. 6.23b, p. 216). Also, the wavelength of the stationary wave equalled that of either progressive wave and was twice the distance between two consecutive nodes or antinodes.

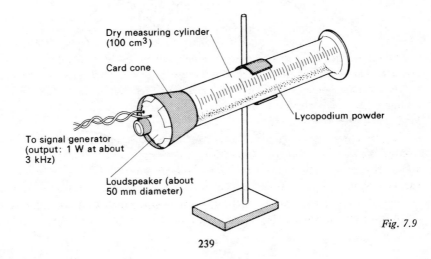

Fig. 7.9

SOUND

Stationary sound waves can be obtained in air using apparatus called *Kundt's tube*, Fig. 7.9. The speaker, driven by a signal generator, produces progressive longitudinal waves which travel through the air to the end of the cylinder where they are reflected to interfere with the incident waves. The frequency of the sound is varied and resonance occurs when it equals one of the natural frequencies of vibration of the air column. Stationary waves are then formed and are revealed by the lycopodium powder which swirls away from the antinodes (where the air is vibrating strongly) and after a short time forms into heaps at the nodes.

(*b*) *Displacement and pressure graphs.* Earlier we discussed the pressure variation in a progressive sound wave (p. 234), we shall now consider how it varies in a stationary sound wave. In Fig. 7.10*a*, curves ① and ② are displacement-distance graphs for a stationary wave for times when the displacements are greatest. As before (Fig. 7.3*b*) displacements to the right are taken to be positive. Consider ①. Air to the left of node N_1 is displaced towards N_1 since this displacement is positive; air to the right has a negative displacement but is also displaced towards N_1. At N_1 the air is thus compressed and the pressure greater than normal. At node N_2 the air is again displaced in opposite directions but away from N_2. In this case the air is less dense and the pressure less than normal. Thus on either side of a node the air is vibrating in antiphase. Approach causes pressure increase and separation makes it decrease. Parts of the air on either side of anti-node A_2, that are equidistant from it, have the same displacement to the left, i.e. they vibrate in phase and the air pressure is normal. The pressure-distance graph corresponding to displacement-distance graph ① is therefore given by ① in Fig. 7.10*b*. By treating displacement graph ② in the same way we obtain pressure-distance graph ②.

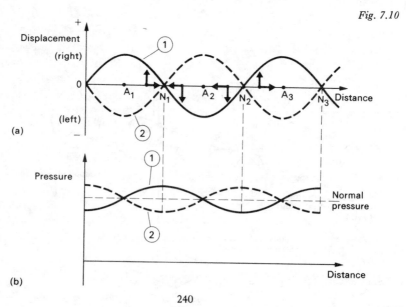

Fig. 7.10

From Fig. 7.10*b* we can conclude that in a stationary longitudinal wave the pressure *variation* is a maximum at a displacement node and is always zero at a displacement antinode. It is sometimes stated that displacement nodes and pressure antinodes coincide as do displacement antinodes and pressure nodes. Adjacent nodes and antinodes are always a quarter of a wavelength apart.

Musical notes

A musical note is produced by vibrations that are regular and repeating, i.e. by periodic motion. Non-periodic motion creates noise which is not pleasant to the ear.

A musical note has three characteristics.

(*a*) *Loudness*. Loudness is a subjective sensation (i.e. it depends on the listener) which is determined by the intensity of the sound and by the sensitivity of the observer's ear. *Intensity*, unlike loudness, is a physical quantity and is defined as the *rate of flow of energy through unit area perpendicular to the direction of travel of the sound at the place in question*. It is measured in watts per square metre (W m^{-2}) and if P is the total energy emitted per second equally in all directions by a point source, then neglecting absorption in the surrounding medium, the intensity I at a distance r from the source is given by

$$I = \frac{P}{4\pi r^2}.$$

This is an *inverse square* relationship and it holds for the intensity at a point due to any source of spherical waves of any type.

The energy of a particle describing simple harmonic motion (s.h.m.) is directly proportional to the square of the amplitude of vibration if other factors are held constant.[1] Now any periodic motion which is not simple harmonic can be resolved into a number of such motions and can conversely be thought of as formed from these separate s.h.m.s. Sound energy is carried by periodic vibrations of the air and so the *intensity* (*and loudness*) *of a sound must also be directly proportional to the square of the amplitude of vibration of the air* and, in turn, of the source.

Amplitudes are alternately positive and negative and their average value is zero. The squares of amplitudes are always positive and their average is not zero, which further confirms their use to measure the rate at which a wave transports energy, i.e. its intensity. The demonstration described on p. 236 using two loudspeakers, a microphone and CRO to show interference with sound waves offers some experimental evidence for this point. An analogous situation occurs with alternating currents where the rate of delivery of energy is proportional to the square of the peak current.

[1] See *Advanced Physics: Materials and Mechanics*, p. 324.

SOUND

(b) *Pitch*. This is also a sensation experienced by a listener. It depends mainly on the frequency of vibration of the air, which of course is the same as that of the source of sound. Frequency is a physical quantity and is measured in hertz (Hz), one vibration per second being one hertz. A high frequency produces a high-pitched note and a low frequency gives a low-pitched note. A note of frequency 500 Hz sounded in front of a microphone connected to a CRO gives twice as many waves (for the same time base setting) as one of frequency 250 Hz. The notes are said to differ in pitch by one *octave*. Pitch is analogous to colour in light.

(c) *Quality or timbre*. The same note played on two different instruments does not sound the same and different waveforms are obtained on a CRO. The notes are said to have different *quality* or *timbre*. This is due to the fact that with one exception (a tuning fork), sounds are never ' pure notes ', i.e. of one frequency; their waveforms are not sinusoidal. They consist of a main or *fundamental* note— which usually predominates—and others, generally with smaller intensities, called *overtones*. The fundamental is the component of lowest frequency and the overtones have frequencies that are multiples of the fundamental frequency. *The number and intensities of the overtones determines the quality of a sound* and depends mostly on the instrument producing the sound. The fundamental is also called the *first harmonic* and if it has frequency *f*, the overtones with frequencies 2*f*, 3*f* etc. are the second, third etc., harmonics. As we shall see shortly (p. 245), with some instruments certain harmonics are not present as overtones in a note.

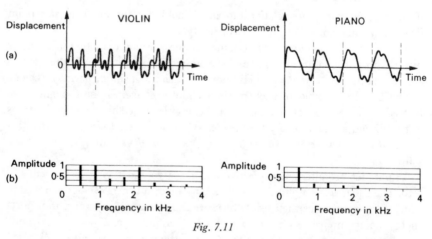

Fig. 7.11

Waveforms for the same note of fundamental frequency 440 Hz played on a violin and a piano are shown in Fig. 7.11a. They are very different and provide visual evidence that the notes are of different ' quality '. Using a mathematical technique known as *Fourier analysis* the frequencies and amplitudes of the harmonics present as overtones can be worked out in each case and sound spectra

242

obtained, Fig. 7.11*b*. Loud higher harmonics (notably the second and fifth) are present in the violin spectrum. The analysis depends on the fact that a periodic waveform, however complex, can be resolved into a number of sine waveforms with frequencies that are multiples of the fundamental. For example, a waveform like that of graph ① in Fig. 6.14 (p. 208) has two components—the fundamental ② and an overtone ③ (the third harmonic) with three times the frequency and half the amplitude of the fundamental.

Vibrating air columns

(*a*) *Wind instruments*. Stationary longitudinal waves in a column of air in a pipe are the source of sound in wind instruments. To set the air into vibration a disturbance is created at one end of the pipe, the other end can be open or closed (stopped). In an organ pipe, Fig. 7.12, a jet of air strikes a sharp edge, in a recorder the player blows across a hole, while in an oboe a reed vibrates when blown. Progressive sound waves travel from the source to the end of the pipe

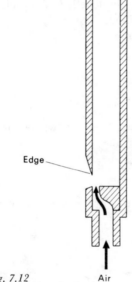

Edge

Fig. 7.12   Air

where they are reflected and interfere with the incident waves. The wavelength of some waves will be such that the length of the pipe produces resonance, i.e. a compression in the reflected wave arrives back at the source just as it is producing another compression. The amplitude of these waves will be large and a stationary longitudinal wave is formed.

The possible modes of vibration of the air in closed and open pipes will now be considered.

(b) *Closed pipe.* In the simplest stationary wave vibration possible, there will be a displacement node N at the closed end of the pipe, since the air there must be at rest, and a displacement antinode A at the open end where the air can vibrate freely, Fig. 7.13a. (Where will the pressure (i) node, and (ii) antinode, be formed?) Fig. 7.13b shows the displacement-distance graph for this vibration

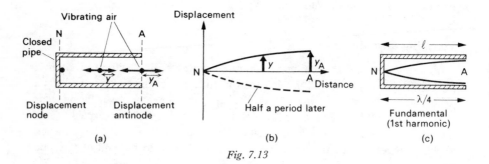

Fig. 7.13

and from Fig. 7.13c we see that the length *l* of the pipe equals the distance between a node and the next antinode. That is NA = $\lambda/4$ where λ is the wavelength of the stationary wave. Hence

$$l = \frac{\lambda}{4}$$

$$\therefore \lambda = 4l.$$

The frequency *f* of the note is given by $v = f\lambda$ where *v* is the speed of sound in air. Therefore

$$f = \frac{v}{\lambda} = \frac{v}{4l}.$$

This is the lowest frequency produced by the pipe, i.e. the fundamental frequency *f* or first harmonic.

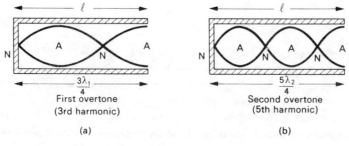

Fig. 7.14

The next two simplest ways in which the air may vibrate in the same pipe are shown in Fig. 7.14a and b. In both cases there are displacement nodes at the closed ends and displacement antinodes at the open ends. In (a), if the wavelength is λ_1, we have

$$l = \frac{3\lambda_1}{4}.$$

The frequency f_1 is therefore given by

$$f_1 = \frac{3v}{4l}.$$

This is the first overtone and since $f_1 = 3f$ it is the third harmonic. Similarly from (b) we find that the frequency f_2 of the second overtone is $5f$, i.e. it is the fifth harmonic and in general a *closed pipe gives only odd-numbered harmonics*.

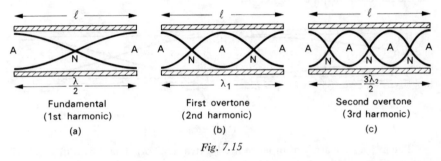

Fundamental	First overtone	Second overtone
(1st harmonic)	(2nd harmonic)	(3rd harmonic)
(a)	(b)	(c)

Fig. 7.15

(c) *Open pipe*. Here, both ends of the pipe are open and are displacement antinodes A. The simplest stationary wave is shown in Fig. 7.15a. There is a displacement node N midway between the two antinodes, hence

$$l = \frac{\lambda}{2}.$$

The fundamental frequency f is given by

$$f = \frac{v}{\lambda} = \frac{v}{2l}.$$

The next simplest mode of vibration, Fig. 7.15b gives the first overtone and we have

$$l = \lambda_1$$

$$\therefore f_1 = \frac{v}{l}.$$

This is the second harmonic since $f_1 = 2f$. The second overtone is obtained from Fig. 7.15c and is the third harmonic. In an open pipe all harmonics are possible as overtones.

(*d*) *Further points.*

(*i*) The fundamental frequency of an open pipe is *twice* that of a closed pipe of the same length.

(*ii*) The note from an open pipe is richer than that from a closed pipe due to the presence of extra overtones.

(*iii*) Higher overtones are encouraged by blowing harder.

(*iv*) The actual vibration of the air in a pipe is the resultant, by superposition, of the various modes that occur.

(*v*) In practice the air just outside the open end of a pipe is set into vibration and the displacement antinode of a stationary wave occurs a distance *c*—called the *end-correction*—beyond the end. The effective length of the air column is therefore greater than the length of the pipe. Hence for the fundamental of a closed pipe, Fig. 7.16*a*, we should write $\lambda/4 = l + c$ and for an open pipe, Fig. 7.16*b*, we have $\lambda/2 = l + 2c$. The value of *c* is generally taken to be 0.6 *r* where *r* is the radius of the pipe.

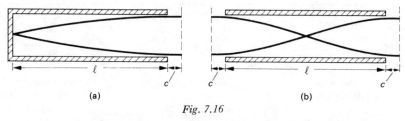

(a) (b)

Fig. 7.16

The impressive array of pipes of the Festival Hall organ is shown in Fig. 7.17.

Fig. 7.17

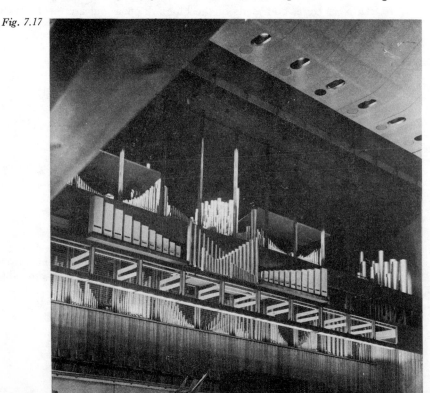

SOUND

Vibrating strings

(a) *String instruments.* The ' string ' is a tightly-stretched wire or length of gut. When it is struck, bowed or plucked, progressive transverse waves travel to both ends, which are fixed, where they are reflected to meet the incid and : waves. A stationary wave pattern is formed for waves whose wavelengths ' fit ' into the length of the string, i.e. resonance occurs. A progressive sound wave (i.e. a longitudinal wave) is produced in the surrounding air with a frequency equal to that of the stationary transverse wave on the string.

The weak sound produced by the vibrating strings is transmitted to a sounding board in a piano and to the hollow body of instruments such as the violin. A larger mass of air is thereby set vibrating and the loudness of the sound increased.

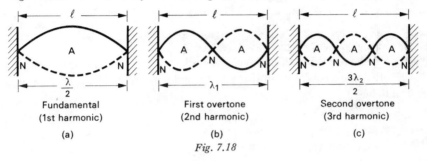

Fundamental	First overtone	Second overtone
(1st harmonic)	(2nd harmonic)	(3rd harmonic)
(a)	(b)	(c)

Fig. 7.18

(b) *Modes of vibration.* The fixed ends must be displacement nodes N. If the string is plucked in the middle the simplest mode of vibration is shown in Fig. 7.18a, A being a displacement antinode. This vibration creates the fundamental note of frequency f, and

$$l = \frac{\lambda}{2}$$

where l is the length of the string. The frequency f is therefore

$$f = \frac{v}{\lambda} = \frac{v}{2l}$$

where v is the speed of the transverse wave along the string.

By plucking the string at a point a quarter of its length from one end, it can vibrate in two segments, Fig. 7.18b. Then, if λ_1 is the wavelength of the resulting stationary wave,

$$l = \lambda_1$$

$$\therefore f_1 = \frac{v}{\lambda_1} = \frac{v}{l} = 2f.$$

This is the first overtone or second harmonic. If the string vibrates in three segments, Fig. 7.18c, the second overtone or third harmonic is obtained.

A string can vibrate in several modes simultaneously, depending on where it is plucked etc., and the frequencies and relative intensities of the overtone produced determine the quality of the note emitted.

Melde's experiment (p. 216) shows the various stationary wave patterns of a vibrating string (or rubber cord). It will be seen that, rather unexpectedly, the driven point of the string is nearly a node; certainly the amplitude of vibration is much less than at an antinode. The motion there is no more than is required for the vibrator to make good the energy dissipated in the system. Also, the frequency of the wave is the same as that of the vibrator.

(c) *Laws of vibration of stretched strings.* It was stated earlier (p. 222) that the speed v of a transverse wave travelling along a stretched string is given by

$$v = \sqrt{\frac{T}{\mu}} \qquad (1)$$

where T is the tension in the string and μ is its mass per unit length. If the string has length l, the frequency f of the fundamental note emitted by it has just been shown to be

$$f = \frac{v}{2l}. \qquad (2)$$

Hence from (1) and (2),

$$f = \frac{1}{2l}\sqrt{\frac{T}{\mu}}. \qquad (3)$$

From (3) it follows that
(i) $f \propto 1/l$ if T and μ are constant
(ii) $f \propto \sqrt{T}$ if l and μ are constant
(iii) $f \propto 1/\sqrt{\mu}$ if l and T are constant.

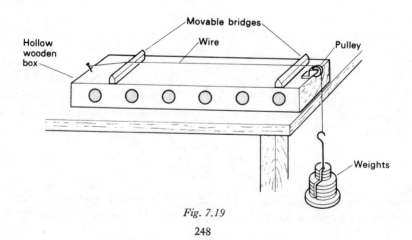

Fig. 7.19

248

These three statements are known as the *laws of vibration of stretched strings*. They indicate that short, thin wires under high tension emit high notes. The laws may be verified experimentally using a *sonometer*. This consists of a wire under tension, arranged on a hollow wooden box as in Fig. 7.19. The vibrations of the wire are passed by the movable bridges to the box and then to the air inside it.

To investigate the relationship between *f* and *l*, the bridges are moved so that different lengths between them emit their fundamental frequencies when plucked at the centre. The frequencies may be found by one or other of the methods outlined on p. 262 or by taking lengths of wire which vibrate in unison (as judged by the ear) with tuning forks of known frequencies. The product $f \times l$ should be constant within the limits of experimental error.

To check that $f \propto \sqrt{T}$, the same length of wire is subjected to different tensions by changing the hanging mass and the frequencies found as before.

The relation between *f* and $1/\sqrt{\mu}$ requires the use of wires of different diameters and materials but of the same vibrating length and under the same tension.

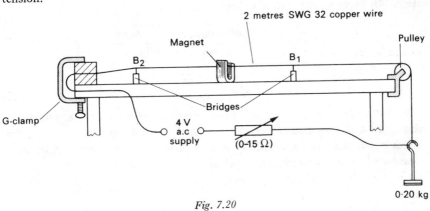

Fig. 7.20

(*d*) *Measuring the frequency of an a.c. supply,* Fig. 7.20. The wire carrying the a.c. experiences an alternating transverse force when the magnet is placed so that the wire passes between its poles. Bridge B_1 is arranged near B_2 at the fixed end of the wire and then slowly separated from it, keeping the magnet approximately midway between them, until the wire vibrates in one loop with maximum amplitude, the current being reduced to a low value. The frequency *f* of the a.c. is given by

$$f = \frac{1}{2l}\sqrt{\frac{T}{\mu}}$$

where $l = B_1B_2$, $T = (0.20 \text{ kg}) \times (10 \text{ N kg}^{-1}) = 2.0 \text{ N}$ and μ is found by weighing a two-metre length of the SWG 32 copper wire. If T is kept constant and the magnet left in position, doubling the distance between the bridges should produce two loops and a second value of f may be calculated (in this case $l = B_1B_2/2$).

A striking variation of the experiment can be shown in a darkened room with a suitable length of nichrome wire connected directly across the 240-V a.c. mains supply—*but great care is necessary not to touch the wire.* If the current is sufficient and several loops are obtained, the nodes become red-hot and show up clearly. (With a ten-metre length of SWG 28 nichrome wire the current is satisfactory and if three metres are allowed to vibrate, four loops are obtained with a load of 0.5 kg.)

Speed of sound

It was shown on p. 221 that the speed of a mechanical wave in a medium depends on 'elasticity' and 'mass' factors and formulae for the speeds of different waves were quoted. An expression for the speed of sound in *any* medium was derived by Newton and may be written

$$\text{speed} = \sqrt{\frac{\text{elastic modulus of medium}}{\text{density of medium}}}.$$

The elastic modulus (i.e. stress/strain) concerned depends on the type of strain produced by the wave in the medium.

(*a*) *Solids.* In a solid rod of small diameter compared with the sound wavelength, the compressions and rarefactions of the sound wave cause a change of length strain and Young's modulus E is appropriate in the above expression. Hence for a solid rod of density ρ in which the speed of sound is v, we have

$$v = \sqrt{\frac{E}{\rho}}.$$

When E is in N m^{-2} (i.e. Pa) and ρ in kg m^{-3}, v is in m s^{-1}. For steel $E = 2.0 \times 10^{11}$ Pa and $\rho = 7.8 \times 10^3$ kg m^{-3} giving $v = 5.1 \times 10^3$ m s^{-1}.

The formula for the speed of sound in a solid is derived in Appendix 5, p. 533.

(*b*) *Gases.* Volume changes occur in the case of gases (and liquids) and the bulk modulus K of the gas is the relevant elastic modulus. It can be shown that under the conditions in which a sound wave travels in a gas, $K = \gamma p$ where p is the pressure of the gas and γ is the ratio of its principal heat capacities, i.e. C_p/C_v (see p. 357). We then have for the speed of sound v in a gas of density ρ

$$v = \sqrt{\frac{\gamma p}{\rho}}.$$

For air at s.t.p., $\rho = 1.3$ kg m^{-3}, $\gamma = 1.4$ and $p = 1.0 \times 10^5$ Pa, therefore $v = 3.3 \times 10^2$ m s^{-1}, which agrees well with the experimental value.

(*c*) *Effect of pressure, temperature and humidity.* For one mole of an ideal gas having volume V and pressure p at temperature T, we can write (p. 334).

$$pV = RT$$

where R is a constant. If M is the mass of gas

$$\rho = \frac{M}{V} = \frac{Mp}{RT}$$

$$\therefore \quad v = \sqrt{\frac{\gamma p}{\rho}} = \sqrt{\frac{\gamma RT}{M}}.$$

The final expression does not contain p and so *the speed of sound in a gas is independent of pressure.* This is not unexpected since changes in density are proportional to changes of pressure, i.e. p/ρ is a constant.

Also, since R has the same value for all gases and γ and M are constants for a particular gas.

$$v \propto \sqrt{T}.$$

That is, *the speed of sound in a gas is directly proportional to the square root of the absolute temperature of the gas* (provided γ is independent of temperature).

For moist air ρ is less than for dry air and γ is slightly greater. The net result is that *the speed of sound increases with humidity.*

(*d*) *Effect of wind.* If the air carrying sound waves is itself moving, i.e. there is a wind, the wind velocity has to be added vectorially to that of the sound to obtain the velocity of the latter with respect to the ground. And since wind velocities usually increase with height, wavefronts become distorted as in Fig. 7.21, resulting in greater audibility down-wind than up-wind.

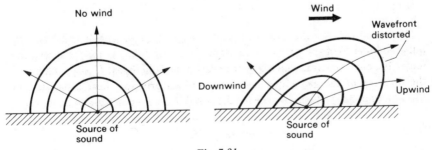

Fig. 7.21

Measuring the speed of sound

(a) *In air by progressive waves using a CRO.* The CRO must have an output terminal from its time base and the latter must be calibrated. The procedure described here refers to the Telequipment S51E (Fig. 11.3b, p. 435).

The VARIABLE control (fine time base) is rotated fully clockwise to the CAL position and the X GAIN turned fully anticlockwise to its minimum setting. These two adjustments ensure that the time base will run at the speed indicated by TIME/CM control. The latter is set to 100 ms. The VOLTS/CM knob is set at the maximum Y-amplifier gain of 0.1 V and the STABILITY and TRIG LEVEL controls rotated to their maximum clockwise positions. The CRO is then switched on at the BRIGHTNESS control and X-SHIFT adjusted so that the start of the trace at the left of the screen is visible.

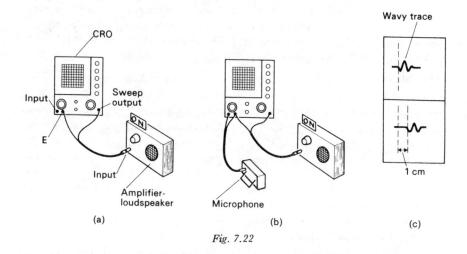

Fig. 7.22

The SWEEP OUTPUT and E terminals on the CRO are connected to the input of an amplifier-loudspeaker, Fig. 7.22a. The method uses the fact that as the time base starts off the trace at the left of the screen, a pulse comes from the sweep output and produces a noise in the loudspeaker. That this is so should be checked.

To make a measurement the time-base speed is increased to 1 ms/cm and a microphone, set close to the loudspeaker, is connected to the INPUT and E terminals on the CRO, Fig. 7.22b. The signal received by the microphone from the loudspeaker causes a ' wavy ' trace on the CRO. (The volume control on the amplifier may have to be turned up and the position of the trace adjusted by the X-SHIFT.) When the microphone is moved away from the speaker the ' wavy ' trace moves to the right on the screen. If we find the distance the microphone has to move to cause the trace to move one centimetre to the right,

SOUND

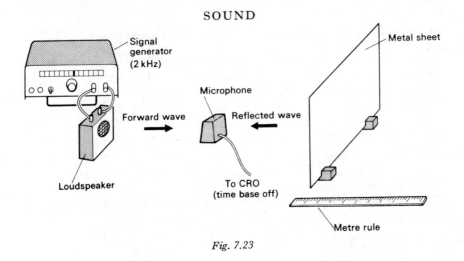

Fig. 7.23

Fig. 7.22c, then we can say sound travels that distance in one millisecond. The speed in air can then be calculated.

(b) *In air by stationary waves.* The apparatus is arranged as in Fig. 7.23. The signal generator delivers a note of known frequency f (2 kHz is suitable) to a loudspeaker which is directed towards a metal sheet. Interference occurs between the forward and reflected progressive sound waves and a stationary wave pattern with nodes and antinodes is established. If the reflector is moved slowly towards or away from the microphone the vertical trace on the CRO varies from a maximum to a minimum and the distance moved by the reflector between two consecutive maxima or minima equals $\lambda/2$ where λ is the wavelength of the sound wave (progressive or stationary). The speed v is obtained from $v = f\lambda$. In practice the reflector is moved through several maxima and minima so that a greater distance is measured.

The stationary wave pattern produced in Kundt's tube (p. 240) and indicated by the lycopodium powder, may also be used to find the speed of sound in air or in different gases.

(c) *In air using a resonance tube.* The arrangement is shown in Fig. 7.24a. The tuning fork is struck and held over the top of the tube whose position in the water is raised or lowered until the note is at its loudest. The fundamental frequency of the air column then equals the frequency of the fork, i.e. there is resonance. A stationary wave now exists in the tube with a displacement node N at the closed end and a displacement antinode A near the open end. We then have

$$l_1 + c = \frac{\lambda}{4} \tag{1}$$

where l_1 is the length from the water level to the top of the tube and c is the end correction.

253

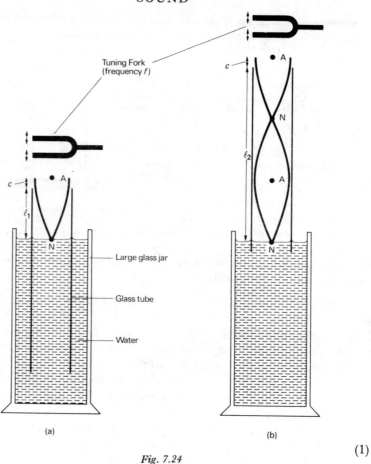

Tuning Fork
(frequency f)

Large glass jar

Glass tube

Water

(a)

(b)

Fig. 7.24

$$(1)$$

A second weaker resonance can be obtained with the same fork by slowly raising the tube out of the water. The air column of greater length l_2 is then producing its first overtone (which equals the frequency of the fork) and from the stationary wave pattern of Fig. 7.24*b* we have

$$l_2 + c = \frac{3\lambda}{4}.$$ $$(2)$$

Subtracting (1) from (2) eliminates c and we get

$$l_2 - l_1 = \frac{\lambda}{2}.$$

Knowing the frequency f of the tuning fork, v can be calculated from $v = f\lambda$.

254

Alternatively using several forks of different, known frequencies f the funda-mental resonance lengths l can be found for each and we have

$$l + c = \frac{\lambda}{4} = \frac{v}{4f}$$

$$\therefore \quad l = \frac{v}{4f} - c.$$

A graph of l against $1/f$ should be a straight line of slope $v/4$ (and intercept $-c$ on the $1/f$ axis).

(d) *In a metal rod.* This method gives an approximate value but allows a rough check to be made on the expression $v = \sqrt{E/\rho}$ (p. 250) for, say, steel. In Fig. 7.25a the output from the signal generator is applied to the input of the CRO when the near end of the suspended rod is hit with the hammer. A compression pulse travels to the far end and is reflected as an expansion (rarefaction) pulse to the near end where *it breaks the contact between the hammer and the rod.* The time of contact is calculated from the length of the trace on the CRO (using its cali-brated time scale of 100 μs cm^{-1}) and is the time for sound to travel *twice* the length of the rod. (*Note.* The stability control on the CRO should be turned only as far anticlockwise as is needed to give a trace when contact is made.)

A demonstration in slow motion with the apparatus of Fig. 7.25b may help to reinforce the idea that the reflected expansion pulse breaks the contact between the hammer and the rod. The four trolleys linked by springs are *all pushed*

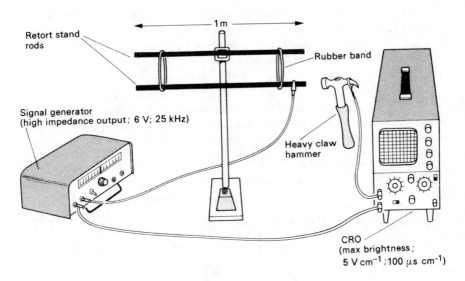

Fig. 7.25a

255

together to the left. The front trolley stops on striking the ' wall ' and a compression pulse travels along the row as each trolley stops in turn. The rear trolley then moves to the right, starting an expansion pulse which travels left to the front trolley and on reaching it contact with the ' wall ' is broken. The disturbance thus travels twice along the row of trolleys during the time the front trolley is in contact with the ' wall '. The scaler records the time of contact and if the speed of the pulses is required the distance from the front of the first trolley to the front of the rear one has to be measured.

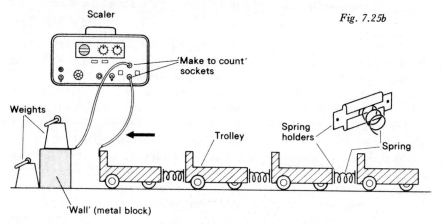

Fig. 7.25b

Doppler effect

The pitch of the note from the siren of a fast-travelling ambulance or police car appears to a stationary observer to drop suddenly as it passes. This *apparent* change in the frequency of a wave motion when there is relative motion between the source and the observer is called the *Doppler effect*. It occurs with electromagnetic waves as well as with sound. The expressions derived below apply to the latter and only to the former if the relative speed of source and observer is small compared with the speed of electromagnetic waves (otherwise relativistic effects have to be considered). The microwave Doppler effect is used in police radar speed checks and in tracking satellites.

(*a*) *Source moving.* In Fig. 7.26*a* S is the source of waves of frequency f and velocity v and O is the stationary observer. If S were at rest, the f waves emitted per second would occupy a distance v and the wavelength would be v/f. When S is moving towards O with velocity u_{s}, f waves are now compressed into the smaller distance $(v - u_{s})$ since S moves a distance u_{s} towards O per second, Fig. 7.26*b*. To O the effect thus appears to be a *decrease of wavelength* to a value λ_{0} given by

$$\lambda_{0} = \frac{(v - u_{s})}{f}.$$

SOUND

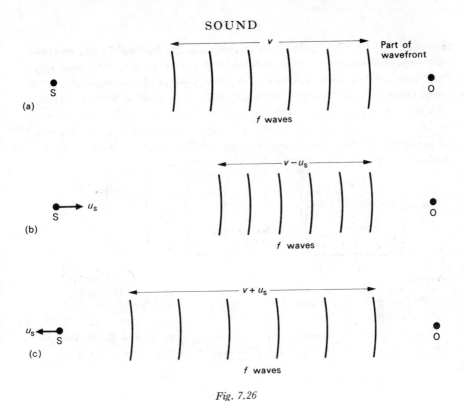

Fig. 7.26

Hence if f_0 is the apparent frequency we have

$$f_0 = \frac{\text{velocity of waves}}{\text{apparent wavelength}}$$

$$= \frac{v}{(v - u_s)/f}$$

$$\therefore \quad f_0 = \left(\frac{v}{v - u_s}\right)f.$$

The apparent frequency is therefore greater than the true frequency since $(v - u_s) < v$.

If S is moving away from the stationary observer O, the apparent wavelength $\lambda_0 = (v + u_s)/f$, Fig. 7.26c, and the apparent frequency f_0 is

$$f_0 = \left(\frac{v}{v + u_s}\right)f.$$

In this case $f_0 < f$ since $(v + u_s) > v$.

It should be noted that if, for example, a source of sound is approaching an observer with *constant velocity*, the apparent pitch of the note heard does not increase as the source gets nearer; it is constant but higher than the true pitch. Similarly, if the source recedes with constant velocity the observer hears a note of constant but lower pitch than the true pitch. The apparent change heard by the observer occurs *suddenly as the source passes.* What expression gives the value of the apparent change?

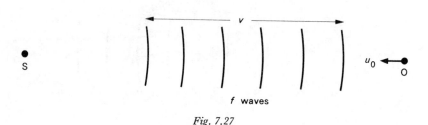

Fig. 7.27

(b) *Observer moving.* In this case the wavelength is unchanged and is given by v/f since the f waves sent out per second by the stationary source S occupy a distance v, Fig. 7.27. If the observer O has velocity u_0 towards S, the velocity of the waves relative to O is $(v + u_0)$. The apparent frequency f_0 is given by

$$f_0 = \frac{\text{relative velocity of waves}}{\text{wavelength}}$$

$$= \frac{v + u_0}{v/f}$$

$$\therefore \ f_0 = \left(\frac{v + u_0}{v}\right) f.$$

Thus $f_0 > f$ since $(v + u_0) > v$.

If O is moving away from the stationary source S, the velocity of the waves relative to O is $(v - u_0)$ and the apparent frequency f_0 is

$$f_0 = \left(\frac{v - u_0}{v}\right) f.$$

Here $f_0 < f$.

(c) *Source and observer moving.* Motion of the source affects the *apparent wavelength* and motion of the observer affects the *velocity of the waves* he receives. If the source and observer are approaching each other with velocities u_S and u_0 respectively then, as before, we can say

velocity of waves relative to O $= v_0 = v + u_0$

apparent wavelength $\lambda_0 \qquad = (v - u_S)/f$

SOUND

For the apparent frequency f_0 we therefore have

$$f_0 = \frac{v_0}{\lambda_0} = \frac{v + u_0}{(v - u_s)/f}$$

$$\therefore \quad f_0 = \left(\frac{v + u_0}{v - u_s}\right)f.$$

When source and observer are moving away from one another

$$\lambda_0 = \frac{v + u_s}{f} \quad \text{and} \quad v_0 = v - u_0.$$

Hence

$$f_0 = \frac{v_0}{\lambda_0} = \frac{v - u_0}{(v + u_s)/f}$$

$$\therefore \quad f_0 = \left(\frac{v - u_0}{v + u_s}\right)f.$$

In general

$$f_0 = \left(\frac{v \pm u_0}{v \mp u_s}\right)f$$

where the upper signs apply to approach and the lower signs to separation.

(*d*) *Doppler effect in light.* Because of the very great speed of light the effect is negligible for most terrestrial sources but it has been used to measure the speeds of stars relative to the earth. The positions of certain wavelengths (i.e. spectral lines), due to an identifiable element in the star's spectrum are compared with their positions in a laboratory produced spectrum of the element. Red shift indicates recession of the star from the earth and from its size the speed may be calculated.

The speed of rotation of the sun has also been found from the apparent difference in wavelength between the Fraunhöfer lines (p. 300) in the spectra of the western and eastern edges of its disc, the former is moving towards the earth and the latter is receding. The value obtained agrees with that deduced from observations of sunspots.

Doppler effect red shift measurements on the light from galaxies (each comprising many millions of stars) suggest that the universe is expanding, with each galaxy retreating from every other at speeds of up to one-third that of light.

One cause of the broadening of spectral lines is due to the Doppler effect. The molecules of a gas or vapour which is emitting light are moving at different angles towards and away from the observer with various very high speeds. As a result the wavelength of a particular spectral line has a range of apparent values. How will the molecules be moving that are responsible for the edges of the line?

SOUND

Sound calculations

1. What are the first two successive resonance lengths of a closed pipe containing air at 27 °C for a tuning fork of frequency 341 Hz? Take the speed of sound in air at 0 °C to be 330 m s⁻¹.

The speed of sound in a gas is directly proportional to the square root of the absolute temperature

$$\therefore \quad \frac{v_1}{v_0} = \sqrt{\frac{273 + 27}{273}} = \sqrt{\frac{300}{273}}$$

where v_1 and v_0 are the speeds of sound at 27 °C and 0 °C respectively,

$$\therefore \quad v_1 = 330\sqrt{\frac{300}{273}} = 346 \text{ m s}^{-1}.$$

For the first resonance length l_1, the closed pipe emits its fundamental frequency f_1 where $f_1 = 341$ Hz. The wavelength of the fundamental note is given by

$$\lambda_1 = \frac{v_1}{f_1} = \frac{346}{341} = 1.02 \text{ m}.$$

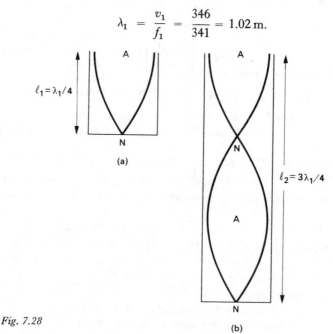

Fig. 7.28

(a)

(b)

The stationary wave pattern in the tube is then as in Fig. 7.28a, hence

$$l_1 = \frac{\lambda_1}{4} = \frac{1.02}{4} = 0.255 \text{ m}.$$

At the second resonance length l_2 the stationary wave is as in Fig. 7.28b. The resonating note has frequency f_1 and wavelength λ_1 as before, but it is now the first overtone for length l_2. Therefore

$$l_2 = \tfrac{3}{4}\lambda_1 = \frac{3 \times 1.02}{4} = 0.765 \text{ m.}$$

2. *A sonometer wire of length 0.50 m and mass per unit length* 1.0×10^{-3} *kg* m^{-1} *is stretched by a load of 4.0 kg. If it is plucked at its mid-point, what will be (a) the wavelength and (b) the frequency, of the note emitted? Take* $g = 10$ *N* kg^{-1}.

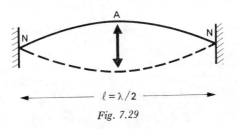

Fig. 7.29

(*a*) The wire vibrates as in Fig. 7.29 and emits its fundamental frequency f of wavelength λ. If l is the length of the wire then

$$\therefore \quad l = \frac{\lambda}{2}$$

$$\therefore \quad \lambda = 2l = 2 \times 0.50 = 1.0 \text{ m.}$$

(*b*) The fundamental frequency f emitted by a wire of length l, mass per unit length μ and under tension T is given by

$$f = \frac{1}{2l}\sqrt{\frac{T}{\mu}}.$$

Now $T = 4.0 \text{ kg} \times 10 \text{ N kg}^{-1} = 40 \text{ N}$, therefore

$$f = \frac{1}{2 \times 0.50}\sqrt{\frac{40}{1.0 \times 10^{-3}}} \text{ Hz}$$

$$= 2.0 \times 10^{2} \text{ Hz.}$$

3. *Calculate the frequency of the beats heard by a stationary observer when a source of sound of frequency 100 Hz moves directly away from him with a speed of 10.0 m* s^{-1} *towards a vertical wall. (Speed of sound in air* $= 340$ *m* s^{-1}.)

The observer hears beats because of interference between the sound coming to him directly from the source and that reflected from the wall. The apparent frequency of the former is less than 100 Hz since the source is moving away from him, while the apparent frequency of the latter is greater than 100 Hz. The motion of the source towards the wall causes the waves incident on it to have a shorter wavelength than they would if the source were at rest and so, to the observer, the reflected waves appear to come from a source moving towards him.

$$\text{Apparent frequency of direct sound} = \left(\frac{340}{340 + 10}\right) 100$$

$$= 97.1 \text{ Hz}.$$

$$\text{Apparent frequency of reflected sound} = \left(\frac{340}{340 - 10}\right) 100$$

$$= 103 \text{ Hz}.$$

$\therefore$ Number of beats per second $= 5.90$.

Measuring the frequency of a source of sound

Three methods will be outlined with instruments that are assumed to be correctly calibrated.

(*a*) *Using a signal generator and a loudspeaker.* The loudspeaker is fed from a signal generator whose frequency is altered until the notes from the speaker and the source of sound are judged to have the same pitch.

(*b*) *Using a microphone and a CRO.* The sound is received by the microphone and fed to the CRO (if need be via a pre-amplifier) which is set on a suitable, known time base range. By noting the number of cycles of a.c. on a certain length of the time scale, the frequency can be worked out.

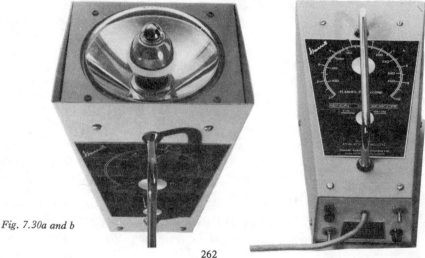

Fig. 7.30a and b

SOUND

(c) *Using a stroboscope.* A stroboscope makes an object which is moving appear to be at rest. A lamp stroboscope, Fig. 7.30, consists of a lamp (containing xenon gas) which emits brief but intense flashes of light from about 2 to 250 times a second according to the setting of the speed control. Thus if, for example, 200 is the *highest* flashing speed which makes a sonometer wire appear to be at rest and in one position, then 200 Hz is the frequency of the stationary wave on the wire and also of the progressive sound wave emitted into the surrounding air by it. What would be seen if the flashing speed were (*i*) 100 Hz, and (*ii*) 400 Hz?

QUESTIONS

1. A point source A emits spherical sound waves. State how the intensity of sound varies with position around the source, assuming that there is no absorption in the medium.

A second identical source B is placed near A, the two sources being in phase. Explain why there will be positions of maximum and minimum intensity near the sources.

If the wavelength for each source is 0.40 m, and the rate at which sound energy is emitted by A is 1.44 times the rate of that emitted by B, explain why the sound intensity is found to be zero at a point 13.2 m from A and 11.0 m from B.

(*J.M.B.*)

2. Explain what is meant by the statement that ' sound is propagated in air as longitudinal progressive waves ', and outline the experimental evidence in favour of this statement. Discuss how changes in atmospheric conditions of pressure, temperature and humidity might be expected to affect the speed of sound.

Two loudspeakers face each other at a separation of about 100 m and are connected to the same oscillator, which gives a signal of frequency 110 Hz. Describe and explain the variation of sound intensity along the line joining the speakers. A man walks along the line with a uniform speed of 2.0 m s^{-1}. What does he hear? (Speed of sound = 330 m s^{-1}.)   (*C.*)

3. Draw a diagram showing the positions of nodes and antinodes in the vibrations of an air column in a pipe closed at one end when it is giving the third harmonic of its fundamental note. Indicate on the diagram the directions and relative magnitudes of motion of the air at several significant points in the pipe at an instant when these motions are at maximum velocity. Calculate the frequency of this third harmonic if the effective length of the closed pipe is 72.0 cm. Speed of sound in air = 330 m s^{-1}.   (*S.*)

4. Distinguish between *free vibrations* and *forced vibrations*, and explain the term *resonance*. Discuss the meanings of these terms by considering a specific vibrating system chosen from any branch of physics other than sound.

A small loudspeaker, actuated by a variable frequency oscillator, is sounded continuously over the open end of a vertical tube 40 cm long and closed at its lower end. At what frequencies will resonance occur as the frequency of the note emitted by the loudspeaker is increased from 200 Hz to 1200 Hz, given that the velocity of sound in air is 3.44×10^4 cm s^{-1}? Neglect the end effect of the tube.

(L.)

5. Two open organ pipes of lengths 50 and 51 cm respectively, give beats of frequency 6.0 Hz when sounding their fundamental notes together. Neglecting end corrections what value does this give for the velocity of sound in air?

(A.E.B. part qn.)

6. Two sonometer wires, A of diameter 7.0×10^{-4} m and B of diameter 6.0×10^{-4} m, of the same material, are stretched side by side under the same tension. They vibrate at the same fundamental frequency of 256 Hz. If the length of B is 0.91 m, find the length of A. Calculate also the number of beats per second which will occur if the length of B is reduced to 0.90 m. *(O. part qn.)*

7. What is meant by a *wave motion*? Define the terms *wavelength* and *frequency* and derive the relationship between them.

Given that the velocity v of transverse waves along a stretched string is related to the tension T and the mass μ per unit length by the equation

$$v = \sqrt{\frac{T}{\mu}}$$

derive an expression for the natural frequencies of a string of length l when fixed at both ends.

Explain how the vibration of a string in a musical instrument produces sound and how this sound reaches the ear. Discuss the factors which determine the quality of the sound heard by the listener. *(O. and C.)*

8. Explain why the note emitted by a stretched string can easily be distinguished from that of a tuning fork with which it is in unison. How would you justify your answer by experiment?

Describe how, using a set of standard forks, you would verify experimentally the relationship between the frequency of the note emitted by a string of fixed length and tension and the mass per unit length of the wire.

A sonometer wire of length 1.0 m emits the same fundamental frequency as a given tuning fork. The wire is shortened by 0.05 m, the tension remaining unaltered, and 10 beats per second are heard when the wire and fork are sounded together. What is the frequency of the fork? If the mass per unit length of the wire is 1.4×10^{-3} kg m^{-1}, what is the tension? *(A.E.B.)*

9. Describe the motion of the particles of a string under constant tension and fixed at both ends when the string executes transverse vibrations of (a) its fundamental frequency, and (b) the first overtone (second harmonic). Illustrate your answer with suitable diagrams.

A horizontal sonometer wire of fixed length 0.50 m and mass 4.5×10^{-3} kg is under a fixed tension of 1.2×10^2 N. The poles of a horse-shoe magnet are arranged to produce a horizontal transverse magnetic field at the midpoint of the wire, and an alternating sinusoidal current passes through the wire. State and explain what happens when the frequency of the current is progressively increased from 100 to 200 Hz. Support your explanation by performing a suitable calculation. Indicate how you would use such an apparatus to measure the fixed frequency of an alternating current. (*J.M.B.*)

10. A source emitting a note of certain frequency f approaches a stationary observer at a constant speed of one-tenth the speed of sound in air. The source is then maintained stationary and the observer moves towards it at the same constant speed. Determine from first principles the frequency of the note heard by the observer in each case. (*J.M.B.*)

11. Show that when a source emitting sound waves of frequency f moves towards a stationary observer with velocity u, the observer hears a note of frequency $fv/(v - u)$, where v is the velocity of sound.

Describe the effect of a steady wind blowing with velocity w directly from source to observer (a) in the case above, and (b) if the source and the observer are both at rest.

A model aircraft on a control line travels in a horizontal circle of 10 m radius, making 1 revolution in 3.0 s. It emits a note of frequency 300 Hz. Calculate the maximum and minimum frequencies of the note heard at a point 20 m from the centre of the circle and in the plane of the path, and find the time interval between a maximum and the minimum that next succeeds it. (Take the velocity of sound in air to be 330 m s⁻¹.) (*O.*)

12. Show that two identical progressive wave-trains travelling along the same straight line in opposite directions in a given medium set up a system of stationary waves. Compare the properties of a stationary wave system in air with those of a progressive wave-train in respect of (a) amplitude, (b) phase, and (c) pressure variation.

An observer moving between two identical sources of sound along the straight line joining them, hears beats at the rate of 4.0 s⁻¹. At what velocity is he moving if the frequency of each source is 500 Hz and the velocity of sound when he makes the observations is 3.40×10^4 cm s⁻¹? (*L.*)

13. A car travelling normally towards a cliff at a speed of 30 m s⁻¹ sounds its horn which emits a note of frequency 100 Hz. What is the apparent frequency of the echo as heard by the driver? Speed of sound in air = 330 m s⁻¹.

14. The velocity of sound v in a gas is given by $v = \sqrt{\dfrac{\gamma p}{\rho}}$, where γ is the ratio of the principal specific heats of the gas, p is its pressure, and ρ is its density. If v is found

to be 400 m s^{-1} for a certain gas under particular conditions, what would be the new value of v if (a) the pressure were reduced by 4 per cent, and (b) the absolute temperature were increased by 4 per cent? Explain how you arrive at your answers.

(S).

15. Compare the mode of propagation of sound with that of light.

How is the speed of sound in a gas affected by (a) an increase in the temperature of the gas, and (b) a decrease in the pressure of the gas?

A generator of ultrasonic waves of frequency 1.00 MHz is set up in a rectangular tank of paraffin facing a detector, which is connected to an amplifier and oscilloscope. It is found that the amplitude of the oscilloscope trace varies periodically as the detector is moved away from the generator. A series of consecutive maxima occurs at 2.99 mm, 3.65 mm, 4.33 mm, 5.00 mm and 5.66 mm from the generator. Explain the variation of the oscilloscope trace, and calculate the speed of sound in paraffin. (C.)

16. Explain the formation of ' beats ' and show that if two notes of frequencies f_1 and f_2 are sounded together the frequency of the beats is given by $f_1 \sim f_2$. How would you determine, other than by ear, which frequency is the higher?

Explain with the aid of diagrams how sound waves may be refracted by (a) a wind gradient, and (b) a temperature gradient. What is the effect of these refractions on the audibility of a sound?

A man standing close to an iron railing consisting of evenly-spaced uprights makes a sharp sound and hears a note of frequency 640 Hz. Calculate the spacing between the uprights. (Speed of sound in air = 330 m s^{-1} at 0 °C. Ambient temperature = 17.0 °C.) (A.E.B.)

17. A vibrating tuning fork, observed by a light which flashes at regular intervals, has the same appearance as it has when at rest. A circular disc with 70 equally-spaced radii drawn on it is rotated from rest with gradually increasing speed. When viewed in the same light the disc first appears at rest when it rotates at one revolution per second. What are the possible values of the frequency of the fork?

(S.)

8 Physical optics

Nature of light

Two apparently contradictory theories of the nature of light were advanced in the seventeenth century.

The *corpuscular theory* regarded light as a stream of tiny particles or corpuscles travelling at high speed in straight lines. Newton supported this view which did account for rectilinear propagation, reflection and refraction—the latter by assuming that on entering an optically denser medium the corpuscles are attracted, thereby causing bending towards the normal. On the other hand, the *wave theory* proposed by Huygens about 1680 considered light to travel as waves. As we have seen (p. 203), a wave model can account satisfactorily for reflection and refraction.

Opponents of the wave theory argued that waves require a transmitting medium and there did not appear to be one for light which was able to travel in a vacuum. Subsequently a medium, called the *ether*, was invented but it defied all attempts at detection. An apparently crucial difference was that whereas the corpuscular theory required light to have a greater speed in a material than in air, the wave theory predicted a lower speed. It was not until 1862 that the wave theory prediction was confirmed when Foucault found the speed of light in water to be less than that in air. In the meantime, interference and diffraction effects had been discovered which were more readily explicable in terms of waves than in terms of corpuscles.

The need for the ether disappeared when Maxwell suggested in 1864 that light was an electromagnetic ' wave ' and consisted of a fluctuating electric field coupled with a fluctuating magnetic field (p. 309). By the end of the nineteenth century Maxwell's electromagnetic wave theory of radiation had established itself as one of the great intellectual pillars of physics, a unifying principle linking electricity, magnetism and light. However, as we shall see later, it did not on its own give a completely satisfactory explanation of all the properties of electromagnetic radiation.

PHYSICAL OPTICS

In this chapter we will deal with physical optics—the subject which is concerned with those effects that make sense if we regard light as having a wave-like nature: such effects are interference, diffraction and polarization.

Speed of light

A knowledge of the speed of light is important for several reasons. First, if the experimental value agrees with the theoretical value predicted by Maxwell for electromagnetic waves (see p. 309) then it is reasonable to assume that (*i*) light is an electromagnetic wave and (*ii*) the electromagnetic theory of radiation is valid. Second, it occurs in certain basic expressions of atomic and nuclear physics, such as Einstein's mass-energy equation $E = mc^2$ (p. 503). Third, the measurement of distance by radar techniques becomes possible (p. 224).

The rotating mirror method is based on one due to Foucault (1862) and in the school laboratory version of the apparatus, Fig. 8.1a, an attempt is made to estimate, in effect, the time taken by light to travel a distance of about four metres. This is of the order of 10^{-8} second, consequently the result obtained is subject to a large error.

Fig. 8.1a

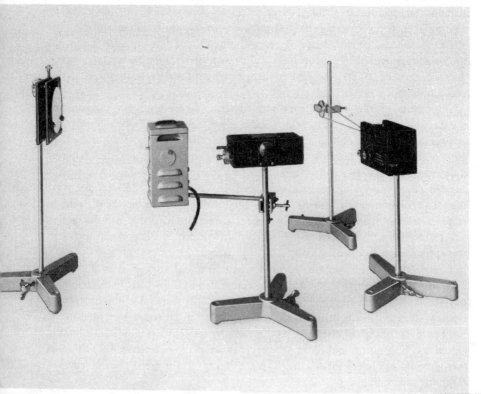

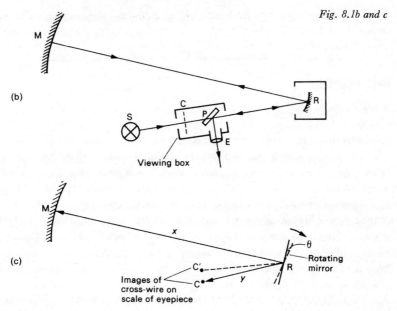

Fig. 8.1b and c

The principle of the method is shown by Fig. 8.1*b*. Light from a source S illuminates a cross-wire C in the viewing box and travels on through the glass plate P to the rotating mirror R where it is reflected to the large fixed concave mirror M, placed so that R is at its centre of curvature. M reflects the light back to R and into the viewing box where it is partially reflected by P into the eyepiece E. An image of C is formed on the eyepiece scale. When R is rotated at high speed (by a small electric motor), the image of C is displaced by a very small distance to C′.

To obtain an expression for the speed of light *c*, consider the simplified diagram Fig. 8.1*c*. Let MR = *x*, CR = *y* and when the rotating mirror makes *n* revolutions per second, suppose the displacement of the image of C is CC′ = *d*. Then

distance travelled by light between reflections at R = $2x$

time for rotating mirror to make one revolution $= 1/n$

time for mirror to rotate angle θ (in rad) $= \theta/(2\pi n)$.

Hence

$$c = \frac{\text{distance travelled by light}}{\text{time taken}} = \frac{2x}{\theta/(2\pi n)}$$

$$= \frac{4\pi n x}{\theta}.$$

269

Also, since the reflected ray turns through twice the angle of rotation θ of the mirror,

angular rotation of image of C (i.e. angle CRC′) = 2θ = d/y.

Substituting for θ,

$$c = \frac{8\pi nxy}{d}.$$

In practice d is extremely small (about 0.05 mm) and the error in measuring it is reduced by noting the displacement $2d$ of the cross-wire when the rotation of the mirror is reversed. Typical values for x and y are 2 m and 0.6 m respectively.

The speed of rotation n can be estimated in various ways. A signal generator connected to a loudspeaker can be adjusted until the note from the latter is judged to be the same frequency as that produced by the rotating mirror. Alternatively, a length of thread is tied to form a loop—say exactly one metre (i.e. one metre from the knot round the loop and back to the knot). It is then placed round the boss of the idler pulley in the rotating mirror unit and run round another pulley on a stand. By finding the time of revolution of the knot in the loop and knowing the diameters of the idler pulley boss, the idler pulley and the mirror shaft, Fig. 8.2, the speed of the mirror can be calculated.

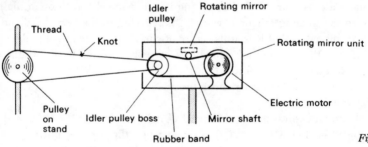

Fig. 8.2

Measurements of the speed of other members of the electromagnetic family confirm that they all—whatever their wavelength—travel in a vacuum with the *same* speed of 2.998×10^8 m s⁻¹. In other media the speed varies with the wavelength; for example, red light travels faster in glass than does blue light.

Interference of light

We saw earlier (p. 209) that interference occurs when waves from two coherent sources cross. Such sources produce waves having the same frequency, equal or comparable amplitudes and a phase difference that does not alter with time.

The wavelength of light must be very small otherwise the diffraction effects it gives would be more evident than they are. It therefore follows from the experiments with water waves in a ripple tank (p. 211) that to obtain nodal and antinodal lines (called *interference fringes*) sufficiently far apart to be seen, we must have

(*i*) the sources very close together

(*ii*) the screen (or eyepiece) as far as possible from the sources.

Also, if the sources are to be coherent they must be derived, in practice, from the same source. If an attempt is made to produce interference with two separate light sources, uniform illumination results instead of regions of light and dark. This is due to the fact that in a light source the phase is constantly changing because light is emitted in short bursts that last about 10^{-9} s when electrons in individual atoms suffer energy changes that occur very quickly and randomly. Phase changes happen abruptly as different atoms come into action and the eye is unable to follow the rapidly-changing interference pattern.

This is also true for light coming from different parts of the same source (with one exception). What is necessary then is that two wave-trains arriving at a given point, say by different paths, should have come from the *same point* of the *same source*. Any phase change occurring in the source then occurs in both wave-trains and stationary interference effects result. To obtain two coherent wave-trains from a point of a single source one of two methods is adopted: (*i*) *division of wavefront*, as is done in Young's double slit, Fresnel's biprism and Lloyd's mirror, and (*ii*) *division of amplitude*, usually by partial reflection and transmission at a boundary, as occurs in wedge fringes (p. 277) and Newton's rings (p. 279).

The exception mentioned above is the *laser* which does produce coherent light because the atoms are made to act in unison and all undergo energy changes simultaneously. Thus if a screen with two small holes is placed in a laser beam so that different parts of the source are used, an interference pattern is obtained.

Interference accounts for the colours of soap bubbles and of thin films of oil on a wet road. It also has practical applications (p. 282).

Young's double slit

One of the first to demonstrate the interference of light was Thomas Young in 1801.

(*a*) *Principle.* The principle of his method is shown in Fig. 8.3*a*. Monochromatic light (i.e. of one colour) from a narrow vertical slit S falls on two other narrow slits S_1 and S_2 which are very close together and parallel to S. S_1 and S_2 act as two coherent sources (both being derived from S) and if they (as well as S) are narrow enough, diffraction causes the emerging beams to spread into the region beyond the slits. Superposition occurs in the shaded area of Fig. 8.3*a* where the diffracted beams overlap. Alternate bright and dark equally-spaced

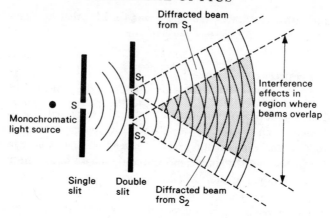

Fig. 8.3a

vertical bands (interference fringes) can be observed on a screen or at the cross-wires of an eyepiece. Fig. 8.3b. If either S_1 or S_2 is covered the bands disappear.

Fig. 8.3b

(b) *Theory.* An expression for the separation of two bright (or dark) fringes can be obtained from Fig. 8.4a.

The path difference between waves reaching O from S_1 and S_2 is zero, i.e. $S_1O = S_2O$, they therefore arrive in phase and so there is a bright fringe at O, in the centre of the pattern. At P, distance x_1 from O, there will be a bright fringe if the path difference is a whole number of wavelengths, that is, if

$$S_2P - S_1P = n\lambda$$

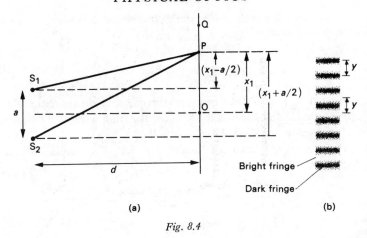

Fig. 8.4

where n is an integer (or zero) and λ is the wavelength of the light. We say the nth bright fringe is formed at P.

In Fig. 8.4a, d is the distance from the screen or cross-wire to the double slit and a is the slit separation. Hence

$$S_2P^2 = d^2 + (x_1 + a/2)^2 = d^2 + x_1^2 + ax_1 + a^2/4$$

$$S_1P^2 = d^2 + (x_1 - a/2)^2 = d^2 + x_1^2 - ax_1 + a^2/4$$

$$\therefore \quad S_2P^2 - S_1P^2 = 2ax_1.$$

But $$S_2P^2 - S_1P^2 = (S_2P - S_1P)(S_2P + S_1P).$$

In practice a is very small (e.g. 0.5 mm) compared with d (e.g. 1 m) and if P is near O then S_2P and S_1P are each just greater than d. Therefore we can say $(S_2P + S_1P) = 2d$. It follows that

$$(S_2P - S_1P)2d = 2ax_1$$

$$\therefore \quad S_2P - S_1P = ax_1/d.$$

For the nth bright fringe at P we have

$$n\lambda = ax_1/d. \tag{1}$$

If the next bright fringe, i.e. the $(n + 1)$th, is formed at Q where $OQ = x_2$ then

$$S_2Q - S_1Q = (n + 1)\lambda$$

$$(n + 1)\lambda = ax_2/d. \tag{2}$$

Subtracting (1) from (2) we get

$$\lambda = a(x_2 - x_1)/d.$$

273

If y is the distance between two adjacent bright (or dark) fringes, called the *fringe spacing*, Fig. 8.4b, then $y = x_2 - x_1$ and so

$$\lambda = \frac{ay}{d}.$$

We see that (i) $y \propto 1/a$ if λ and d constant (therefore the slit separation should be small), (ii) $y \propto d$ if λ and a constant (therefore the fringes should be viewed from a distance) and (iii) $y \propto \lambda$ if a and d constant.

If a dark fringe were formed at P then $(S_2P - S_1P)$ would equal an odd number of half wavelengths.

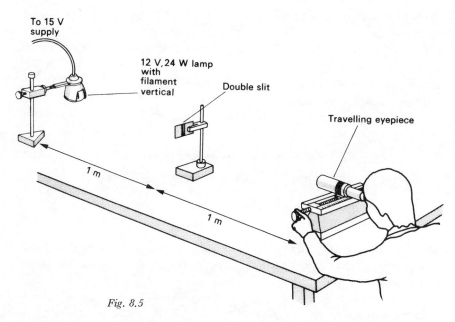

Fig. 8.5

(c) *Measurement of* λ. One experimental arrangement is shown in Fig. 8.5 in which the lamp filament acts as the single slit. (The double slit may be made by ruling two slits 0.5 mm apart on a microscope slide coated with Aquadag, using a blunt needle as described in Appendix 6, p. 534, or it can be purchased ready-made from a laboratory supplier). The fringes are viewed in a blacked-out room at the cross-wire of a travelling eyepiece (e.g. a travelling microscope with the objective removed). Filters are used to obtain coloured light from a white light source.

The average fringe spacing y is found by measuring across as many fringes as possible with the travelling eyepiece. A metre rule is used to measure d and the slit separation a is measured directly with a travelling microscope. The value of λ obtained is approximate; for violet light it is about 4×10^{-7} m (0.4 μm) and for red light about 7×10^{-7} m (0.7 μm).

(d) Further points.

(i) The overlapping beams which interfere are produced by diffraction at S_1 and S_2. Since the fringes can be observed anywhere in the overlapping region they are called *non-localized* fringes.

(ii) Point sources would give the same fringe system but slits (i.e. a line of point sources) give brighter fringes by reinforcing the pattern.

(iii) The fringes are really the intersections with a vertical plane of hyperboloids of revolution having S_1 and S_2 as foci. (These correspond in three dimensions to the two-dimensional hyperbolic nodal and antinodal lines, see Fig. 6.16, p. 211.)

(iv) The interference is incomplete because for all fringes except the central bright one, the amplitudes of the two wave-trains are not exactly equal. Why?

(v) Using white light fewer fringes are seen and each colour produces its own set of fringes which overlap. Only the central one is white, its position being the only one where the path difference is zero for all colours. The first coloured fringe is bluish near the central fringe and red at its far side, Fig. 8.6. Why?

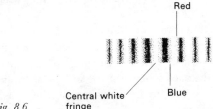

Fig. 8.6

Red

Central white fringe

Blue

(vi) The fringe spacing for red light is greater than for blue light. Red light must therefore have a greater wavelength than blue light since $y \propto \lambda$ (if a and d are constant).

(vii) The number of fringes obtained depends on the amount of diffraction occurring at the slits and this in turn depends on their width (see pp. 290 and 534). The narrower the slits, the greater will be the number of fringes due to the increased diffraction but the fainter will they be, since less light gets through. In practice, to give easily seen fringes, the slits have to be many wavelengths wide; the case is similar to that shown for water waves in Fig. 6.20*b* (p. 214).

Fresnel's biprism: Lloyd's mirror

Two other ways of producing interference fringes from two coherent sources that are derived by division of the wavefront from a single source will now be outlined.

(a) *Fresnel's biprism.* Monochromatic light from a narrow slit S falls on a double glass prism arranged as in Fig. 8.7. Two virtual images S_1 and S_2 are formed of S, one by refraction at each half of the prism, and these act as coherent

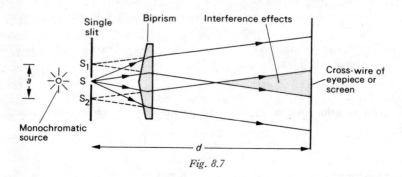

Fig. 8.7

sources which are close together because of the small refracting angles (about 0.5°) of the prisms. An interference pattern, similar to that given by the double slit but brighter, is obtained in the shaded region where the two refracted beams overlap. The fringes can be observed as before using a travelling eyepiece focused on the fringes at its cross-wire.

The theory and the expression for the fringe spacing y are the same as for Young's method, i.e. $y = d\lambda/a$. To obtain a, a convex lens is moved between the biprism and the eyepiece until a real magnified image of the virtual 'objects' S_1 and S_2 is in focus. The distance b between the images of S_1 and S_2 (on the cross-wire of the eyepiece) is measured, then knowing the object and image distances u and v respectively we can say that the magnification m is given by $m = v/u = b/a$. The wavelength λ can then be calculated if y and d are also determined, but this is not one of the most accurate ways of measuring the wavelength of light.

(b) *Lloyd's mirror.* A plane glass plate (acting as a mirror) is illuminated at almost grazing incidence by light from a slit S_1, parallel to the plate, Fig. 8.8. A virtual image S_2 of S_1 is formed close to S_1 by reflection and these two act as

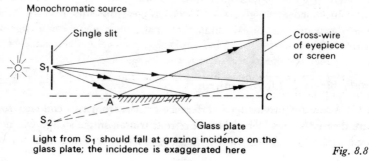

Light from S_1 should fall at grazing incidence on the glass plate; the incidence is exaggerated here

Fig. 8.8

coherent sources. In the shaded area direct waves from S_1 cross reflected waves which appear to come from S_2 and interference fringes can be seen.

The expression giving the fringe spacing is the same as for the double slit and the biprism experiments, but the fringe system differs in one important respect. In Lloyd's mirror, if the point P, for example, is such that the path difference $(S_1A + AP) - S_1P$ (or $S_2P - S_1P$) is a whole number of wavelengths, the fringe at P is *dark*, not bright. This is due to the 180° phase change which occurs when light is reflected at a rare-dense boundary. This is equivalent to adding an extra half-wavelength to the path of the reflected wave and is similar to the phase change a pulse on a spring undergoes at a fixed end, or to that when microwaves are reflected at a metal plate (p. 224). At grazing incidence a fringe is formed at C, where the geometrical path difference between the direct and reflected waves is zero and it follows that it will be dark rather than bright.

Wedge fringes

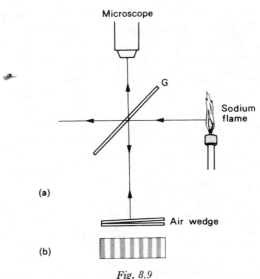

Fig. 8.9

Interference fringes are produced by a thin wedge-shaped film of air, the thickness of which gradually increases from zero along its length. The wedge can be formed from two microscope slides clamped at one end and separated by a thin piece of paper at the other so that the wedge angle is very small. In the arrangement of Fig. 8.9a monochromatic light from an extended source (e.g. a sodium lamp or flame) is partially reflected vertically downwards by the glass plate G. When the microscope is focused on the wedge, light and dark equally-spaced fringes are seen, parallel to the edge of contact of the wedge, Fig. 8.9b.

Some of the light falling on the wedge is reflected upwards from the bottom surface of the top slide and some of the rest, which is transmitted through the air wedge, is reflected upwards from the top surface of the bottom slide. Both wave-trains have arisen from the same point P, Fig. 8.10, by *division of the amplitude*. They are therefore coherent and when brought together (by the eye or a micro-scope) they can interfere.

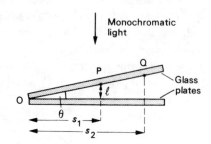

Fig. 8.10

If l is the thickness of the air wedge at P then, since the incidence is nearly normal, the path difference between the rays at P is $2l$. At O where the path difference is zero, we would expect a bright band but a *dark* band is observed. This is due to the 180° phase change which occurs when the wave-train *in* the air wedge is reflected at the top surface of the bottom slide, i.e. at a denser medium. The phase change, in effect, adds an extra path of half a wavelength (i.e. a crest is reflected as a trough). The path difference between the two wave-trains at P is thus $(2l + \lambda/2)$ where λ is the wavelength of the light. A bright fringe is formed at P if

$$2l + \lambda/2 = n\lambda$$

or
$$2l = (n - \tfrac{1}{2})\lambda$$

where $n = 1$ gives the first bright fringe, $n = 2$ gives the second bright fringe, etc. ($n = 0$ is impossible). A dark fringe is formed at P if

$$2l = n\lambda$$

where $n = 0$ gives the first dark fringe, $n = 1$ gives the second dark fringe, $n = 2$ gives the third dark fringe etc.

The microscope (or the unaided eye) has, for normal incidence, to be focused on the top surface of the air wedge. This ensures that superposition of the two interfering wave-trains then occurs at a particular point in the retina. The fringes in this case are called *localized* fringes. Each fringe is the locus of points of equal air wedge thickness (i.e. same path difference) and they are often referred to as ' fringes of equal thickness '. There are also reflections from the other surfaces; these are usually out of focus and since they involve large path differ-ences they do not spoil the fringes.

The angle θ of the wedge can be found if the reading on a travelling microscope is taken when the cross-wire is on a dark fringe at P, say the nth from O. We then have from Fig. 8.10 that

$$2l = n\lambda.$$

But $l = s_1\theta$ (if θ is in radians), hence

$$2s_1\theta = n\lambda. \tag{1}$$

If the microscope is now moved until it is on the $(n + k)$th dark fringe, say at Q, and the reading again taken, then

$$2s_2\theta = (n + k)\lambda. \tag{2}$$

Subtracting (1) from (2)

$$2\theta(s_2 - s_1) = k\lambda$$

$$\therefore \quad \theta = \frac{k\lambda}{2(s_2 - s_1)}$$

where $(s_2 - s_1)$ is the distance moved by the microscope and λ is the wavelength of the light.

Newton's rings

This system of interference fringes, also produced by division of amplitude, was discovered by Newton but it was Young who gave a satisfactory explanation of their formation in terms of waves.

(*a*) *Principle.* The arrangement is shown in Fig. 8.11. Monochromatic light (e.g. from a sodium lamp or flame) is reflected by the glass plate G so that it falls

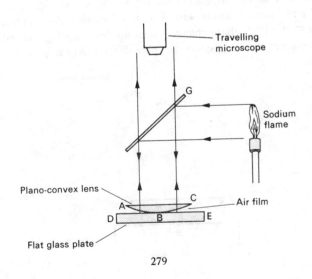

Fig. 8.11

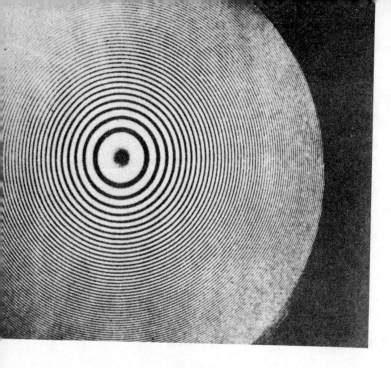

Fig. 8.12

normally on the air film formed between the convex lens of long focal length (about one metre) and the flat glass plate. The thickness of the air film gradually increases outwards from zero at the point of contact B but is the same at all points on any circle with centre B.

Interference occurs between light reflected from the lower surface ABC of the lens and the upper surface DBE of the plate. A series of bright and dark rings is seen through G when a travelling microscope is focused *on the air film*, Fig. 8.12. The rings are fringes of ' equal thickness ', ' localized ' in the air film (like those formed by a wedge) and as their radii increase, the separation decreases. At the centre B of the fringe system where the geometrical path difference between the two wave-trains is zero, there is a *dark* spot. As with wedge fringes, this is due to the 180° phase change which occurs when light is reflected at an optically denser medium. In effect, the path difference at B is not zero but half a wavelength.

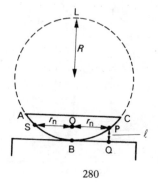

Fig. 8.13

(b) *Radius of a ring.* In Fig. 8.13 a complete circular section of a sphere of radius R is shown, R being the radius of curvature of the lower surface ABC of the lens. Let r_n be the radius OP of the ring at P, where the thickness PQ of the film is l. If BO is produced to meet the circle at L then BOL is a diameter. By the theorem of intersecting chords we have

$$SO \cdot OP = LO \cdot OB$$

$$\therefore \qquad r_n \cdot r_n = (2R - l)l \quad \text{(since OB = PQ = } l)$$

$$\therefore \qquad r_n{}^2 = 2Rl - l^2.$$

But l^2 is very small compared with $2Rl$ since R is large, hence

$$r_n{}^2 = 2Rl.$$

The path difference between the two interfering wave-trains at P is $2l$ for light incident normally on the air film. If a dark ring is formed at P then

$$2l = n\lambda$$

where n is an integer and λ is the wavelength of the light. Even though the geometrical path difference is a whole number of wavelengths, the $\lambda/2$ phase change at Q results in destructive interference. Hence from the two previous expressions we can write for the radius r_n of a *dark* fringe

$$r_n{}^2 = Rn\lambda.$$

Thus $n = 0$ gives the central dark spot, $n = 1$ gives the first dark ring, $n = 2$ gives the second dark ring and so on.

A *bright* ring would be formed at P if $2l$ equalled $\frac{1}{2}\lambda$, or $1\frac{1}{2}\lambda$, or $2\frac{1}{2}\lambda$ etc., since the phase change at the glass plate in effect increases the path length of the light reflected there by $\lambda/2$. Hence if the nth bright ring is formed at P, then

$$2l = (n - \tfrac{1}{2})\lambda.$$

The radius r_n of a *bright* ring is therefore given by

$$r_n{}^2 = R(n - \tfrac{1}{2})\lambda$$

where $n = 1$ gives the first bright ring, $n = 2$ gives the second bright ring and so on.

(c) *Measurement of λ.* It is better to measure the diameters of rings rather than their radii because of the uncertainty of the position of the centre of the system. Hence if d_n is the diameter of the nth *dark* ring we have

$$r_n{}^2 = (d_n/2)^2 = Rn\lambda$$

$$\therefore \quad d_n{}^2 = 4Rn\lambda.$$

A graph of d_n against n is thus a straight line of slope $4R\lambda$. If it does not pass through the origin, n may have been miscounted each time *or* the contact between lens and plate is poor (and may give a central bright spot). Neither effect, however, affects the slope of the graph. A travelling microscope is used to measure d_n and R can be found by Boys' method.

Using interference

(*a*) *Testing of optical surfaces.* Fringes of equal thickness are useful for testing optical components. For example, in the making of optical ' flats ', the plate under test is made to form an air wedge with a standard plane glass surface. Any uneven parts of the surface which require more grinding will show up as irregularities in what should be a parallel, equally-spaced, straight set of fringes.

The grinding of a lens surface may be checked if it is placed on an optical flat and Newton's rings observed in monochromatic light. The rings should be exactly circular if the lens is spherical.

(*b*) *Non-reflecting glass.* In optical instruments containing lenses or prisms light is lost by reflection at each refracting surface and results in reduced brightness of the final image. There will also be a loss of contrast against such a background of stray light.

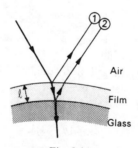

Fig. 8.14

The amount of light reflected at a surface can be appreciably reduced by coating it (by evaporation in a vacuum) with a film of transparent material (e.g. magnesium fluoride) to a thickness of *one quarter of a wavelength of light in the film*, Fig. 8.14. Light reflected from the top (ray 1) and bottom (ray 2) surfaces of the film then interfere destructively since the latter has to travel twice the thickness of the film, i.e. the path difference is $2l$. The refractive index of the film is less than that of glass and so each reflection, being at a rare-dense boundary, suffers a 180° phase change. The net effect of the phase changes on the path difference is thus zero. The refractive index of the film should be as nearly as possible the mean of the refractive indices for air and glass so that the amounts of light reflected at the two surfaces are almost equal.

For light of wavelength λ_a in air and λ_f in the film, the condition for destructive interference, at normal incidence, is

$$2l = \tfrac{1}{2}\lambda_f$$

$$\therefore \quad l = \tfrac{1}{4}\lambda_f.$$

When light passes from one medium into another, its speed and wavelength change but not its frequency f. Therefore if v_a and v_f are the speeds of light in air and the film respectively, we have

$$v_a = f\lambda_a \quad \text{and} \quad v_f = f\lambda_f.$$

Also if $_an_f$ is the refractive index of the material of the film relative to that of air then

$$_an_f = \frac{v_a}{v_f} \qquad \text{(see p. 206)}$$

$$= \frac{\lambda_a}{\lambda_f}$$

Hence
$$\lambda_f = \frac{\lambda_a}{_an_f} \; .$$

The previous expression for l may then be written more conveniently as

$$l = \frac{\lambda_a}{4 \, _an_f} \; .$$

The interference is complete for one wavelength only, usually taken to be that at the centre of the visible range (i.e. yellow-green). For red and blue light the reflection is weakened but not eliminated and a coated or ' bloomed ' lens appears purple in white light. Energy which would have been wasted as reflected light increases the amount of transmitted light when destructive interference occurs. If $\lambda_a = 550$ nm and $_an_f = 1.38$, then $l = 100$ nm.

(c) *Measurement of length.* The SI unit of length, the metre, is defined as a certain number of wavelengths (1 650 763.73) in a vacuum of a particular line in the spectrum of an isotope of the inert gas krypton. The measurement required to set up this standard involves forming an interference pattern between two mirrors and counting the number of fringes which cross the field of view as one mirror is moved to the other. The movement of one-tenth of a fringe can just be detected and so the accuracy attainable is about 1 part in 10^7.

A similar technique is used to measure the very small expansion of a crystal when it is heated, the motion of interference fringes again being observed.

Everyday examples of interference

(*a*) *Colours of oil films on water.* Interference occurs between two wave-trains—one reflected from the surface of the oil and the other from the oil-water interface, Fig. 8.15. When the path difference gives constructive interference for light of one wavelength, the corresponding colour is seen in the film. The path difference varies with the thickness of the film and the angle of viewing, both of which affect the colour produced at any one part.

Fig. 8.15

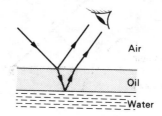

(*b*) *Pulsing of the picture on a television receiver.* This occurs when an aircraft passes low overhead. The signal travelling directly from the transmitting to the receiving aerial interferes with that reflected from the aircraft, Fig. 8.16. Usually the reflected wave is much weaker and so the interference is never completely destructive.

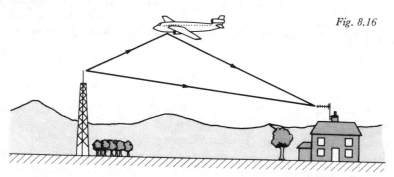

Fig. 8.16

Optical path length

We can show that a.length l in a medium of refractive index n is equivalent to a length nl in a vacuum. Let PQ in Fig. 8.17 be a plane wavefront falling obliquely on the surface separating a vacuum and the medium, in which the speeds of light are c and v respectively. Let t be the time for Q to reach R, i.e. QR $= ct$. In this time suppose P travels a distance PS in the medium, then PS $= vt$. If n is the absolute refractive index of the medium then

$$n = \frac{c}{v} \quad \text{and} \quad v = \frac{c}{n}.$$

Hence

$$PS = \frac{ct}{n} = \frac{QR}{n}$$

$$\therefore \quad QR = n \cdot PS.$$

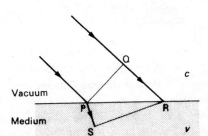

Fig. 8.17

That is, if light travels a distance PS in the medium then in the same time it would travel a distance $n \cdot PS$ in a vacuum. In general,

length l in medium of refractive index n is
optically equivalent to length nl in a vacuum.

nl is called the *optical path length* of distance l in the medium. It follows that if a thickness l of transparent material of refractive index n is placed in the path of a beam of light, the path difference between the new and the previous paths is $nl - l = (n - 1)l$.

Interference calculations

1. In a Young's double slit experiment the distance between the slits and the screen is 1.60 m and using light of wavelength 5.89×10^{-7} m the distance between the centre of the interference pattern and the fourth bright fringe on either side is 16.0 mm. What is the slit separation?

From $\lambda = ay/d$ (see p. 274) we have

$$a = \frac{d\lambda}{y}$$

where $\lambda = 5.89 \times 10^{-7}$ m, $d = 1.60$ m and y = fringe spacing = 16.0/4 = 4.0 mm = 4.0×10^{-3} m. Hence

$$a = \frac{1.60 \times 5.89 \times 10^{-7}}{4.0 \times 10^{-3}} \quad \frac{m \times m}{m}$$

$$= 0.236 \times 10^{-3} \, m$$

$$= 0.236 \, mm.$$

2. *An air wedge is formed between two glass plates which are in contact at one end and separated by a piece of thin metal foil at the other end. Calculate the thickness of the foil if 30 dark fringes are observed between the ends when light of wavelength 6.0×10^{-7} m is incident normally on the wedge, Fig. 8.18.*

Fig. 8.18

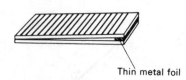

Thin metal foil

The *first* dark fringe is formed at the end where the plates are in contact and the thickness l of the air wedge is zero.

The *second* dark fringe occurs where $l = \lambda/2$; the geometrical path difference is then $2l = \lambda$ and the $\lambda/2$ phase change at the bottom plate causes destructive interference.

The *third* dark fringe occurs where $l = 2(\lambda/2)$, the *fourth* where $l = 3(\lambda/2)$ and the *thirtieth* where $l = 29(\lambda/2)$.

$$\therefore \quad \text{Thickness of foil} = 29 \times \frac{6.0 \times 10^{-7}}{2} \text{ m}$$

$$= 8.7 \times 10^{-3} \text{ mm}.$$

3. *Calculate the radius of curvature of a plano-convex lens used to produce Newton's rings with a flat glass plate if the diameter of the tenth dark ring is 4.48mm, viewed by normally reflected light of wavelength 5.00×10^{-7} m. What is the diameter of the twentieth bright ring?*

For the nth dark ring of radius r_n

$$r_n{}^2 = Rn\lambda \qquad \text{(see p. 281)}$$

where R is the radius of curvature of the lens surface and λ is the wavelength of the light

$$\therefore \quad R = \frac{r_n{}^2}{n\lambda}$$

$$= \frac{(4.48 \times 10^{-3}/2)^2}{10 \times 5.00 \times 10^{-7}} \frac{\text{m}^2}{\text{m}}$$

since $n = 10$ for the tenth dark ring

$$\therefore \quad R = 1.00 \text{ m}.$$

The radius r_n of the twentieth bright ring is given by

$$r_n{}^2 = R(n - \tfrac{1}{2})\lambda$$

where $n = 20$ and $R = 1.00$ m. Hence

$$r_n^2 = 1(20 - \tfrac{1}{2})5.00 \times 10^{-7}$$

$$\therefore \ r_n = \sqrt{9.75} \times 10^{-3} \, \text{m}$$

$$= 3.12 \, \text{mm}$$

$$\therefore \ d_n = 6.24 \, \text{mm}.$$

Diffraction of light

Light can spread round obstacles into regions that would be in shadow if it travelled exactly in straight lines, i.e. it exhibits the typical wave-like property of diffraction. Thus the edges of shadows are not sharp. Behind the obstacles or apertures at which diffraction occurs, a diffraction pattern of dark and bright fringes is formed which, under the right conditions, may be seen on a screen or at the cross-wire of an eyepiece.

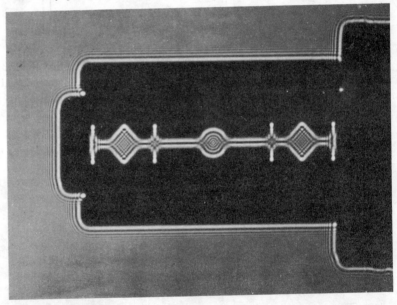

Fig. 8.19

A striking example of a diffraction pattern is shown in Fig. 8.19 and was produced by placing a razor blade midway between an illuminated pinhole and a photographic film. In most cases, small sources (e.g. a pinhole or a slit) are needed to observe diffraction effects and these are more evident if obstacles and apertures have linear dimensions comparable with the wavelength of light. With a large source each point gives rise to a diffraction pattern and there is uniform illumination when these overlap.

PHYSICAL OPTICS

Diffraction is regarded as being due to the superposition of secondary wavelets from coherent sources on the unrestricted part of a wavefront that has been obstructed by an obstacle or aperture (p. 215). Thus, whereas interference involves the superposition of waves on two different wavefronts, in diffraction there is superposition of waves from different parts of the *same* wavefront.

Before considering the important case of diffraction at a single slit, we shall briefly explain in general terms the diffraction patterns due to various other obstacles.

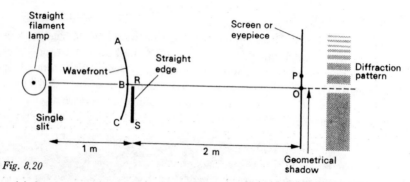

Fig. 8.20

(*a*) *Straight edge* (e.g. a razor blade). It can be shown that the fringes at points such as P in Fig. 8.20, which are close to the region of geometrical shadow, are due to the superposition of secondary wavelets from point sources on the unrestricted wavefront near R.

(*b*) *Circular obstacle* (e.g. a ball-bearing). In this case there is, rather surprisingly, a *bright* spot at the centre of the geometrical shadow, Fig. 8.21.

Fig. 8.21

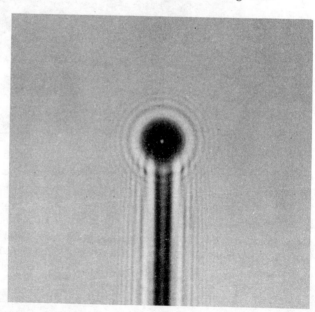

(c) *Straight narrow obstacle* (e.g. a pin). Between P and Q in Fig. 8.22 the fringes arise largely from the superposition of secondary wavelets from points near R, i.e. the pattern is a diffraction one due to coherent sources on the same wavefront. Similarly, secondary wavelets from points near S cause the pattern between L and M. The evenly-spaced fringes inside the geometrical shadow LP are formed by the superposition of secondary wavelets from points around R and S acting as *two* coherent sources; this is an interference pattern similar to that obtained with Young's double slit.

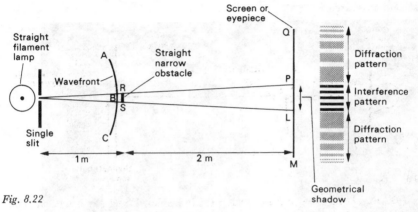

Fig. 8.22

Diffraction at a single slit

(a) *Experimental arrangement*, Fig. 8.23. The lamp has a straight filament, arranged to be vertical and the slit, whose width can be adjusted by a screw, is mounted with its length parallel to the filament. Initially the slit is opened wide and the lens moved to give a sharp image of the filament on the translucent screen. The shield and the large stop cut out stray light.

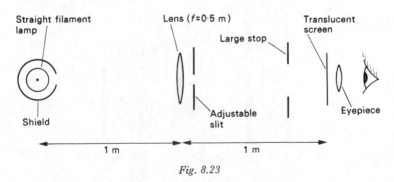

Fig. 8.23

The diffraction pattern depends on the slit width and is observed from behind the screen, preferably through a magnifying glass. For widths of a few millimetres a rectangle of light is produced, the result of near-rectilinear propagation. As the slit is narrowed, a pattern is obtained with a white central band having dark bands either side, fringed with colour. Inserting a red filter between lamp and screen gives red and black bands and the effect is as in Fig. 8.24a. In the diagrammatic representation of Fig. 8.24b, bright red bands are formed at A, B, C, D and E. With blue light the bands are closer together. Just before the slit closes the pattern dims and the central bright band widens causing illumination well into the geometrical shadow of the slit, i.e. the diffraction is very marked as the width of the slit approaches the wavelength of light and the slit acts like a secondary source.

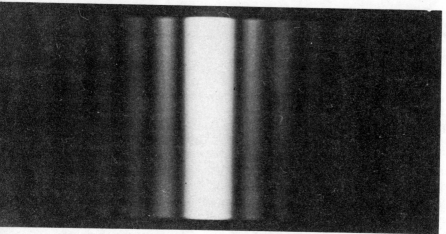

Fig. 8.24a and b

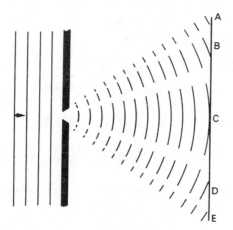

(b) *Theory.* To account for the maxima and minima of the diffraction pattern we use Huygens' construction and consider each point of the slit as a source of secondary wavelets. The slit is imagined to consist of strips of equal width, parallel to the length of the slit. The total effect in a particular direction is then found by adding the wavelets emitted in that direction by all the strips, using the superposition principle. In practice this operation presents mathematical difficulties too complex to be dealt with here, but a simplified two-dimensional treatment based on dividing the slit into strips is possible.

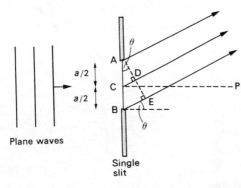

Fig. 8.25

Suppose plane waves (e.g. from a distant light source) fall normally on a narrow rectangular slit of width a, Fig. 8.25. Consider the first dark band (i.e. the first minimum) where there is no light. It will be formed at an angle θ to the incident beam if the path difference for the secondary wavelets from the strip just below A and the strip just below C (the midpoint of the slit) is $\lambda/2$, where λ is the wavelength of the light. Destructive interference will then occur for wavelets from this pair of strips since a crest from one strip reaches the observer with a trough from the other. This happens for all pairs of *corresponding* strips in AC and CB because the same path difference of $\lambda/2$ exists. Hence there is no light in direction θ when

$$CD = \lambda/2 \quad (\text{or } BE = \lambda).$$

But $\sin \theta = \sin CAD = CD/(a/2)$, therefore

$$CD = \frac{a \sin \theta}{2} \quad (\text{or } BE = a \sin \theta).$$

That is, the direction of the first minimum is given by

$$a \frac{\sin \theta}{2} = \frac{\lambda}{2} \quad (\text{or } a \sin \theta = \lambda)$$

$$\therefore \quad \sin \theta = \frac{\lambda}{a}.$$

It can be shown by similar ' pairing ' processes that other minima occur when

$$\sin \theta = \frac{n\lambda}{a}$$

where $n = \pm 1, \pm 2, \pm 3$, etc.; the $\pm$ signs indicate that, for example, there are two ' first order ' minima, one on each side of the original direction of the incident beam. If θ is small and in radiens we can write $\theta \simeq n\lambda/a$.

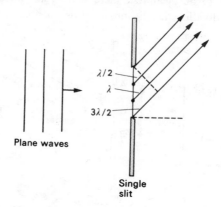

Plane waves

Fig. 8.26

Single slit

At P in Fig. 8.25, where the central bright band is formed, wavelets from all the imaginary strips in the slit arrive in phase since they have the same path length and the intensity of light is greatest. Other maxima occur roughly half-way between the minima at angles such that $\sin \theta$ has values $\pm 3\lambda/2a$, i.e. $(\lambda/a + 2\lambda/a)/2$, $\pm 5\lambda/2a$, etc. The first maximum is explained if the slit is divided into three equal parts and a direction considered in which the path differences between their ends are $\lambda/2$, Fig. 8.26. Wavelets from strips in two adjacent parts then cancel (as above), leaving only wavelets from one part to give a much less bright band. A graph of the relative intensity distribution for a single slit diffraction pattern is shown in Fig. 8.27.

Fig. 8.27

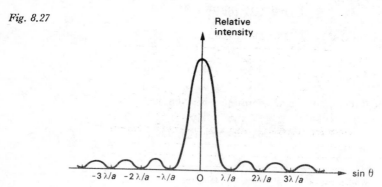

Relative intensity

$-3\lambda/a \quad -2\lambda/a \quad -\lambda/a \qquad O \qquad \lambda/a \quad 2\lambda/a \quad 3\lambda/a$

$\sin \theta$

(c) *Further points.* From $\sin \theta = \lambda/a$ we see that if the slit is wide, λ is much less than a and so $\sin \theta$, and therefore θ, are very small. The directions of the first (and all other) minima are thus extremely close to the middle of the central maximum. Most of the light emerging from the slit is in the direction of the incident light and there is little diffraction. Propagation is almost rectilinear and the laws of geometrical optics are applicable. On the other hand, if the slit is one wavelength wide, $\lambda = a$, $\sin \theta = \sin 90 = 1$ and the central bright band spreads completely into the geometrical shadow. In this case the behaviour of light requires a wave model.

The width of the central bright band can be seen from Fig. 8.27 to be twice that of any other bright band. (Also see Fig. 8.24a, p. 290.)

Diffraction at multiple slits

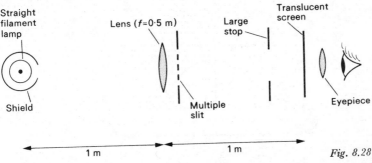

Fig. 8.28

Effects similar to those given by Young's double slit are obtained if more than two slits are used. They may be produced with the apparatus of Fig. 8.28, set up as for observing diffraction at a single slit (p. 289). It can be seen that as the number of slits is increased, the bright bands become both brighter and sharper. With equally-spaced slits each pattern is in effect the same as a two-slit one.

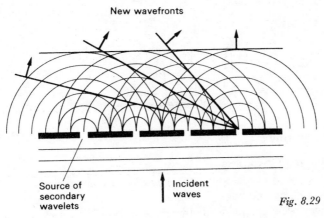

Fig. 8.29

Diffraction occurs at the slits and these, being very narrow, act as sources of secondary wavelets (semicircular in two dimensions) which superpose beyond the slits, Fig. 8.29. In certain directions the wavelets interfere constructively

293

to form a new, straight wavefront (a bright band), whilst in others they interfere destructively (a dark band).

As well as interference occurring between secondary wavelets from *different* slits, it also occurs between secondary wavelets from the *same* slit. Each slit therefore produces its own diffraction pattern; these are similar, in the same direction and will coincide exactly if focused by a lens. The diffraction pattern due to a single slit is thus superimposed on the interference pattern and determines the variations of intensity of the bright bands in the latter, as shown by the dotted curves in Fig. 8.30 and by the double slit interference fringes in the photograph of Fig. 8.3*b* (p. 272).

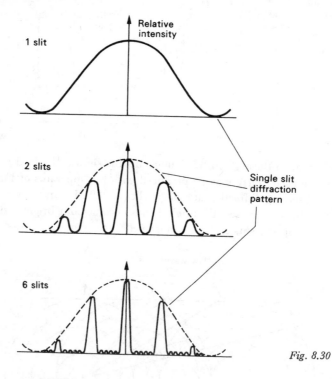

Fig. 8.30

Diffraction grating

A diffraction grating consists of a large number of fine, equidistant, closely-spaced parallel lines of equal width, ruled on glass or polished metal by a diamond point. In *transmission* gratings glass is used; the lines scatter the incident light and are more or less opaque while the spaces between them transmit light and act as slits. Such gratings are very expensive and cheaper plastic replicas are made. In *reflection* gratings the lines, ruled on metal, are again opaque but the unruled parts reflect light regularly. This type has the advantage that radiation absorbed by transmission grating material can be studied and if it is ruled on a concave spherical surface it focuses the radiation as well as diffracting it and no lenses are needed.

Diffraction gratings are used to produce spectra and for measuring wavelengths accurately. They have replaced the prism in much modern spectroscopy. Their usefulness arises from the fact that they give very sharp spectra, most of the incident light being concentrated in certain directions.

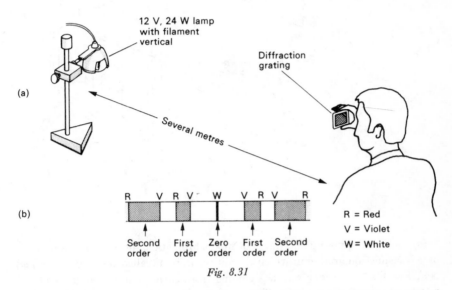

Fig. 8.31

(*a*) ' *Fine* ' *and* ' *coarse* ' *gratings*. A low-voltage lamp with its filament vertical is viewed at a distance of a few metres through a transmission grating held near the eye and with its lines parallel to the filament, Fig. 8.31*a*. A ' fine ' grating (e.g. 300 lines per mm) and a ' coarse ' grating (e.g. 100 lines per mm) should be tried in turn.

A typical pattern for a ' fine ' grating is shown in Fig. 8.31*b*. The central bright band, called the *zero order image*, is white (W) but on either side of it are brilliant bands of colour, called *first-* and *second order spectra*, like those given by a prism but having red light (R) deviated more than violet (V) and the dispersion increasing with order. Each spectrum consists of a series of adjacent images of the filament formed by the constituent colours of white light. A ' coarse ' grating forms many more orders of spectra, closer together. The effect of allowing a narrower band of wavelengths to fall on the grating can be observed by placing red and green filters in turn in front of the lamp.

A 'very coarse' grating (e.g. 10 lines per mm) gives a pattern that resembles Young's fringes.

The result is very striking and beautiful if white light from the lamp is viewed through two ' coarse ' gratings (100 lines per mm) with their rulings ' crossed '. An arrangement for projecting grating spectra on to a screen is shown in Fig. 8.32.

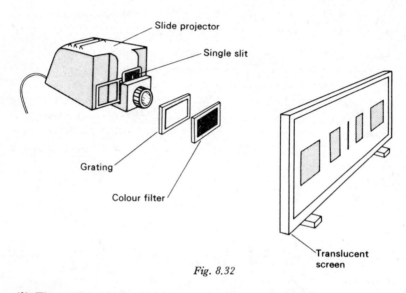

Slide projector

Single slit

Grating

Colour filter

Translucent screen

Fig. 8.32

(*b*) *Theory*. Suppose plane waves of monochromatic light of wavelength λ fall on a transmission grating in which the slit separation (called the *grating spacing*) is d, Fig. 8.33. Consider wavelets coming from corresponding points A and B on two successive slits and travelling at an angle θ to the direction of the incident beam. The path difference AC between the wavelets is $d \sin \theta$, as it is for all pairs

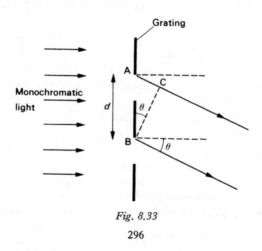

Grating

Monochromatic light

Fig. 8.33

296

of wavelets from other corresponding points in these two slits and in all pairs of slits in the grating. Hence if

$$d \sin \theta = n\lambda$$

where n is an integer giving the order of the spectrum, then reinforcement of the diffracted wavelets occurs in direction θ and a maximum will be obtained when the wavelets are brought to a focus by a lens.

When $n = 0$, $\theta = 0$ and we observe in the direction of the incident light the central bright maximum, i.e. the zero order image, for which the path difference of diffracted wavelets is zero. First, second etc. order spectra are given by $n = 1$, 2 etc., but these are much less bright than the $n = 0$ maximum.

As an example, consider a grating with 500 lines per mm on which yellow light of wavelength 6×10^{-7} m falls normally. Since there are 500 lines and 500 spaces per mm of the grating, the grating spacing (i.e. 1 line + 1 space) $d = 1/500$ mm $= 10^{-3}/500$ m $= 2 \times 10^{-6}$ m (this is about three times the wavelength of yellow light). For the first order, $n = 1$

$$\therefore \quad \sin \theta_1 = \frac{\lambda}{d} = \frac{6 \times 10^{-7}}{2 \times 10^{-6}}$$

$$= 0.3$$

$$\therefore \quad \theta_1 = 17°.$$

For the second order, $n = 2$

$$\therefore \quad \sin \theta_2 = \frac{2\lambda}{d} = 0.6$$

$$\therefore \quad \theta_2 = 37°.$$

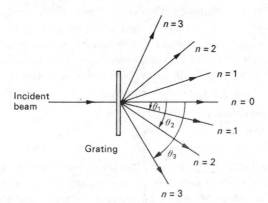

Fig. 8.34

With this grating a third order spectrum is obtained, Fig. 8.34, but not a fourth. Why?

PHYSICAL OPTICS

(c) *Measurement of wavelength.* The wavelength of monochromatic light, e.g. from a sodium lamp or flame, can be measured to four-figure accuracy using a spectrometer and a transmission diffraction grating having 600 lines per mm (which is typical of many gratings).

The usual adjustments are first made on (*i*) cross-wires, (*ii*) telescope and (*iii*) collimator of the spectrometer[1] so that the collimator produces parallel light and the telescope focuses it at the cross-wires.

The telescope is then turned through 90° exactly, from the position T_1 in which it is directly opposite the illuminated collimator slit, to position T_2, Fig. 8.35*a*.

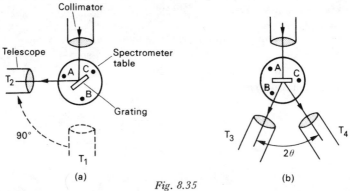

(a)   (b)

Fig. 8.35

The grating is placed on the spectrometer table, at right angles to the line joining two of the levelling screws, say A and B, and the table turned until the image of the slit reflected from the grating is in the centre of the field of view. Adjustment of A or B may be necessary to achieve this. The plane of the grating is now parallel to the axis of rotation of the telescope.

The table (and grating) are next turned through 45° exactly so that the incident light falls on the grating normally. The telescope is rotated, say to T_3, Fig. 8.35*b*, where the first order image is seen. If the lines of the grating are parallel to the axis of rotation of the telescope the image will be central, otherwise levelling screw C will need altering. The first order reading on the other side of the normal, at T_4, should also be taken. Half the angle between these two telescope settings gives θ, whence λ can be calculated from $\lambda = d \sin \theta$ where d is the grating spacing.

Optical spectra

Optical spectra fall into two basic groups, as do the spectra of all types of electromagnetic radiation when analysed by an appropriate spectrometer. As we shall see later (p. 419) the study of spectra, known as *spectroscopy*, provides information about the structure of atoms.

[1] See *Advanced Physics: Materials and Mechanics*, p. 215.

(*a*) *Emission spectra*. These are obtained when the light from a luminous source undergoes dispersion (formerly by a prism, nowadays often by a diffraction grating) and is observed directly. There are three types.

(*i*) *Line spectra* consist of quite separate bright lines of definite wavelengths on a dark background and are given by luminous gases and vapours at low pressure. Sodium vapour emits two bright yellow lines which are very close together (wavelengths 0.5896 and 0.5890 μm), Fig. 8.36*a*. Hydrogen emits red, blue-green and violet lines, Fig. 8.36*b*. Each line is an image of the slit (on the collimator) of the spectrometer on which the light falls. No two elements give the same line spectrum and spectroscopic methods are so sensitive that they can identify and reveal the presence of the most minute quantities of materials. In general line spectra are due to the individual *atoms* concerned since only in a gas (especially at low pressure) are the atoms far enough apart not to interact. They are also called *atomic* spectra.

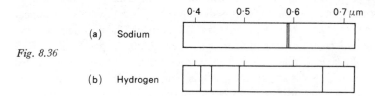

Fig. 8.36

A convenient source for producing line spectra is the discharge tube. This comprises a glass tube containing a gas (e.g. hydrogen, helium or neon) or vapour (e.g. mercury or sodium) at low pressure and two metal electrodes, across which a p.d. of several thousand volts is applied. The gas or vapour conducts and a luminous discharge occurs—an effect used in neon advertising signs and certain types of street lighting.

(*ii*) *Band spectra* have several well-defined groups or bands of lines. The lines are close together at one side of each band, making this side sharper and brighter than the other, Fig. 8.37. Band spectra are complex and are obtained from the molecules of glowing gases or vapours, heated or excited electrically. They arise from interaction between atoms in each molecule. The blue inner cone of a Bunsen burner flame gives a band spectrum as do nitrogen and oxygen in a discharge tube.

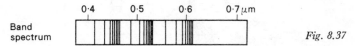

Fig. 8.37

(*iii*) *Continuous spectra* are emitted by hot solids and liquids and also hot gases at high pressures. The atoms are then so close that interaction is inevitable, and all wavelengths are emitted, Fig. 8.38. A continuous spectrum is not characteristic of the source and is conveniently produced by the white-hot tungsten filament of an electric lamp.

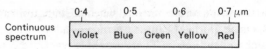

Fig. 8.38

(b) *Absorption spectra.* These form the second basic group of spectra and are observed when part of the radiation emitted is absorbed by a material between the source and observer. Line, band and continuous spectra are again obtained.

A line absorption spectrum occurs when white light passes through a cooler gas or vapour. Dark lines occur, against the continuous spectrum of white light, exactly at some of those wavelengths which are present in the line emission spectrum of the gas or vapour. The absorption spectrum of an element is thus the same as its emission spectrum except that the latter consists of bright lines on a dark background and the former of dark lines on a bright background. The atoms of the cooler gas absorb light of the wavelengths which they can emit, and then re-radiate the same wavelengths almost immediately but in all directions. Consequently, the parts of the spectrum corresponding to those wavelengths appear dark by comparison with other wavelengths not absorbed. The production of the line absorption spectrum of iodine vapour is described in Appendix 9, p. 537. The presence of a layer of relatively cooler gas round the sun causes the so-called *Fraunhöfer* lines in the solar spectrum, which is thus an example of a line absorption spectrum. The lines indicate the presence of hydrogen, helium, sodium, etc. in the sun's atmosphere (i.e. the chromosphere).

Band absorption spectra are formed when a continuous emission spectrum is observed through a material which itself can emit a band spectrum. In a continuous absorption spectrum the brightness of one region of the spectrum may be reduced but not another.

Resolving power

This is the ability of an optical system to form separate images of objects that are very close together and so reveal detail.

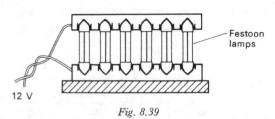

Fig. 8.39

(a) *Factors affecting it.* It can be shown that resolving power depends on the size of the slit or hole through which the objects are viewed and on the colour, i.e. wavelength, of the light used. Thus if the multiple light source of Fig. 8.39, with a green filter in front of it, is observed through an adjustable slit it will be found

300

that if the slit is narrow or the viewing distance large the lamps are not seen separately but appear to be continuous.

If the distance and the slit width are arranged so that the lamps can *just* be resolved, replacing the green filter by a blue one improves the resolution whilst a red one worsens it. Evidently the shorter the wavelength of the light the easier is it to see detail.

(*b*) *Resolving power of the eye.* The smallest angle θ which two points can subtend at the eye and still be seen as separate is taken as a measure of the resolving power of the eye. For example, if two well-lit black lines 2 mm apart on a card can just be distinguished separately by a certain observer at a distance of 5 m, then from Fig. 8.40 and since θ will be small, we have $\theta = a/d =$ 2×10^{-3} m/5 m $= 4 \times 10^{-4}$ rad. The resolving power of his eye is 4×10^{-4} rad. The smaller θ is, the *greater* the resolving power.

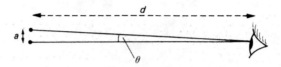

Fig. 8.40

(*c*) *Resolving power and diffraction.* A limit on the resolving power of an optical system is set by diffraction at the slit or hole through which observation occurs. A point object does not give a point image (even in an aberration-free system as we assume in geometrical optics); instead a diffraction pattern is obtained with its centre where the point image would be formed.

Rayleigh suggested that we should consider objects to be *just* resolved when the first minimum of the diffraction pattern of one falls on the central maximum of the other, Fig. 8.41. In the case of a slit of width a, the angular separation of

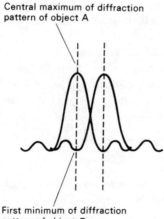

Central maximum of diffraction pattern of object A

Fig. 8.41   First minimum of diffraction pattern of object B

301

these two fringes is λ/a (see p. 291) and in general we can take this as being roughly true for other diffraction apertures of width a, e.g. a circle of diameter a.

Assuming the pupil of the eye has a diameter of two millimetres and that the average wavelength of light is 6×10^{-7} m, we might expect the eye to have a resolving power of $6 \times 10^{-7}/(2 \times 10^{-3}) = 3 \times 10^{-4}$ rad. This agrees with experimental values—see (b) above.

However, it must be pointed out that in many cameras, for example, the resolving power is limited more by aberrations than by diffraction and so the lens system has to be ' stopped down ' for finer resolution.

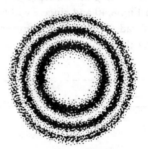

Fig. 8.42

(d) *Resolving power and magnifying power of optical instruments.* When a lens forms an image of a small object, it acts as a circular aperture and the image is a diffraction pattern like that in Fig. 8.42. From (c) it follows that the resolving power (λ/a) of a lens is improved by increasing its diameter a or by decreasing the wavelength λ of the light used. By having a large diameter, a telescope objective gathers a large amount of light and forms a small diffraction pattern of a distant star, i.e. the image is bright, sharp and detailed. Some microscopes obtain greater resolution by using ultraviolet radiation, which has a smaller wavelength than light. In an electron microscope moving electrons behave as waves with wavelengths about 10^5 times less than those of light and so extremely small objects can be resolved.

The resolving power of an instrument sets a limit on its useful magnifying power. For example, the objective of the Mount Palomar telescope has a diameter of 5 m, giving an approximate resolving power of $6 \times 10^{-7}/5 = 10^{-7}$ rad. The resolving power of the eye is about 3×10^{-4} rad, therefore the magnifying power required to increase the angular separation from 10^{-7} to 3×10^{-4} rad is given by

$$\frac{\text{resolving power of the eye}}{\text{resolving power of telescope}} = \frac{3 \times 10^{-4}}{10^{-7}} = 3 \times 10^3.$$

Any magnification beyond this reveals no further detail and is like stretching a rubber sheet on which there is a picture. The size of the picture increases but not the resolution.

In an electron microscope with a resolving power of 10^{-12} rad, a magnifying power of the order of 10^5 could be usefully employed. In any type of microscope the magnifying (and the resolving) power is limited basically by the wavelength of the ' illumination ' used in the instrument.

Polarization of light

Interference and diffraction require a wave model of light but they do not show whether the waves are longitudinal or transverse. Polarization suggests they are transverse in character.

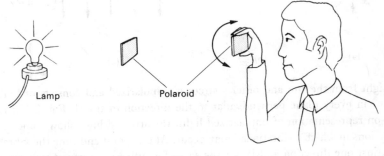

Lamp   Polaroid

Fig. 8.43

If a lamp is viewed through a piece of Polaroid (used for example in sunglasses), apart from it appearing slightly less bright there is no effect when the Polaroid is rotated about an axis perpendicular to itself. However, using two pieces, one of which is kept at rest and the other rotated slowly, Fig. 8.43, the light is cut off more or less completely in one position. The Polaroids are then said to be ' crossed '. Rotation through a further 90° allows maximum light transmission.

Fig. 8.44

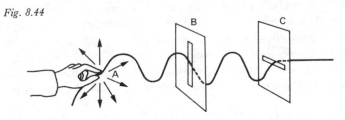

This experiment is analogous to that shown in Fig. 8.44 where transverse waves are sent along a spring to two slots B and C. The waves are generated by moving end A of the spring to and fro in *all* directions perpendicular to the direction of travel. Slot B passes only waves due to vibrations in a vertical plane. The transmitted wave is said to be *plane polarized* in a vertical plane and is unable to pass through the horizontal slot C. In this position slots B and C are ' crossed '. A longitudinal wave would emerge from both slots whatever their relative position, i.e. it cannot be polarized.

The optical experiment can be explained if light is regarded as a transverse wave motion which the first Polaroid plane polarizes by transmitting only the light ' vibrations ' in one particular plane and absorbing those in a plane at right angles. The second Polaroid transmits or absorbs the plane polarized light incident on it depending on its orientation with respect to the first Polaroid. It thus acts as a detector of polarized light.

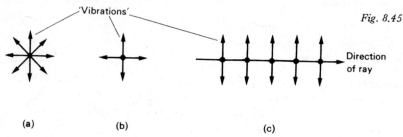

Fig. 8.45

(a) (b) (c)

Light from the sun and other sources is unpolarized and consists of ' vibrations ' in every plane perpendicular to the direction of travel. Fig. 8.45*a* is an end-on representation of unpolarized light, the arrowed lines show some of the directions in which ' vibrations ' may occur. At one point and time the vibration is in just one direction and is not the same for different points at that time. It also changes rapidly (about 10^9 times a second) and randomly at each point. Any vibration can be resolved into two perpendicular components and another end-on picture of unpolarized light is given in Fig. 8.45*b*. A side view of an unpolarized ray is shown in Fig. 8.45*c*, the dots representing ' vibrations ' at right angles to the paper and the arrowed lines ' vibrations ' in the plane of the paper.

Like other forms of electromagnetic radiation light is regarded as a varying electric field (E) coupled with a varying magnetic field (B), at right angles to each other and to the direction of travel. Fig. 8.46 is an attempt to indicate diagrammatically an electromagnetic wave. Experiment shows that the coupling of light with matter is more often through the electric field, e.g. affecting a photographic film. For this reason light ' vibrations ' are taken to be the variations of the electric field strength E. If light is vertically polarized we mean that the plane containing E and the direction of travel of the wave, called the *plane of vibration*, is vertical.

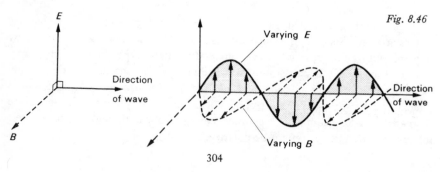

Fig. 8.46

304

PHYSICAL OPTICS

Producing polarized light

(a) *By Polaroid*. Polaroid is made from tiny crystals of quinine iodosulphate all lined up in the same direction in a sheet of nitrocellulose. Crystals which transmit light vibrations, i.e. electric field variations, in one particular plane and absorb those in a mutually perpendicular plane are said to be *dichroic*. They produce polarization by *selective absorption*. The effect is analogous to the absorption of vertically polarized microwaves by a grid of vertical wires, Fig. 8.47a, and transmission when the wires are horizontal, Fig. 8.47b.

Fig. 8.47

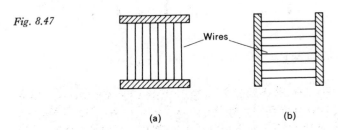

(a)   (b)

(b) *By reflection*. When unpolarized light falls on glass, water and some other materials the reflected light is, in general, partially plane polarized. But at one particular angle of incidence, called the *polarizing angle* i_p, the polarization is complete. At this angle, the reflected ray and the refracted ray in a transparent medium are found to be at right angles to each other. The vibrations in the reflected ray are parallel to the surface.

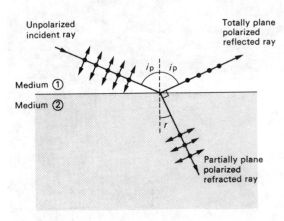

Fig. 8.48

Applying Snell's law to Fig. 8.48 we have

$$n_1 \sin i_p = n_2 \sin r$$

where n_1 and n_2 are the absolute refractive indices of media ① and ②. Also

$$i_p + 90° + r = 180°$$

$$\therefore \quad r = 90 - i_p.$$

Hence

$$n_1 \sin i_p = n_2 \sin (90 - i_p)$$

$$= n_2 \cos i_p$$

$$\therefore \quad \frac{n_2}{n_1} = \frac{\sin i_p}{\cos i_p} = \tan i_p.$$

If medium ① is air or a vacuum $n_1 = 1$ and so

$$n_2 = \tan i_p.$$

This is *Brewster's law*. For glass of refractive index 1.5, $i_p = 57°$.

(c) *By double refraction.* If a crystal of calcite (a form of calcium carbonate and also called ' Iceland spar ') is placed over, say, a triangle drawn on a piece of paper, two images are seen, Fig. 8.49. Such crystals exhibit double refraction and an incident unpolarized ray is split into two rays, called the *ordinary* and *extraordinary* rays, which are plane polarized in directions at right angles to each other. The former obeys Snell's law, but the latter does not because it travels with different speeds in different directions in the crystal.

Fig. 8.49

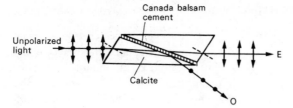

Canada balsam cement

Unpolarized light

E

Calcite

O

Fig. 8.50

In a *Nicol prism*, Fig. 8.50, which is made by cutting a calcite crystal in a certain way, only the extraordinary ray (E) is transmitted. The prism may therefore be used to produce plane polarized light, as Polaroid does. If the angle of incidence exceeds the critical angle, total internal reflection of the ordinary ray (O) occurs at the transparent layer of Canada balsam cement because it is optically less dense than the crystal for the ordinary ray but not for the extraordinary ray.

(*d*) *By scattering.* This may be shown by sending a narrow beam of unpolarized light through a tank of water in which scattering particles are obtained by adding one drop of milk. Polarization of the scattered light is detected by holding a piece of Polaroid in different positions, shown by the dotted lines in Fig. 8.51.

Fig. 8.51

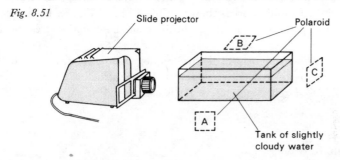

Slide projector

Polaroid

B

C

A

Tank of slightly cloudy water

The polarization arises from the transverse wave nature of light. That scattered along PA in Fig. 8.52 must be polarized in a direction parallel to PB (otherwise light would be longitudinal in character) and along PB the direction of polarization must be parallel to PA. What will happen if a piece of Polaroid is placed between the tank and the lamp?

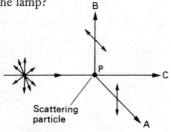

B

P

C

Scattering particle

A

Fig. 8.52

Using polarized light

(a) *Reducing glare.* Glare caused by light reflected from a smooth surface can be reduced by using polarizing materials since the reflected light is partially or completely polarized. Thus Polaroid discs, suitably orientated, are used in sunglasses and also in photography as ' filters ' where they are placed in front of the camera lens, thereby enabling detail to be seen that would otherwise be hidden by glare. Blue light from the sky, in a direction at right angles to the sun, is polarized and the brightness of the sky may be reduced in a colour photograph by a polarizing ' filter '.

(b) *Optical activity.* Certain crystals (e.g. quartz) and liquids (e.g. sugar solutions) rotate the plane of vibration of polarized light passing through them and are said to be optically active. For a solution the angle of rotation depends on its concentration and in an instrument known as a *polarimeter* this is used to measure concentration.

(c) *Stress analysis.* When glass, Perspex, polythene and some other plastics are under stress (e.g. by bending, twisting or uneven heating) they become doubly refracting and if viewed in white light between two ' crossed ' Polaroids, coloured fringes are seen round the regions of strain.[1] The effect is called *photoelasticity* and is used to analyse stresses in plastic models of various structures. A modern polariscope for such work is shown in Fig. 8.53 and Fig. 8.54 is a photograph of the strain pattern for the model of the crane-hook under stress in the polariscope.

In a more recent development of the technique, actual structures are coated with a photoelastic plastic to which strains created in the structure are transmitted directly. In this case a reflection polariscope is used.

Electromagnetic waves

The idea that fields of force exist in the space surrounding electric charges, current-carrying conductors and magnets is useful for ' explaining ' electrical and magnetic effects. The flow of current in a conductor may also be considered to be due to the existence of an electric field in the conductor and since current can be induced by a changing magnetic field it follows that *a changing magnetic field creates an electric field* in the medium where the change occurs (whether it be free space or a material medium). This effect is electromagnetic induction and was discovered by Faraday in 1831 as we have seen (p. 106).

The converse effect, i.e. *a changing electric field sets up a magnetic field*, was proposed by Maxwell in 1864 on the grounds that if this assumption is made, the mathematical equations expressing the laws of electricity and magnetism (in

---

[1] See *Advanced Physics: Materials and Mechanics*, Fig. 2.17a.

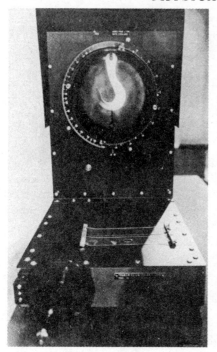

Fig. 8.53

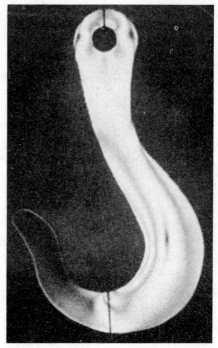

Fig. 8.54

terms of Faraday's ideas of fields of force) become both simple and symmetrical. There is no easy way of demonstrating Maxwell's assumption directly on account of the difficulty of detecting a weak magnetic field, but one consequence is that any electric field produced by a changing magnetic field must inevitably create a magnetic field and vice versa, i.e. one effect is coupled with the other.

On this basis Maxwell predicted that when a moving electric charge is oscillating it should radiate an electromagnetic wave consisting of a fluctuating electric field accompanied by a fluctuating magnetic field of the same frequency and phase, the fields being at right angles to each other and to the direction of travel of the wave. The intensities (E and B) of the fields thus vary periodically with time *like* the amplitude of a wave motion of the transverse type but, in fact, no medium seems to be involved in the transmission. A diagrammatic representation of an electromagnetic wave was given in Fig. 8.46 (p. 304).

Maxwell also showed that the speed c of all electromagnetic waves in free space, irrespective of their wavelength or origin, should be given by

$$c = \frac{1}{\sqrt{\mu_0\epsilon_0}}$$

where μ_0 and ϵ_0 are the permeability and permittivity respectively of free space.

Substituting numerical values in this expression gives $c = 3.0 \times 10^8$ m s^{-1}, i.e. the speed of light. This led Maxwell to suggest that light might be one form of electromagnetic wave motion.

The search for other forms of electromagnetic radiation was taken up and resulted in the discovery by H. Hertz in 1887 of waves having the same speed as light but with wavelengths of several metres compared with a very small fraction of a millimetre for light. These waves, now called *radio waves*, were generated by Hertz using a spark gap transmitter which produced bursts of high frequency a.c.

A whole family of electromagnetic radiations is now known, extending from gamma rays of very short wavelength to very long radio waves. Fig. 8.55 shows

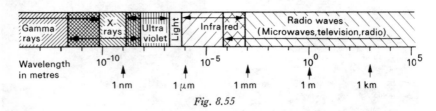

Fig. 8.55

the wavelength ranges of the various types but there is no rapid change of properties from one to the next. Although methods of production and detection differ from member to member, all exhibit reflection, refraction, interference, diffraction and polarization. When they fall on a body partial reflection, transmission or absorption occurs; the part absorbed becomes internal energy.

For any wave motion, $v = f\lambda$. For all electromagnetic waves in a vacuum $v = c = 3.0 \times 10^8$ m s^{-1} and so if f or λ for a particular radiation is known, the other can be calculated. Thus for violet light of $\lambda = 0.40$ μm $= 4.0 \times 10^{-7}$ m we have $f = c/\lambda = 3.0 \times 10^8/(4.0 \times 10^{-7}) = 7.5 \times 10^{14}$ Hz. By contrast a 1-MHz radio signal has a wavelength of 300 m. Electromagnetic waves may be distinguished either by their wavelength or frequency but the latter is more fundamental since unlike wavelength (and speed) it does not change when the wave travels from one medium to another. The production of various types of electromagnetic radiation will be considered later.

Whilst a wave model accounts for many of the properties of electromagnetic radiation it cannot explain some other aspects of its behaviour (see p. 392).

Infrared radiation

The wavelength of light varies from 0.4 μm for violet light to about 0.7 μm for red light. This is the ' visible ' region of the electromagnetic spectrum. Longer wavelength radiation, extending from 0.7 μm to about one millimetre, is called *infrared* and is not detected by the eye. It has all the properties of electromagnetic waves.

(a) *Sources.* The surfaces of all bodies emit infrared radiation in a continuous range of wavelengths. The relative amount of each wavelength depends mainly on the temperature but also on the nature of the surface of the body. At low temperatures, long wavelength infrared is emitted; as the temperature of the body rises the emission is more copious and shorter and shorter wavelengths are present. At about 500 °C, red light is emitted as well as long and short wave infrared, i.e. the body is red-hot. Increasing the temperature adds orange, yellow, green, blue and violet light in turn and at around 1000 °C the body becomes white-hot. Eventually ultraviolet radiation (p. 314) is emitted as well.

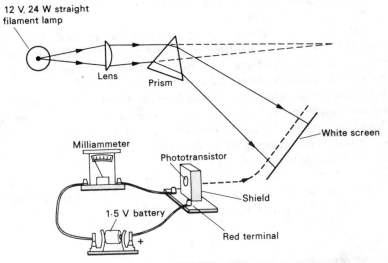

12 V, 24 W straight filament lamp

Lens

Prism

White screen

Milliammeter

Phototransistor

Shield

1·5 V battery

Red terminal

Fig. 8.56

The presence of infrared in the radiation from a white-hot filament lamp may be shown with the arrangement of Fig. 8.56 in which a phototransistor is used as a detector. When this is moved slowly through the spectrum produced by the glass prism, the milliammeter gives a reading beyond the visible red region, i.e. in the infrared.

(b) *Absorption.* When absorbed by matter, infrared—like all other types of electromagnetic radiation—causes an increase of internal energy which usually results in a temperature rise. Thus, infrared falling on the skin produces the sensation of warmth. Infrared lamps are used for the treatment of muscular complaints and to dry the paint on cars during manufacture. Infrared cooking is also practised.

Except for wavelengths near to that of red light, infrared is absorbed by glass but transmitted by rock salt. The action of a greenhouse depends on the fact that sunlight passes through the glass, is absorbed by and warms up the plants, soil etc., which reradiate long wavelength infrared (because of their comparatively low temperature). Most of this infrared cannot penetrate the glass and is trapped in the greenhouse.

Water vapour and carbon dioxide in the lower layers of the atmosphere exhibit the same ' selective absorption ' effect and prevent infrared emitted by the earth from escaping. It has been estimated that if the average temperature of the earth rose by 3.5 °C due, for example, to the combustion of fossil fuels and the resulting increase of carbon dioxide in the atmosphere, dramatic climatic and geographical changes could occur. Some scientists believe that ' thermal ' pollution is as great a threat as any other form.

(*c*) *Detection*. There are three types of detector—photographic, photoelectric (e.g. phototransistors) and thermal.

Special *photographic films* sensitive to infrared may be used. They enable pictures to be taken in the dark (like that of Fig. 8.57 of a car; the white parts are hottest) or in hazy conditions since infrared is scattered less than light by particles in the atmosphere (because of its longer wavelength).

Very sensitive *photoelectric devices* have been developed which enable the infrared emitted by, for example, a distant rocket to be detected and early warning of its launching obtained. Similar detectors are used in weather satellites to obtain cloud formation patterns over the whole of the earth's surface and are especially useful for detecting hurricanes.

Thermal detectors include the thermometer, the thermopile and the bolometer.

A *thermopile* consists of many thermocouples in series. In a thermocouple an e.m.f. is produced between the junction of two different metals or semiconductors

Fig. 8.57

if one junction is at a higher temperature than the other, Fig. 8.58*a*. The hot junctions of a thermopile, Fig. 8.58*b*, are blackened to make them good absorbers of the incident radiation to be measured whilst the cold junctions are shielded from it. The hot junction temperature rises until the rate of gain of heat equals the rate of loss of heat to the surroundings and then the e.m.f. is measured with a

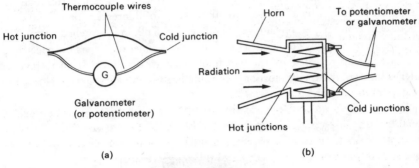

Fig. 8.58

potentiometer. Alternatively it may be enough to observe the reading produced on a galvanometer connected across the thermopile.

A *bolometer* is a device such as a thermistor in which a small temperature change causes a large change of electrical resistance. In practice two thermistors are used in a Wheatstone bridge circuit, Fig. 8.59, radiation being allowed to fall

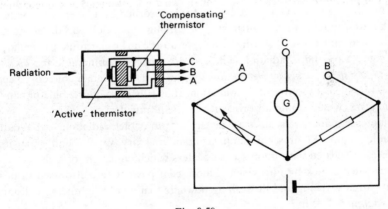

Fig. 8.59

on the ' active ' one but not on the ' compensating ' one. The resulting tempera-ture change creates an out-of-balance p.d. and the bridge has to be rebalanced. Greater sensitivity is achieved if the incident radiation is ' chopped ', i.e. inter-rupted by a rotating disc with slots, so that the output from the bridge is an alternating p.d. which can be amplified.

313

PHYSICAL OPTICS

Ultraviolet radiation

Shortly after infrared was discovered in 1800, radiation beyond the violet end of the sun's visible spectrum was found. This radiation, which has typical electromagnetic wave properties, has a shorter wavelength than light and covers the range 0.4 μm to about 1 nm.

The existence of ultraviolet in the spectrum formed from an overrun filament lamp may be shown by allowing it to fall on strips of fluorescent paper and non-fluorescent white paper, one above the other, Fig. 8.60, in a darkened room. The fluorescent paper glows outside (as well as at) the violet of the visible spectrum, the latter being revealed by the non-fluorescent white paper. If an ultraviolet filter is inserted it cuts off most of the visible light.

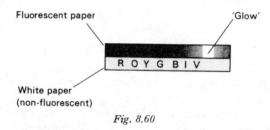

Fig. 8.60

The sun emits ultraviolet but much of it is absorbed by a layer of ozone in the earth's atmosphere. This is fortunate, for although ultraviolet produces vitamins in the skin and causes ' sun-tan ', an overdose can be harmful, particularly if it is short-wavelength radiation. The eyes are especially vulnerable and *an ultraviolet lamp should never be viewed directly.* Such a lamp contains mercury vapour and is a convenient source of the radiation. It usually has a quartz bulb or window —not glass, which would absorb much of the ultraviolet.

Fluorescent materials absorb ' invisible ' ultraviolet radiation and reradiate ' visible ' light. Fluorescent lamps contain mercury vapour and their inner surfaces are coated with fluorescent powders which emit light of a characteristic colour when struck by ultraviolet. Fluorescent powders are also used in poster paints for advertising and in washing powders that add ' brightness to cleanness and whiteness '. Why?

Photoelectric devices and photographic films can also detect ultraviolet.

Thermal radiation

Thermal radiation is energy which travels as electromagnetic waves, having been produced by a source because of its temperature. Its chief component is usually infrared radiation but light and ultraviolet may also be present. Being energy in the process of transfer it can be regarded as heat.[1]

At the start of the twentieth century certain aspects of the emission of thermal radiation were responsible, along with the photoelectric effect to be discussed later (p. 390), for a revolutionary change of view about the nature of light and other types of electromagnetic radiation. This resulted in the emergence of the *quantum theory*. For the present we shall deal mainly with the facts concerning the emission and absorption of thermal radiation and with other related ideas.

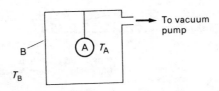

Fig. 8.61

(a) *Prévost's theory of exchanges.* In Fig. 8.61 the small body A, at temperature T_A, is suspended by a non-conducting thread inside the box B whose walls are at a constant, different temperature T_B. If B is then evacuated, energy exchange between A and B can occur only by radiation. If $T_A > T_B$, A's temperature falls until it is also T_B, but if $T_A < T_B$, it rises to T_B. In either case A acquires B's temperature and it might appear that energy exchange then ceases.

Prévost suggested in 1792 that, on the contrary, *when a body is at the same temperature as its surroundings its rate of emission of radiation to the surroundings equals its rate of absorption of radiation from the surroundings*. That is, there is dynamic equilibrium, and energy exchange continues and at a rate depending on the temperature. This view is now generally accepted.

It follows that a body which is a good absorber of radiation must also be a good emitter of radiation otherwise its temperature would rise above that of its surroundings. Conversely a good emitter must be a good absorber. Experiments confirm these conclusions and indicate that a dull, black surface is the best absorber and, as we might anticipate, is also the best emitter.

(b) *Black body radiation.* The theoretical concept of a perfect absorber, called a *black body, which absorbs all the radiation of every wavelength falling on it*, is, like the idea of an ideal gas in kinetic theory (p. 335), a useful standard for judging the performance of other bodies. At normal temperatures it must appear black since it does not reflect light—hence the term ' black body '. We would expect a black body to be the best possible emitter at any given temperature. The radiation emitted by it is called *black body radiation*, or *full radiation* or *temperature radiation*; the latter term is used because the relative intensities of the various wavelengths present depend only on the temperature of the black body.

[1] See *Advanced Physics: Materials and Mechanics*, p. 115.

In practice an almost perfect black body consists of an enclosure, such as a cylinder, with dull black interior walls and a small hole, Fig. 8.62. Radiation entering the hole from outside has little chance of escaping since any energy not absorbed when the wall is first struck, will be subsequently. The *hole* in the enclosure thus acts as a black body because it absorbs all the radiation falling on it.

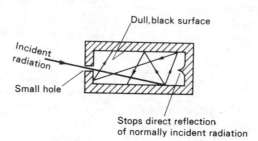

Fig. 8.62

When the enclosure is heated to a certain temperature, say by an electric heating coil wrapped round it, *black body* or *cavity* radiation emerges from the hole (which nonetheless may appear red or yellow or even white if the temperature is high enough). A spectrum of the radiation can be formed if it falls on the slit of a suitable spectrometer and in general infrared, light and ultraviolet will be present. All three have a heating effect when absorbed and can therefore be detected by a thermopile or a bolometer. If a narrow slit is placed in front of one of these instruments and moved along the spectrum, the energy carried by the small band of wavelengths passing through the slit at different mean wavelengths

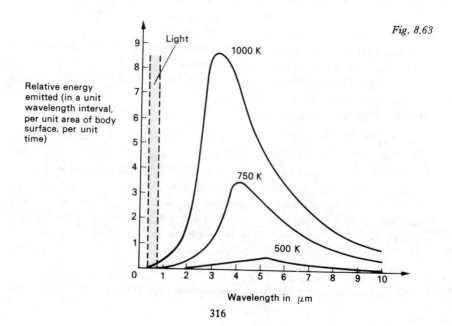

Fig. 8.63

can be obtained. Repeating this for different temperatures enables a family of curves to be drawn showing how the energy in the spectrum of a black body is distributed among the various wavelengths.

The curves have the form shown in Fig. 8.63; the following points emerge.

(*i*) As the temperature rises the energy emitted in each band of wavelengths increases, i.e. the body becomes ' brighter '.

(*ii*) Even at 1000 K only a small fraction of the radiation is light. (The maximum does not lie in the visible region until about 4000 K.)

(*iii*) At each temperature T the energy radiated is a maximum for a certain wavelength λ_{max} which decreases with rising temperature; in fact $\lambda_{max} \propto 1/T$, i.e. $\lambda_{max}T = $ constant—a statement known as *Wien's displacement law*. This explains why a body appears successively red-hot, yellow-hot and white-hot, etc. Sirius (the Dog Star) looks blue, not white. Why?

It is interesting to compare the spectral emission curve of a tungsten filament lamp (a non-black body) with that of a black body at the same temperature (2000 K), Fig. 8.64.

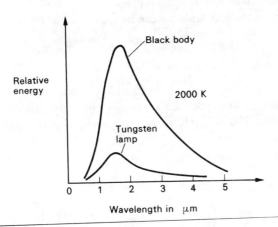

Fig. 8.64

(*c*) *Stefan's law*. This states that if E is the *total* energy radiated of all wavelengths per unit area per unit time by a *black body* at thermodynamic temperature T (see page 333), then

$$E \propto T^4$$

or

$$E = \sigma T^4$$

where σ is a constant, called *Stefan's constant*. If E is in W m^{-2} (i.e. J s^{-1} m^{-2}) and T is in K (kelvins) then σ has units W m^{-2} K^{-4}. Its value is

$$\sigma = 5.7 \times 10^{-8} \text{ W m}^{-2} \text{ K}^{-4}.$$

The area under each spectral emission curve, Fig. 8.64, represents the total energy radiated per unit area per unit time at temperature T and is found to be proportional to T^4—agreeing with Stefan's law.

A black body at temperature T in an enclosure at a lower temperature T_0 loses energy by emission but also gains some from the enclosure. The net loss of energy E from the black body per unit area per unit time is thus given by

$$E = \sigma T^4 - \sigma T_0^4$$

$$= \sigma(T^4 - T_0^4).$$

If $T \gg T_0$ then T_0^4 can be neglected and we have

$$E = \sigma T^4.$$

If $(T - T_0)$ is small then we can show that

$$E \simeq 4\sigma T_0^3(T - T_0).$$

Thus, let $x = T - T_0$ and so $E = \sigma(T^4 - T_0^4) = \sigma[(T_0 + x)^4 - T_0^4] = \sigma[T_0^4 + 4T_0^3x + 6T_0^2x^2 + 4T_0x^3 + x^4 - T_0^4] \simeq 4\sigma T_0^3x$ since x^2, x^3 and x^4 are negligible. The rate of loss of energy by radiation is therefore proportional to the temperature excess of the body over its surroundings if this is small.

For a non-black body Stefan's law can be applied in the form

$$E = \epsilon\sigma T^4$$

where ϵ is called the *total emissivity* of the body and has a value between 0 and 1.

(*d*) *Optical pyrometer.* A pyrometer is an instrument for measuring high temperatures and certain types use all or part of the thermal radiation emitted by the hot source. The principle of the *disappearing filament* optical pyrometer,

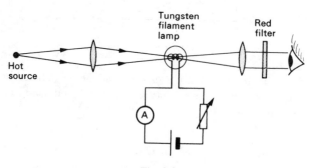

Fig. 8.65

which responds to light only, is shown in Fig. 8.65. Light from the source and the tungsten filament lamp passes through a red filter of a known wavelength range before reaching the eye; both appear red. The current through the filament is adjusted until it appears as bright as the background of light from the source. The temperature is then read off from the ammeter, previously calibrated in K.

Calibration may be done against a gas thermometer up to 1800 K. The range of a pyrometer can be extended up to about 3000 K by having a sectored disc rotating in front of the source which cuts down the radiation in a known proportion.

Pyrometers are used to measure the temperatures of the interiors of furnaces—without getting too close—and of the surfaces of molten metals.

QUESTIONS

Speed of light: interference

1. A plane mirror rotating at 35 revolutions per second reflects a narrow beam of light to a stationary mirror 200 m away. The stationary mirror reflects the light normally so that it is again reflected from the rotating mirror. The light now makes an angle of 2.0 minutes of arc with the path it would travel if both mirrors were stationary. Calculate the velocity of light.

Give *two* reasons why it is important that an accurate value of the velocity of light should be known.
(*J.M.B.*)

2. A beam of monochromatic light of wavelength 6.0×10^{-7} m in air passes into glass of refractive index 1.5. If the speed of light in air is 3.0×10^8 m s^{-1}, calculate (*a*) the speed of the light in glass, (*b*) the frequency of the light, and (*c*) the wavelength of the light in glass.

3. Describe an experiment by means of which both the wave nature of light can be demonstrated and the wavelength of the light can be determined. Draw and label a diagram of the apparatus and give the theory of the experiment.

Plane waves in air are incident obliquely on a plane boundary between the air and glass. Use Huygens' construction to derive the sine law of refraction.

What test of the wave theory is suggested by the result? (*J.M.B.*)

4. Monochromatic light illuminates a narrow slit which is 4.0 m away from a screen. Two very narrow parallel slits 0.50 mm apart are placed midway between the single slit and the screen so that interference fringes are obtained. If the spacing of five fringes is 10 mm, calculate the wavelength of the light.

What will be the effect of (*a*) halving the distance between the double slit and the screen, (*b*) halving the slit separation, (*c*) covering one of the double slits, and (*d*) using white light?

Discuss whether interference, diffraction or both are involved in the formation of the fringes.

5. Describe the experimental arrangement for the production of interference fringes by Young's two-slit method. Describe and explain the appearance of the fringes if a white light source is used.

Explain how you would take the necessary measurements in order to determine the wavelength of light from a monochromatic source. Give the theory which enables you to calculate your result.

What would be the effect of introducing a microscope slide in the path of one of the interfering beams?

(A.E.B.)

6. Draw and label diagrams to illustrate the apparatus you would use to demonstrate optical interference giving:

(a) straight line fringes, and

(b) circular fringes.

In a Young's double-slit experiment the fringe separation observed using yellow light was found to be 0.275 mm. The yellow lamp, giving a wavelength 5.50×10^{-7} m is replaced by a purple one giving wavelengths of 4.00×10^{-7} m in the violet and 6.00×10^{-7} m in the red. The remainder of the apparatus is undisturbed. Calculate

(i) the distance between the fringes formed by the violet light,

(ii) the distance between the fringes formed by the red light, and

(iii) the distance from the purple fringe on the axis to the next purple fringe observed.

Hence draw a diagram of the appearance of the new fringe system, indicating the colours, and extending as far as 1 mm from the axis.

(C.)

7. Explain what is meant by the term *path-difference* with reference to the interference of two wave-motions.

Why is it not possible to see interference where the light beams from the head-lamps of a car overlap?

Interference fringes were produced by the Young's slits method, the wavelength of the light being 6.0×10^{-5} cm. When a film of material 3.6×10^{-3} cm thick was placed over one of the slits, the fringe pattern was displaced by a distance equal to 30 times that between two adjacent fringes. Calculate the refractive index of the material. To which side are the fringes displaced?

(When a layer of transparent material whose refractive index is n and whose thickness is d is placed in the path of a beam of light, it introduces a path difference equal to $(n - 1)d$.)

(O. and C.)

8. Give an account of the evidence that leads you to believe that light is propagated as a transverse wave motion.

In an interferometer, two coherent beams are made to follow separate paths, one traversing the length of a glass tube A, the other the length of a glass tube B, before being brought together to interfere. The tubes, each 20 cm long, are closed by plane glass plates, and initially contain air at atmospheric pressure. Fringes are obtained with light of wavelength 5.6×10^{-5} cm. Explain why the fringes move across the field of view while one of the tubes is being evacuated, and calculate the number which cross a fixed point in the field of view before a good vacuum is finally reached.

(Take the refractive index of air to be 1.000 28 at atmospheric pressure.)   (O.)

PHYSICAL OPTICS

9. State the conditions necessary for the effects of interference in optics to be observed.

Two microscope slides, each 7.5 cm in length are placed with two of their faces in contact. A cover glass is then inserted between them at one extreme end to enclose a wedge-shaped layer of air. When illuminated normally by sodium light and the reflected light viewed, parallel interference bands are seen at a uniform distance apart of 0.11 mm. Explain the formation of these bands and calculate the thickness of the cover glass. (Wavelength of sodium light may be taken as 5.9×10^{-5} cm.) *(L.)*

10. Draw a diagram to illustrate how you would set up apparatus to view Newton's rings, formed by normal reflection of sodium light. Explain the formation of the rings.

If the diameter of the fifth bright ring is 2.55×10^{-1} cm and the diameter of the fifteenth bright ring is 4.27×10^{-1} cm, calculate an approximate value for the wavelength of the sodium light. The radius of curvature of the lens employed is 50.0 cm. *(W.)*

11. Describe how you would set up the necessary apparatus to observe and measure Newton's rings. Derive an expression for the difference in the squares of the radii of two consecutive bright rings in the pattern.

A set of Newton's rings was produced between one surface of a biconvex lens and a glass plate, using green light of wavelength 5.46×10^{-5} cm. The diameters of two particular bright rings, of orders of interference p and $p + 10$, were found to be 5.72 mm and 8.10 mm respectively. When the space between the lens surface and the plate was filled with liquid the corresponding values were 4.95 mm and 7.02 mm. Determine the radius of curvature of the lens surface and the refractive index of the liquid. *(L.)*

Diffraction: spectra: polarization

12. Calculate the angular separation of the red and violet rays of the first order spectrum when a parallel beam of white light is incident normally on a diffraction grating with 5000 lines per cm. Take the wavelengths of red and violet light to be 7×10^{-7} m and 4×10^{-7} m respectively.

13. When a source of monochromatic light of wavelength 6.0×10^{-7} m is viewed 2.0 m away through a diffraction grating two similar lamps are ' seen ' one on each side of it, each 0.30 m away from it. Calculate the number of lines in the grating.

14. Explain the action of a plane diffraction grating. Giving the necessary theory and experimental detail, explain how it may be used to measure the wavelength of light from a monochromatic source. It may be assumed that the initial adjustments to the spectrometer have been made.

With a given grating used at normal incidence a yellow line, $\lambda = 6.0 \times 10^{-7}$ m, in one spectrum coincidences with a blue line, $\lambda = 4.8 \times 10^{-7}$ m, in the next order spectrum. If the angle of diffraction for these two lines is $30°$ calculate the spacing between the grating lines. *(A.E.B.)*

PHYSICAL OPTICS

15. Describe *two* experiments to illustrate the diffraction of light. How, in general, do the effects produced depend on the size of the obstacle or aperture relative to the wavelength of the light used?

A diffraction grating consists of a number of fine equidistant wires each of diameter 1.00×10^{-3} cm with spaces of width 0.75×10^{-3} cm between them. A parallel beam of infrared radiation of wavelength 3.00×10^{-4} cm falls normally on the grating, behind which is a converging lens made of material transparent to the infrared. Find the angular positions of the second order maxima in the diffraction pattern.

(L.)

16. Draw a diagram to show plane monochromatic waves falling normally on a grating and being diffracted. Using the diagram, explain in which directions diffracted maxima will be seen.

A diffraction grating is fitted on a spectrometer table with monochromatic light incident normally on it. The instrument is adjusted to observe the spectra produced. State two ways in which the spectra would be affected if a grating of greater spacing were used.

The slit of the collimator is now illuminated with light of wavelengths 5.89×10^{-5} cm and 6.15×10^{-5} cm. The grating has 6.00×10^3 lines per cm and the telescope objective has a focal length of 20.0 cm. Calculate (*a*) the angle between the first-order diffracted waves leaving the grating, and (*b*) the separation between the centres of the two first-order lines formed in the focal plane of the objective.

(*J.M.B.*)

17. Distinguish between emission spectra and absorption spectra. Describe the spectrum of the light emitted by (*i*) the sun, (*ii*) a car headlamp fitted with yellow glass, and (*iii*) a sodium vapour street lamp.

What are the approximate wavelength limits of the visible spectrum? How would you demonstrate the existence of radiations whose wavelengths lie just outside these limits?

(*O. and C.*)

18. Describe and explain the action of one device for producing a beam of plane polarized light.

How would you make an experimental determination of the refractive index of a polished sheet of black glass using this device?

Mention two practical applications of polarized light. (*L. part qn.*)

19. Explain what is meant by *plane polarized light.*

Describe how a beam of plane polarized light may be produced by (*a*) reflection, and (*b*) double refraction. Give an account of two uses of polarized light.

Why is it not possible to polarize sound waves? (*A.E.B.*)

Thermal radiation

20. The cylindrical element of a 1.0-kW electric fire is 30.0 cm long and 1.0 cm in diameter. Taking the temperature of the surroundings to be 20 °C and Stefan's constant to be 5.7×10^{-8} W m^{-2} K^{-4}, estimate the working temperature of the element.

Give *two* reasons why the actual temperature is likely to differ from your estimate.

(*J.M.B.*)

21. State *Stefan's* law and explain what is meant by *Prévost's theory of exchanges*.

What is meant by a *black body radiator*? Draw labelled curves which show how the energy in the spectrum of the radiation from such a body varies with wavelength at various temperatures.

A copper sphere, at a temperature of 727 °C, is suspended in an evacuated enclosure the walls of which are maintained at 27 °C. If the radius of the sphere is 5.0 cm, and if conditions are such that Stefan's law is obeyed, calculate the initial rate of fall of temperature of the sphere. (Density of copper $= 9.0 \times 10^3$ kg m^{-3}; specific heat capacity of copper $= 400$ J kg^{-1} K^{-1}; Stefan's constant $= 5.8 \times 10^{-8}$ W m^{-2} K^{-4}.)

(*A.E.B.*)

22. What do you understand by a *black body*? In what respects does the radiation from a black body at 2000 K differ from that from a black body at 1000 K? How would you devise a black body to radiate at 1000 K?

A blackened sphere, of radius 2.0 cm, is contained within a hollow evacuated enclosure the walls of which are maintained at 27 °C. Assuming that the sphere radiates like a black body and that Stefan's constant is 5.7×10^{-8} W m^{-2} K^{-4}, calculate the rate at which the sphere loses heat when its temperature is 227 °C.

(*J.M.B.*)

23. Discuss the nature of the processes by which a hot body may lose heat to its surroundings.

A blackened platinum strip of area 0.20 cm^2 is placed at a distance of 200 cm from a white-hot iron sphere of diameter 1.0 cm, so that the radiation falls normally on the strip. The radiation causes the temperature, and hence the resistance, of the platinum to increase. It is found that the same increase in resistance can be produced under similar conditions, but in the absence of radiation, when a current of 3.0 mA is passed through the platinum strip, the potential difference between its ends being 24 mV. Estimate the temperature of the iron sphere.

(Stefan's constant $= 5.7 \times 10^{-8}$ W m^{-2} K^{-4}.)

(*C.*)

Part 3 | ATOMS

9 Kinetic theory: thermodynamics

Introduction

The behaviour of matter in bulk can be described in terms of quantities such as density, pressure and temperature which are measurable and are often capable of perception by the senses. One of the aims of modern science is to relate these macroscopic (i.e. large-scale) properties of matter to the masses, speeds, energies etc. of the constituent atoms and molecules that cannot be perceived directly. An attempt is made to explain the macroscopic in terms of the microscopic. If the atomic model of matter is valid this should be possible since the same thing is being considered from two different points of view. We thereby hope to obtain deeper understanding of the phenomenon under investigation.

Here we will pursue this approach by studying *thermodynamics* and *kinetic theory* (a branch of statistical mechanics). In thermodynamics, thermal effects are considered using macroscopic quantities like pressure, temperature, volume and internal energy. Kinetic theory covers similar ground but assumes the existence of atoms and molecules and applies the laws of mechanics to large numbers of them, using simple averaging techniques.

We will be concerned mainly with gases. In these, conditions are simpler and the mathematical problems less difficult as a result.

Gas laws

Experiments have established three laws describing the thermal behaviour of gases.

(*a*) *Boyle's law.* This relates the pressure and volume of a gas and arose from work done by Boyle about 1660.

The pressure of a fixed mass of gas is inversely proportional to its volume if the temperature is constant.

In symbols, if p is the pressure and V the volume then

$$p \propto \frac{1}{V} \text{ or } p = \frac{\text{constant}}{V}$$

$$\therefore \quad pV = \text{constant}.$$

The value of the constant depends on the mass of gas and the temperature. Graphically the law may be represented by plotting (i) p against $1/V$, Fig. 9.1a, which is a straight line through the origins, (ii) pV against p, Fig. 9.1b, giving a straight line parallel to the p-axis, or (iii) p against V, Fig. 9.1c, to obtain a rectangular hyperbola. If a fixed mass of gas is investigated at different temperatures, a series of graphs, each called an *isothermal* (i.e. a curve of constant temperature), can be drawn as shown.

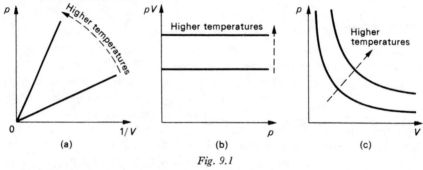

Fig. 9.1

Measurements made since Boyle's time have indicated that the law is obeyed only when the density of the gas is low. Thus, gases such as oxygen, nitrogen, hydrogen and helium (known as ' permanent ' gases) follow the law well at normal temperatures and pressures but they deviate from it at high pressures (e.g. above several hundred atmospheres) when their density is high. Their p-V

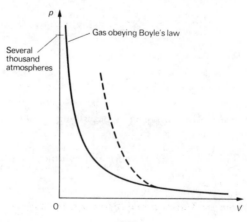

Fig. 9.2

graphs become more like that for a liquid and the volume decrease is quite small for large pressure increases, as shown by the dotted curve in Fig. 9.2. At high pressures carbon dioxide and some other gases and vapours are close enough to being liquid at ordinary temperatures to show marked deviations from the law.

(b) *Charles' law*. The connection between the volume change with temperature of a fixed mass of gas kept at constant pressure was published by Charles in 1787 (and independently by Gay-Lussac in 1802). It states that the volume V of the gas, of fixed mass and pressure, measured at temperature θ by, say, a mercury-in-glass thermometer is related approximately to its volume V_0 at the ice point (O °C) by

$$V = V_0(1 + \alpha\theta)$$

where α is the *cubic expansivity of the gas at constant pressure* and has roughly the same value of 0.003 66 °C^{-1} (1/273) for all gases at low pressure. This is a linear relationship and a graph of V against θ would be a straight line but not passing through the volume origin. How would you find α from such a graph?

(c) *Pressure law*. A similar relationship is found to exist between the pressure p of a fixed mass of a gas, kept at constant volume, and the temperature θ, again measured by the mercury-in-glass thermometer. It is

$$p = p_0(1 + \beta\theta)$$

where p_0 is the pressure at the ice point and β is a constant, known as the *pressure coefficient* of the gas. It has practically the same value as α.

Temperature scales

(a) *Disagreement between thermometers*. The early experimenters measured temperatures with mercury thermometers and often found that different instruments gave different readings for the same temperature. Later, when thermometers based on other temperature-dependent properties of matter were used (e.g. electrical resistance of a wire, pressure of a gas at constant volume) there was disagreement between the different kinds as well (except at the fixed points where they had to agree by definition).[1]

It became clear that one type of thermometer based on one scale of temperature would have to be chosen as a standard. Gas thermometers, either constant volume Fig 9.3, or constant pressure, showed the closest agreement over a wide range of temperatures, especially at low pressures. They were also very sensitive, accurate and highly reproducible. Although the readings depended on the nature of the gas used and on its pressure, all indicated the same reading as the pressure was lowered and approached zero.

[1] See *Advanced Physics: Materials and Mechanics*, pp. 110–114.

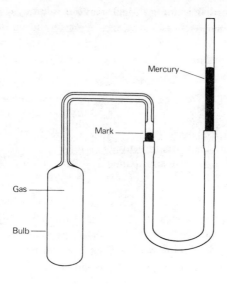

Mercury

Mark

Gas

Bulb

Fig. 9.3

(b) *Ideal gas scale.* Kelvin suggested that the standard scale of temperature should be based on an imaginary ideal gas with properties that were those of real gases at very low pressure, i.e. it obeyed Boyle's law. He proposed that the product of the pressure and the volume of this ideal gas be used as the thermometric property and the simplest procedure is then to say that if p_1 and p_2 are the pressures and V_1 and V_2 are the corresponding volumes of a fixed mass of the gas at two temperatures T_1 and T_2 on this scale, then these temperatures are defined by

$$\frac{T_1}{T_2} = \frac{p_1 V_1}{p_2 V_2}.$$

A linear relationship between pV and temperature is thus assumed and means we have *chosen* to make equal changes of pV represent equal changes of temperature. That is, the ideal gas scale of temperature has been *defined* so that pV for a fixed mass of ideal gas is directly proportional to the temperature of the gas measured on this scale. The measurements can be imagined to be made either by a constant volume or a constant pressure gas thermometer or by an experiment in which neither pressure nor volume remain constant. In all cases the ratio T_1/T_2 will be the same.

If the volume of the gas is kept constant we have

$$\frac{p_1}{p_2} = \frac{T_1}{T_2}$$

which is the pressure law for an ideal gas and is a consequence of the way temperature is defined on the ideal gas scale. Similarly, at constant pressure we can say

$$\frac{V_1}{V_2} = \frac{T_1}{T_2}.$$

(c) *Fixed points.* The unit of temperature on this scale has to be defined so that thermometers can be calibrated. This is done by choosing two fixed points and assigning numbers to them. Before 1954 the ice and steam points were used and on the Celsius method of numbering, given the values 0 °C and 100 °C respectively.

For practical reasons (among which was the extreme sensitivity of the steam point to pressure change) the upper fixed point is now taken as the only temperature at which ice, liquid water and water vapour coexist in equilibrium and is called the *triple point* of water. It is given the value of 273.16 K (which is nearly 0.01 °C), K being the symbol for the post-1954 unit of temperature, i.e. the kelvin. This number was chosen so that there were still 100 kelvins between the ice and steam points. The temperatures of the ice, triple and steam points in °C and in K are given below.

	°C	K
Steam point	100	373.15
Triple point	0.01	273.16
Ice point	0	273.15

The lower fixed point, to which a value 0 K is assigned, is called *absolute zero* and more will be said about it presently (0 K = −273.15 °C).

If we use a constant volume gas thermometer, and in the equation $T_1/T_2 = p_1/p_2$ we put

$T_1 = T$ (the unknown temperature)
$T_2 = 273.16$ (triple point)
$p_1 = p$ (pressure of the ideal gas in the constant volume thermometer at temperature T)

and $p_2 = p_{tr}$ (pressure at the triple point),

we get

$$T = 273.16 \left(\frac{p}{p_{tr}}\right).$$

(d) *Measuring temperature on the ideal gas scale.* To find an unknown temperature T using a constant volume gas thermometer, the pressure p_{tr} at the triple point has to be found when the bulb is surrounded by water at 273.16 K in a

331

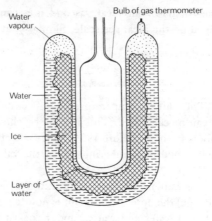

Fig. 9.4

triple point cell, Fig. 9.4. The pressure p at the unknown temperature must also be determined (keeping the volume constant) and then the value calculated of 273.16 (p/p_{tr}).

Some of the gas in the thermometer is then removed so reducing the pressure and the new values for p_{tr} and p measured. The value of 273.16 (p/p_{tr}) is again obtained and will be different. This process is repeated for smaller and smaller amounts of gas in the bulb and a graph of the calculated value of 273.16 (p/p_{tr}) plotted against p_{tr}.

Graphs for different gases at the temperature of steam in equilibrium with water at standard pressure are shown in Fig. 9.5. When extrapolated to $p_{tr} = 0$, they all indicate the same value, which is the temperature T of the steam point on the ideal gas scale. A constant volume thermometer containing a *real* gas, e.g.

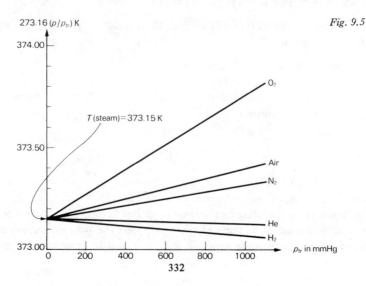

Fig. 9.5

helium, hydrogen or nitrogen, can thus be used to measure temperatures on the ideal gas scale.

(e) *Absolute thermodynamic scale.* Whilst the ideal gas scale is independent of the properties of any particular gas, it does depend on the properties of gases in general. The absolute thermodynamic or Kelvin scale is quite independent of the properties of any substance and is based on the efficiency of an ideal reversible heat engine (p. 363). It is a theoretical scale and the SI unit of temperature, i.e. the kelvin, is defined in terms of it, as is absolute zero (0 K).

The ideal gas scale and the Kelvin scale can be shown in a more advanced treatment, to be theoretically identical at all temperatures. We are therefore justified in writing ' K ' after an ideal gas scale temperature as we have done. The symbol for temperatures in kelvins is T.

Theory also predicts that temperatures below absolute zero do not exist and certainly so far all attempts to reach 0 K have failed.

(f) *International Practical Temperature Scale* (IPTS). Measuring ideal gas scale temperatures is a tedious task and the IPTS has been adopted as a more practical scale for general use. It consists of eleven primary fixed points (ranging from the triple point of hydrogen at 13.81 K to the freezing point of gold at 1337.58 K), which have been accurately determined on the ideal gas scale by a constant volume gas thermometer, and also some other secondary fixed points. Particular types of thermometer are specified for measuring temperatures over a certain range using agreed formulae. At the fixed points there is agreement between the IPTS and the Kelvin scale; the differences at intervening temperatures are usually negligible.

Ideal gas equation

The equation defining temperature on the ideal gas scale is (p. 330),

$$\frac{T_1}{T_2} = \frac{p_1 V_1}{p_2 V_2}$$

or

$$\frac{p_1 V_1}{T_1} = \frac{p_2 V_2}{T_2}.$$

Hence for an *ideal* gas we can say

$$\frac{pV}{T} = \text{constant}$$

where p and V represent any pair of simultaneous values of the pressure and volume of a given mass of ideal gas at the ideal gas scale temperature T.

This equation is *not* based on experiment but incorporates *two definitions*. The first is of an ideal gas as one which obeys Boyle's law ($pV = \text{constant}$) and the second of temperature on the ideal gas scale as a quantity proportional to pV.

The value of the constant depends on the mass of gas and experiments with real gases at low enough pressures show that it has the same value R for all gases if one mole is considered. R is called the *universal molar gas constant* and has a value 8.31 J mol^{-1} K^{-1}. (Check that the units are correct.) We can now write for one mole

$$pV = RT.$$

This is called the *equation of state for an ideal gas*, the state being determined by the values of pressure and temperature. (In view of this use of the term ' state ', confusion is avoided if we refer on other occasions to the three ' phases ' of matter, i.e. solid, liquid and gas.) The ideal gas does not exist of course, but since real gases behave almost ideally at low pressures the concept is sufficiently close to actual fact to be useful.

Notes. 1. The *mole* is the amount of substance which contains the same number of elementary entities as there are atoms in 12 grams of carbon 12. Experiment shows this to be about 6.02×10^{23}—a value denoted by N_A and called the *Avogadro constant*, i.e. $N_A = 6.02 \times 10^{23}$ particles per mole where the particles may be atoms, molecules, electrons etc.

2. The *relative molecular mass* M_r (known previously as the *molecular weight*) of a substance is defined by

$$M_r = \frac{\text{mass of a molecule of substance}}{\text{mass of carbon 12 atom}} \times 12.$$

The values for hydrogen, carbon and oxygen are 2, 12 (by definition) and 32 respectively.

3. The *molar mass* M_m of a substance is the mass of one mole and if expressed in kg mol^{-1} it is related (in SI units) to M_r by the equation

$$M_m = M_r \times 10^{-3}.$$

M_m may thus be the mass of a substance (in kg) containing N_A molecules; its value in SI units for oxygen is 32×10^{-3} kg mol^{-1}. (' Molar ' means ' per unit amount of substance ' and terms such as ' gram-molecule ' are now obsolescent.)

4. In $pV = RT$, V is the molar volume given by

$$V = \frac{\text{molar mass (in kg mol}^{-1}\text{)}}{\text{density (in kg m}^{-3}\text{)}}$$

$$= \frac{M_m}{\rho} \text{ (in m}^3 \text{ mol}^{-1}\text{)}.$$

Hence we can also write $R = \dfrac{pM_m}{T\rho}$

$$= \frac{pM_r}{T\rho} \times 10^{-3}.$$

Also for n moles of an ideal gas

$$pV = nRT$$

where $n =$ mass of gas (in kg)/molar mass (in kg mol^{-1}).

5. One mole of an ideal gas at s.t.p. has a volume of 22.4×10^{-3} m^3 as can be shown by substituting $p = 1.01 \times 10^5$ Pa, $R = 8.31$ J mol^{-1} K^{-1}, $T = 273$ K and $n = 1$ in $pV = nRT$. (1×10^{-3} m$^3 = 1$ litre.)

Kinetic theory of an ideal gas

So far, the ideal gas has been considered from the macroscopic point of view and treated in terms of quantities such as pressure, volume and temperature, which are properties of the gas as a whole. The kinetic theory, on the other hand, attempts to relate the macroscopic behaviour of an ideal gas with the microscopic properties (e.g. speed, mass) of its molecules. A theoretical model is constructed by attributing certain properties to the molecules and the laws of mechanics are applied to it. Any results obtained must then be studied to see if the theory throws any light on the macroscopic behaviour of ideal and real gases.

(a) *Assumptions.* The main assumption, which in effect defines an ideal gas on the microscopic scale, is that the range of intermolecular forces (both attractive and repulsive) is small compared with the average distance between molecules. From this it follows that

(i) the intermolecular forces are negligible except during a collision.

(ii) the volume of the molecules themselves can be neglected compared with the volume occupied by the gas (see p. 349),

(iii) the time occupied by a collision is negligible compared with the time spent by a molecule between collisions, and

(iv) between collisions a molecule moves with uniform velocity.

We also assume that there are a large number of molecules even in a small volume and that a large number of collisions occur in a small time.

(b) *Calculation of pressure.* An expression for the pressure exerted by an ideal gas due to molecular bombardment can be derived and the model made quantitative.

Suppose a gas in a cubical box of side l contains N molecules each of mass m. Consider one molecule moving with velocity c as shown in Fig. 9.6. We can resolve c into components u, v and w in the three directions Ox, Oy and Oz respectively, parallel to the edges of the box. First consider motion along Ox. The molecule has momentum mu to the right and on striking the shaded wall X of the box, we can assume for mathematical simplicity that on average (since the system is in equilibrium) it rebounds with the same velocity reversed. Its

Fig. 9.6

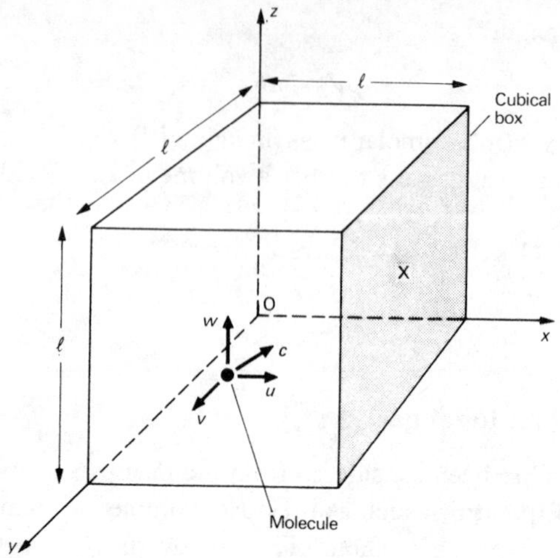

momentum is now *mu* to the left, i.e. $-mu$ and so, by providing an impulse (i.e. a force for a short time), the wall has caused a change of momentum of

$$mu - (-mu) = 2\,mu.$$

The momentum components *mv* and *mw* have no effect on the momentum exchanges at wall X since they are directed at right angles to O*x*.

If the molecule travels to the wall opposite X and rebounds back to X again without striking any other molecule on the way, it covers a distance 2*l* in time $2l/u$. This is the time interval between successive collisions of the molecule with wall X. Hence

$$\text{rate of change of momentum at X} = \frac{\text{change of momentum}}{\text{time}}$$

$$= \frac{2mu}{2l/u} = \frac{mu^2}{l}.$$

Newton's second law of motion states that force equals rate of change of momentum. Therefore the *total* force on wall X due to impacts by all *N* molecules is given by

$$\text{force on X} = \frac{m}{l}\,(u_1{}^2 + u_2{}^2 + \ldots + u_N{}^2)$$

where u_1, u_2 etc., are the different O*x* components of the velocities of molecules 1, 2 etc.

Since pressure is force per unit area, we can say that the pressure *p* on wall X of area l^2 is given by

$$p = \frac{m}{l^3}\,(u_1{}^2 + u_2{}^2 + \ldots + u_N{}^2).$$

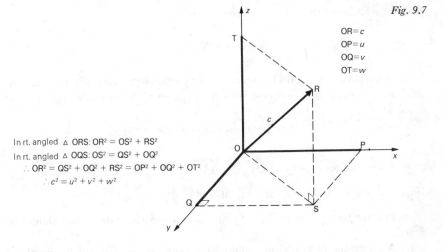

Fig. 9.7

$OR = c$
$OP = u$
$OQ = v$
$OT = w$

In rt. angled $\triangle$ ORS: $OR^2 = OS^2 + RS^2$
In rt. angled $\triangle$ OQS: $OS^2 = QS^2 + OQ^2$
$\therefore OR^2 = QS^2 + OQ^2 + RS^2 = OP^2 + OQ^2 + OT^2$
$\therefore c^2 = u^2 + v^2 + w^2$

If $\overline{u^2}$ represents the mean value of the squares of all the velocity components in the Ox direction, then

$$\overline{u^2} = \frac{u_1^2 + u_2^2 + \ldots + u_N^2}{N}$$

and

$$N\overline{u^2} = u_1^2 + u_2^2 + \ldots + u_N^2$$

$$\therefore \quad p = \frac{Nm\overline{u^2}}{l^3}.$$

For any molecule, the application of Pythagoras' theorem twice to Fig. 9.7 shows that

$$c^2 = u^2 + v^2 + w^2.$$

This will also hold for the mean square values

$$\therefore \quad \overline{c^2} = \overline{u^2} + \overline{v^2} + \overline{w^2}.$$

But since N is large and the molecules move randomly (i.e. they show no preference for moving parallel to any edge of the cube, see (c) (v) below), it follows that the mean values of $\overline{u^2}$, $\overline{v^2}$ and $\overline{w^2}$ are equal. Thus

$$\overline{c^2} = 3\overline{u^2}$$

$$\therefore \quad \overline{u^2} = \frac{\overline{c^2}}{3}.$$

Hence

$$p = \frac{Nm\overline{c^2}}{3l^3}.$$

But l^3 = volume of the cube = volume of gas = V and so

$$pV = \frac{1}{3} Nm\overline{c^2}.$$

This important result links the macroscopic property pressure with the number, mass and speed of the molecules. We will see presently what further information it yields.

Alternatively, if ρ is the density of the gas we can write

$$p = \frac{1}{3} \rho \overline{c^2}$$

since $\rho = Nm/V$, Nm being the total mass of gas.

(c) *Further points.* The following points about the above derivation should be noted.

(i) A container of any shape could have been chosen but a cube simplifies matters.

(ii) Intermolecular collisions were ignored but these would not affect the result because the average momentum of the molecules on striking the walls is unaltered by their collisions with each other.

(iii) The mean square speed $\overline{c^2}$ is not the same as the square of the mean speed. Thus if five molecules have speeds of 1, 2, 3, 4, 5 units, their mean speed is

Fig. 9.8

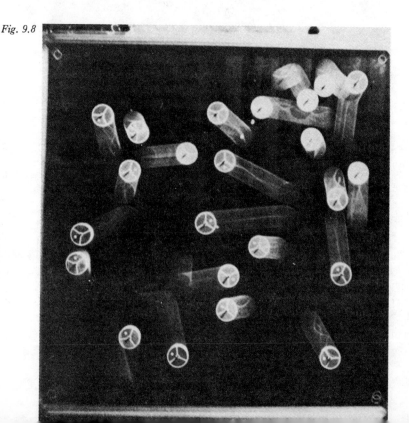

$(1 + 2 + 3 + 4 + 5)/5 = 3$ and its square is 9. The mean square speed on the other hand is $(1^2 + 2^2 + 3^2 + 4^2 + 5^2)/5 = 55/5 = 11$.

(*iv*) The pressure on one wall only was found but it is the same on all sides (neglecting the weight of the gas).

(*v*) Fig. 9.8 shows a horizontal table with lots of small holes through which air is blown and forms a cushion on which lightweight pucks can move almost without friction. A vibrating wire around the ' air table ' keeps the pucks moving randomly and compensates for the inelastic collisions they have with each other and with the sides. The apparatus is a two-dimensional model of a gas.

The photograph was taken from above by opening a camera shutter about one second before the air supply was cut off and closing it one second after. The pucks stop immediately the air supply stops and ' tracks ' of their direction of motion are obtained. We see from them that *roughly* half the pucks are moving east or west if we include those that are going more nearly east or west than north or south. The other half are going more nearly north or south.

(*vi*) The derivation is a simplified version of a more complex piece of theory which makes similar assumptions and arrives at the same result.

Temperature and kinetic theory

We have found from the kinetic theory that for a mole of an ideal gas,

$$pV = \frac{1}{3} N_A m\overline{c^2}$$

where N_A is the Avogadro constant. This may be rewritten as

$$pV = \frac{2}{3} N_A (\tfrac{1}{2} m\overline{c^2}).$$

The ideal gas equation of state, also for a mole, is

$$pV = RT$$

where R is the universal molar gas constant. If we combine these last two equations we get

$$\frac{2}{3} N_A (\tfrac{1}{2} m\overline{c^2}) = RT$$

$$\therefore \quad \tfrac{1}{2} m\overline{c^2} = \frac{3}{2} \frac{R}{N_A} T.$$

Now $\tfrac{1}{2} m\overline{c^2}$ is the average kinetic energy (k.e.) of translational motion[1] per mole-

[1] This is motion in which every point of the body has the same instantaneous velocity and there is no rotational or vibrational motion.

cule and from the above equation we see, since R and N_A are constants, that *for an ideal gas, it is proportional to the thermodynamic temperature T.*

The kinetic theory thus offers a microscopic basis for the macroscopic quantity of temperature. It suggests that an increase in the temperature of a gas is a manifestation of an increase of molecular speed and so of translational k.e.

The ratio R/N_A is called *Boltzmann's constant* and is denoted by k. It is the gas constant per molecule and has the value 1.38×10^{-23} J K^{-1}. Hence we can say

$$\tfrac{1}{2}m\overline{c^2} = \frac{3}{2}kT.$$

The total translational k.e. per mole of an ideal gas is given by

$$\tfrac{1}{2}N_A m\overline{c^2} = \frac{3}{2}RT.$$

Whilst the above expression indicates that at absolute zero the molecular k.e. of an *ideal gas* is zero, this is not so for real substances. In their case, the molecular k.e. can be shown to have a *minimum* value, called the ' zero-point energy ' and it is wrong to think that all molecular motion stops at 0 K.

Deductions from kinetic theory

Three well-established laws relating to real gases will be deduced.

(*a*) *Avogadro's law (or hypothesis).* Consider two ideal gases 1 and 2. We can write

$$p_1 V_1 = \frac{1}{3}N_1 m_1 \overline{c_1^2}$$

$$p_2 V_2 = \frac{1}{3}N_2 m_2 \overline{c_2^2}.$$

If their pressures, volumes and temperatures are the same, then

$$p_1 = p_2$$
$$V_1 = V_2$$
$$\tfrac{1}{2} m_1 \overline{c_1^2} = \tfrac{1}{2} m_2 \overline{c_2^2}$$
$$\therefore \quad N_1 = N_2.$$

Hence, *equal volumes of ideal gases existing under the same conditions of temperature and pressure contain equal numbers of molecules.*

This statement is Avogadro's law. It cannot be proved directly for real gases but it has to be assumed to explain the volume relationships which exist between combining gases (and are expressed by Gay-Lussac's law).

(*b*) *Dalton's law of partial pressures.* The kinetic theory attributes gas pressure to bombardment of the walls of the containing vessel by molecules. In a mixture of ideal gases we might therefore expect the total pressure to be the sum of the partial pressures due to each gas. This statement is Dalton's law of partial pressures, which holds for real gases when they behave ideally. Fig. 9.9 illustrates the law.

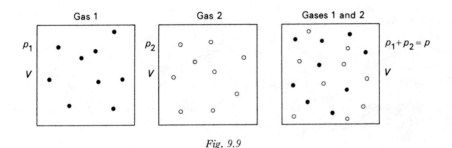

Fig. 9.9

(*c*) *Graham's law of diffusion.* It is reasonable to assume that in a closed container, the rate of diffusion of a gas through a small hole is directly proportional to the rate at which each molecule hits the side with the hole, i.e. the rate is proportional to $\sqrt{\overline{c^2}}/2l$, i.e. to $\sqrt{\overline{c^2}}$, where l is the length of the container. Hence if two gases, 1 and 2, diffuse through the same hole under the same conditions of pressure and temperature,

$$\frac{\text{rate of diffusion of 1}}{\text{rate of diffusion of 2}} = \sqrt{\frac{\overline{c_1^2}}{\overline{c_2^2}}}.$$

But, from $p = \dfrac{1}{3}\rho\overline{c^2}$ we can say

$$\rho_1\overline{c_1^2} = \rho_2\overline{c_2^2}$$

$$\therefore \quad \frac{\overline{c_1^2}}{\overline{c_2^2}} = \frac{\rho_2}{\rho_1}.$$

Hence

$$\frac{\text{rate of diffusion of 1}}{\text{rate of diffusion of 2}} = \frac{\sqrt{\rho_2}}{\sqrt{\rho_1}}.$$

341

That is, the *rate of diffusion of a gas is inversely proportional to the square root of its density*. This is the law which Graham discovered by experiment.

Molecular magnitudes

Information about gas molecules can be obtained from the kinetic theory if use is also made of knowledge obtained from macroscopic measurements on gases.

(*a*) *Speed of air molecules*. For a gas of density ρ, exerting pressure p we have.

$$p = \frac{1}{3}\rho\overline{c^2}$$

where $\overline{c^2}$ is the mean square speed of the molecules. Assuming we can apply this equation to air, for which $\rho = 1.29$ kg m^{-3} at s.t.p. (i.e. 273 K and 1.01×10^5 Pa) then

$$\overline{c^2} = \frac{3p}{\rho} = \frac{3 \times 1.01 \times 10^5}{1.29} \frac{\text{Pa}}{\text{kg m}^{-3}}$$

$$= 2.35 \times 10^5 \text{ m}^2 \text{ s}^{-2}.$$

The square root of $\overline{c^2}$ is called the *root mean square* (r.m.s.) *speed*. It is denoted by $\sqrt{\overline{c^2}}$ or $c_{\text{r.m.s.}}$ (is greater than the mean speed, see p. 339) and for air has the value

$$\sqrt{\overline{c^2}} = \sqrt{2.35 \times 10^5} = 485 \text{ m s}^{-1}.$$

This is a surprisingly high speed. Some molecules will have greater speeds, others smaller, and these will change due to energy exchanges in intermolecular collisions.

The speeds of gas molecules have been measured directly and the results are in good agreement with calculated values. The fact that the speed of sound in a gas is of the same order as the speed of its molecules offers some confirmation. We would not expect the speed with which the energy of a sound wave is passed on from one air molecule to the next to be greater than the speed of the air molecules themselves. In fact, it could well be less due to the random nature of molecular motion. A ' message ' carried by a ' runner ' can only be transmitted as fast as he can ' run ' and may even travel more slowly if there is ' traffic ' congestion. The speed of sound in air at 0 °C is about 330 m s^{-1}, which compares with the 485 m s^{-1} we calculated for the r.m.s. speed of air molecules at 0 °C.

The very high speed with which molecules of brown bromine vapour travel into a vacuum may be demonstrated, but *care is necessary as the liquid and vapour are both dangerous*. (For precautions see Appendix 7, p. 535.) The air is removed from the tube of Fig. 9.10*a*, the tap closed and another tube containing a bromine capsule connected, Fig. 9.10*b*. The neck of the capsule is broken by squeezing the rubber tubing with pliers, Fig. 9.10*c*, and the tap opened.

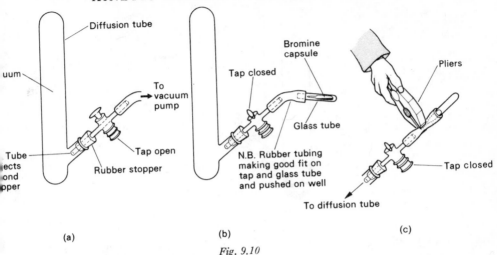

(a)   (b)   (c)

Fig. 9.10

(b) *Mean free path.* Despite their high average molecular speeds, gases diffuse very slowly into air as we know from the fact that a ' smell ' (even when greatly aided by convection) takes time to travel. This suggests that gas molecules, because of their finite size, suffer frequent collisions with other molecules. At each collision the direction of motion changes and the path followed is a succession of zig-zag steps—called a *random walk*. The progress of a molecule is slow, although the sum of all the separate steps it has taken between collisions is a very large distance.

There is a statistical rule (which you can verify by a game, see Appendix 8, p. 536) which states that the average distance travelled in a ' random walk ' of n equal steps taken in succession in random directions will be $\sqrt{n}$ steps. Thus in Fig. 9.11, if A and B are the starting and finishing points respectively and 25 equal random steps are taken, then the distance from A to B as the crow flies is five steps.

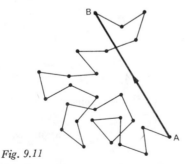

Fig. 9.11

Now consider the diffusion of bromine into air. Using the same apparatus as before the average distance bromine has diffused up the tube (containing air)

Fig. 9.12

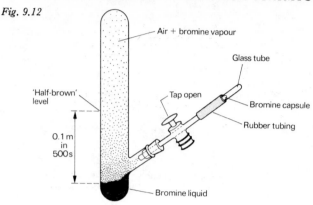

Air + bromine vapour

Glass tube

'Half-brown' level

Tap open

Bromine capsule

Rubber tubing

0.1 m in 500 s

Bromine liquid

in, say, 500 seconds can be estimated by guessing where the vapour looks ' half-brown ', Fig. 9.12. At the bottom it is ' full-brown ' and at the top it is almost clear. Suppose the ' half-brown ' level is 0.1 m up from the original, fairly distinct bromine-air boundary. In 500 seconds the average progress of a bromine molecule is, from start to finish, 0.1 m. If the average length of a step between collisions (called the *mean free path*) in this random walk of, say, n steps is λ then, applying the statistical rule, we have

$$0.1 = \sqrt{n} \cdot \lambda. \tag{1}$$

But it is also known from calculations like the one we did for air molecules that bromine molecules have an r.m.s. speed of 200 m s^{-1} and so in 500 seconds they will travel $200 \times 500 = 10^5$ m. This will be the length of the *straightened-out* track of a bromine molecule and since it makes n steps (and therefore n collisions) in 500 seconds, we can say that the mean free path λ is also given by

$$\lambda = \frac{10^5}{n}. \tag{2}$$

Substituting for n in (1),

$$0.1 = \sqrt{\frac{10^5}{\lambda}} \cdot \lambda = \sqrt{10^5 \lambda}.$$

Squaring both sides gives

$$\lambda = 10^{-7} \text{ m}.$$

How many collisions does a bromine molecule make (mostly with air molecules) in 500 seconds?

The mean free path of an air molecule in air at atmospheric pressure is also about 10^{-7} m $= 100 \times 10^{-9}$ m $= 100$ nm.

Mean free path λ is not to be confused with the average separation D of the gas molecules. At atmospheric pressure λ is appreciably greater than D, Fig. 9.13, as we shall see presently.

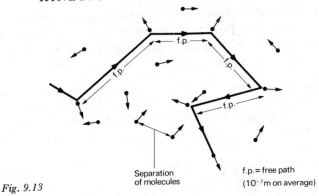

Separation
of molecules

f.p. = free path
(10^{-7} m on average)

Fig. 9.13

(c) *Size of an air molecule.* The volume swept out on average by a molecule between two consecutive collisions is the volume of a cylinder of radius d and length λ where λ is the mean free path and d is the *diameter* of a molecule. This can be seen from Fig. 9.14; a collision will occur when the centres of the two molecules involved are up to distance d apart. In other words there is, on average, one gas molecule in a volume $\pi d^2 \lambda$ and this is therefore the volume 'taken up' by a gas molecule.

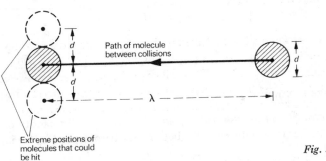

Path of molecule
between collisions

Extreme positions of
molecules that could
be hit

Fig. 9.14

When liquid air changes to atmospheric air the volume increases about 750 times and so we may assume that the mean volume per gas molecule is 750 times the mean volume per liquid molecule. Assuming that the latter is d^3 (i.e. we suppose liquid molecules are close-packed like spheres in a box), then the former will be $750\,d^3$. Hence

$$750\,d^3 = \pi d^2 \lambda$$

$$\therefore \quad d = \frac{\pi \lambda}{750}.$$

Substituting $\lambda = 10^{-7}$ m, we get

$$d \simeq 4 \times 10^{-10} \text{ m.}$$

The diameter of an air molecule is thus 0.4 nm, approximately. More precise measurements can be made but there is no point in trying to obtain too exact values since molecules do not have hard, definite surfaces. The value obtained depends to some extent on the method of measurement and there will be an element of doubt when this involves collisions between molecules.

(d) *Separation of air molecules.* Gas molecules lack any kind of regular spacing such as exists in the case of the molecules of a solid. However, we can obtain a value for their ' average separation ' D for air using the fact that in the change from liquid air to atmospheric air, the volume increases 750 times. We assume as above that each liquid molecule occupies a volume d^3, where d is the diameter of a molecule (in whatever phase). If we *imagine* a given volume of gaseous air as divided up into tiny cubes each of side D and each having one molecule at its centre, then the volume of a gaseous air molecule will be D^3. Hence

$$750 \, d^3 = D^3$$

$$\therefore \quad \frac{D}{d} = \sqrt[3]{750} \simeq 9.$$

The average separation of molecules in air is roughly 9 molecular diameters. Taking $d = 0.4$ nm, $D \simeq 3.6$ nm which is much smaller than the mean free path. We now have the following *rough* estimates for air molecules at s.t.p.

Diameter (d)	Separation (D)	Mean free path (λ)	$\sqrt{\overline{c^2}}$ or $c_{\text{r.m.s.}}$
0.4 nm	4 nm	100 nm	500 m s⁻¹

Properties of vapours

(a) *Saturated and unsaturated vapours.* If a very small quantity of a volatile liquid like ether is introduced at the bottom of a barometer, Fig. 9.15a, it rises to

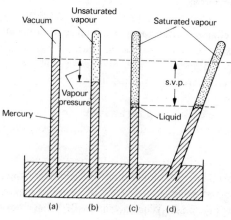

Fig. 9.15

(a) (b) (c) (d)

the top of the mercury column. Here it evaporates into the vacuum and the pressure exerted by the vapour causes the mercury level to fall, Fig. 9.15b. When more ether is injected, the level drops further but stops as soon as liquid ether appears on top of the column. The ether vapour is then said to be *saturated* and exerting its *saturation vapour pressure* (s.v.p.) at the temperature of the surroundings, Fig. 9.15c. Before liquid ether gathers above the mercury, the ether vapour is *unsaturated*.

The s.v.p. of a substance is the pressure exerted by the vapour in equilibrium with the liquid, e.g. the vapour is in a closed vessel above the liquid. If the space above the mercury in Fig. 9.15a had contained some air initially, the s.v.p. would be the same but the evaporation would take place more slowly. At 15 °C the s.v.p. of water is 1.3 cmHg and of ether 35 cmHg.

(*b*) *Vapours and the gas laws.* Unsaturated vapours obey Boyle's law roughly up to near saturation point, i.e. along AB in the *p-V* graph of Fig. 9.16a. At B condensation of the vapour starts, liquid and saturated vapour exist together along BC and, since the mass of vapour is changing, Boyle's law is no longer

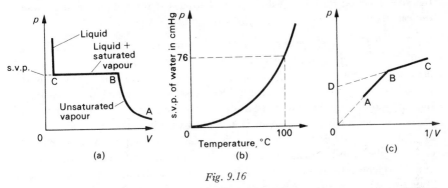

Fig. 9.16

relevant. The pressure along BC is the s.v.p. at the temperature concerned and we see that it does not change as the volume of saturated vapour decreases. This may be shown by tilting a barometer tube containing saturated vapour as in Fig. 9.15d; the vertical depth of the mercury surface below its level in Fig. 9.15a indicates the s.v.p. and remains the same as the volume of saturated vapour diminishes.

Unsaturated vapours obey Charles' law and the pressure law approximately: saturated vapours do not. The s.v.p. of a vapour does increase with temperature but much more rapidly than is required by Charles' law, as Fig. 9.16b shows for water.

A mixture of gas and unsaturated vapour obeys the gas laws fairly well, each exerting the same pressure as it would exert if it alone occupied the total volume (Dalton's law of partial pressures, p. 341). A graph of *p* against 1/*V* for such a mixture is shown in Fig. 9.16c. What occurs at B? What does OD represent?

(c) *Kinetic theory explanations*. In an unsaturated vapour the rate at which molecules leave the liquid surface exceeds that at which they enter it from the vapour, i.e. evaporation occurs more rapidly than condensation. In a saturated vapour the rates are equal and *dynamic equilibrium* exists. The rate of leaving depends on the average k.e. of the molecules and so on the temperature; the rate of entering is determined by the temperature and density of the vapour.

The s.v.p. increases when the temperature of the liquid is raised because the average k.e. of the molecules increases and more are able to escape from the surface. The rate of evaporation becomes greater, thereby increasing the density of the vapour and so also the rate of condensation. Eventually equilibrium and saturation are re-established at a greater s.v.p. than before.

The s.v.p. is not affected by changes of volume (at a constant temperature). If the volume available to the vapour decreases, its density momentarily increases and more molecules return to the liquid in a given time than previously. The rate at which molecules leave the surface remains steady, however, and so the rate of condensation now exceeds the rate of evaporation until the density of the vapour falls to its original value and dynamic equilibrium is restored once more, with the s.v.p. having its initial value. What happens if the volume of the vapour increases?

(d) *Boiling*. Whereas evaporation occurs from the surface of a liquid at all temperatures, boiling takes place at a temperature determined by the external pressure and consists in the formation of bubbles of vapour throughout the liquid. The pressure inside such bubbles must at least equal the pressure in the surrounding liquid. The pressure inside is the s.v.p. at the temperature of the boiling point (since the vapour is in contact with liquid), that outside may be taken as practically equal to atmospheric pressure for a liquid in an open vessel (if we neglect the hydrostatic pressure of the liquid itself and surface tension effects). Hence, from this qualitative treatment it would seem that *a liquid boils when its s.v.p. equals the external pressure.*

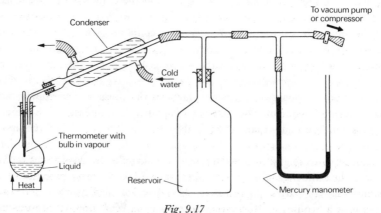

Fig. 9.17

348

(*e*) *Measuring s.v.p.* The apparatus for the 'dynamic method' is shown in Fig. 9.17 and is suitable for s.v.p.s above 50 mmHg (for water from about 50 °C upwards). The boiling point of the liquid is found at different external pressures and these equal the s.v.p.s.

The vacuum pump or compressor enables the pressure to be set to any desired value below or above atmospheric pressure. With cold water circulating through the condenser, the liquid in the flask is heated until it boils. The temperature of the saturated vapour is then given by the thermometer and the difference between the pressure inside and outside the apparatus is registered by the manometer. The reservoir helps to prevent large pressure fluctuations by acting as a buffer.

Van der Waals' equation

The equation of state for an ideal gas $pV = RT$ holds fairly well for real gases (and unsaturated vapours) at low pressures but not at moderate to high pressures. An accurate knowledge of the deviations of real gases from ideal behaviour is important for two reasons. First, they are required to make the necessary corrections to constant volume gas thermometer readings so that temperatures can be expressed on the thermodynamic scale. Second, they provide information about the nature of intermolecular forces and the structure of molecules.

The ideal gas equation is consistent with the kinetic theory. As a step towards obtaining an equation of state for real gases and explaining their deviations, we might consider if any of the assumptions of the theory are invalid for real gases. Van der Waals assumed that when real gases are *not* at low pressure, the range of intermolecular forces is not small compared with the average distance between the molecules (see assumptions, p. 335). There are then two points to consider.

(*i*) *Effect of repulsive intermolecular forces.* These act at very short range, i.e. when molecules approach each other very closely and in effect cause a reduction in the volume in which the molecules can move. This is often expressed by saying that the molecules themselves have a certain volume which is not negligible in relation to the volume V occupied by the gas. A factor called the 'co-volume', denoted by b is introduced, making the 'free-volume' $(V - b)$.

(*ii*) *Effect of attractive intermolecular forces.* These act over slightly greater distances and may cause two or more molecules to form loose associations or 'complexes', thereby reducing the number of particles in the gas. The observed pressure p is less than it would be if this did not happen. (The 'complexes' do not have greater momentum than single molecules because, being in thermal equilibrium with the latter they have the same average translational k.e.) It can be shown in a more advanced treatment that to allow for this 'pressure defect', p, which still represents the pressure, should be replaced by $(p + a/V^2)$ where a, like b, is a 'constant' that varies from gas to gas and is found by experiment.

Van der Waals' equation of state for a mole of a real gas is

$$\left(p + \frac{a}{V^2}\right)(V - b) = RT.$$

The quantities a/V^2 and b become important at high pressures when the molecules are close together.

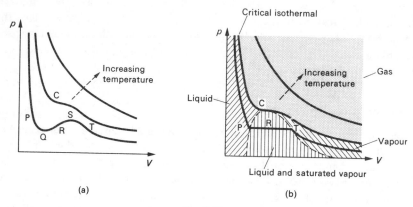

Fig. 9.18

The isothermals (i.e. graphs of p against V at different, constant temperatures) predicted by van der Waals' equation for three temperatures are shown in Fig. 9.18a. Those in Fig. 9.18b were obtained experimentally by Andrews (in 1863) for carbon dioxide. The top curve is a rectangular hyperbola, the middle one (which represents a lower temperature) has a point of inflexion C and occurs at the *critical temperature*, i.e. the temperature above which a gas cannot be liquefied by increasing the pressure (see *Note* below). The upper two van der Waals curves are very similar to the corresponding Andrews curves but the third one has a curved section PQRST whereas the lowest Andrews curve has a straight section PRT (which represents the change from liquid to vapour), although parts PQ and TS have been realized in practice.

The van der Waals equation thus describes fairly well the behaviour of real gases above, but not below, their critical temperature. Evidently the modifications we have made to the kinetic theory are still over-simplified. Other equations of state have been tried but there is no simple one which applies to all gases under all conditions.

Note. The critical temperature of carbon dioxide is 304 K (31 °C) and of oxygen 154 K (−119 °C). Gases such as oxygen were wrongly called 'permanent' gases because attempts to liquefy them by compression without cooling were unsuccessful. A distinction is sometimes made between a gas and a vapour; the former is above its critical temperature and the latter below.

KINETIC THEORY: THERMODYNAMICS

Thermodynamics, heat and work

Thermodynamics deals with processes which cause energy changes as a result of heat flow to or from a system and/or of work done on or by a system. A thermodynamic system consists of a fixed mass of matter, often a gas, separated from its surroundings, perhaps by a cylinder and a piston. Heat engines such as a petrol engine, a steam turbine and a jet engine all contain thermodynamic systems designed to convert heat into mechanical work. Heat pumps and refrigerators are thermodynamic devices for transferring heat from a cold body to a hotter one.

Heat and work are terms used to describe *energy in the process of transfer*.

Heat is energy that flows by conduction, convection or radiation from one body to another because of a temperature difference between them.

The energy is only heat as it flows. The notion that heat is something in a body is contrary to experimental fact. For example, an unlimited amount of heat can be obtained by rubbing two surfaces together and the system remains the same during the process. No definite meaning can therefore be given to the expression ' the heat in the system '.

Work is energy that is transferred from one system to another by a force moving its point of application in its own direction.

The force may have a mechanical, electrical, magnetic, gravitational etc., origin but no temperature difference is involved. Again, experience shows that an indefinite amount of work can be put into a system without it changing and so the phrase ' work in a body ' is also meaningless.

Energy may be transferred to or from a system as heat or work and afterwards it is impossible to tell which form it took. Thus if the air in a bicycle pump is hot, it could either have been compressed by the piston or held near a body at a higher temperature. Again there is no point in talking about the ' heat in the air ' since heat may not have been transmitted to it. Unfortunately everyday usage militates against this use of the term ' heat '.

The energy in a system, whether transferred to it as heat or work, is called *internal energy* and will be considered shortly.

Laws of thermodynamics

(a) *Zeroth law and thermal equilibrium.* This basic law was proposed after the first law had been stated. It stems from the fact that all energy exchange appears to stop after a time when hot and cold bodies are brought into contact. The bodies are then said to be in ' thermal equilibrium ' and we call the common property which they have *temperature.*

The zeroth law states that if bodies A and B are each separately in thermal equilibrium with body C, then A and B are in thermal equilibrium with each other.

For example, if C is a thermometer and reads the same when in contact with two bodies A and B, then A and B are at the same temperature. This apparently simple fact is not altogether obvious since in the realm of human relations Black and White may both know Brown but they may not know each other. In effect, the law says that there exists a useful quantity called 'temperature' and if it was not true, taking thermometer readings would be a pointless exercise.

(b) *First law and internal energy.* Heat supplied to a gas (or a liquid or a solid) may (i) raise its *internal energy*, and (ii) enable it to expand and thereby do *external work* by pushing back the atmosphere or if it is in a cylinder, by moving a piston against a force.

In general, the internal energy of a gas consists of two components.

(i) Kinetic energy due to the translational, rotational and vibrational motion of the molecules, all of which depend only on the temperature. (In a monatomic gas there is only translational motion.)

(ii) Potential energy due to the intermolecular forces; this depends on the separation of the molecules, i.e. the volume of the gas.

In an ideal gas only the kinetic component is present (Why?) and the kinetic theory shows that the translational part of it equals $3\,RT/2$ per mole. For real gases, both components are present with the kinetic form predominating and their internal energy depends on the temperature and the volume of the gas. (In a solid k.e. and p.e. are present in roughly equal amounts.)

If δQ is the heat supplied to a mass of gas and if δW is the external work done by it then the increase of internal energy δU equals $(\delta Q - \delta W)$ if energy is conserved. Hence

$$\delta Q = \delta U + \delta W.$$

This equation is taken as the first law of thermodynamics and is a particular case of the principle of conservation of energy. δQ is taken as positive if heat is supplied to the gas and negative if heat is transferred from it. δW is positive when external work is done by the gas (expansion) and negative if work is done on it (compression).

(c) *Second law, heat engines and heat pumps.* Experience shows that in a heat engine, of the total heat Q_1 absorbed at the high-temperature reservoir (at T_1) only part is converted into work W, the rest Q_2 being rejected at the low-tempera-ture reservoir (at T_2) or exhaust of the engine, Fig. 9.19a. A reservoir in thermo-dynamics is a source or sink of infinite heat capacity whose temperature remains constant however much heat it supplies or receives. In practice a source may take the form of a continuous combustion of fuel and a sink may be the atmosphere. The ratio W/Q_1 for one cycle of operations is defined as the *efficiency* of the engine and it can be shown that it increases as the temperature ratio T_1/T_2 increases.

In a modern power station steam enters the turbines at about 830 K (560 °C) but, even so, the efficiency attained is just over 30 per cent which means that about two-thirds of the heat absorbed is rejected to the atmosphere usually via the cooling towers. In a jet engine the turbine blades, Fig. 9.20, 'glow' during operation.

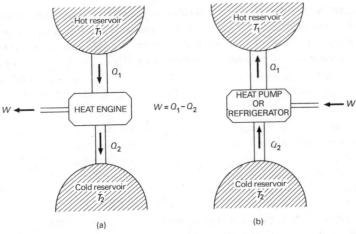

Fig. 9.19

The heat engine version of the second law may be stated as follows.

No continually working heat engine can take heat from a source and convert it completely into work.

Whilst all the work done on a system may become heat, the conversion of heat to work can only occur to a limited extent no matter how good the engine design may be.

The action of heat pumps and refrigerators is the reverse of that occurring in a heat engine. In these devices work W is done on the system which enables it to transfer heat Q_2 from a low-temperature reservoir at T_2 to a high-temperature reservoir at T_1, Fig. 9.19b. The function of a heat pump is to supply as

Fig. 9.20

much heat as possible to the hot body, i.e. to make Q_1 large; its *coefficient of performance* is measured by Q_1/W (which is greater than 1). A refrigerator, on the other hand, is designed to remove heat from the cold body, i.e. to make Q_2 large; its *coefficient of performance* is given by Q_2/W.

Heat naturally flows from a higher temperature to a lower one and the heat pump-refrigerator statement of the second law may be expressed in the following way.

Heat cannot be transferred continually from one body to another at a higher temperature unless work is done by an external agent.

The two statements of the second law are equivalent and state the same thing in different ways. They cannot be proved directly but their consequences can. The law has been applied to a very wide range of phenomena.

(d) *Degradation of energy.* All energy ultimately becomes internal energy of the surroundings through being used to overcome friction. This is the most unavailable form of energy and is useless unless we have something cooler to which to transfer it. Whilst energy is always conserved, it does become inaccessible and we say it is *degraded* as internal energy. The oceans of the world are an enormous energy reservoir but their use requires another reservoir at a lower temperature and this we unfortunately do not have.

Work done by an expanding gas

(a) *Expression for work done.* Consider a mass of gas enclosed in a cylinder by a frictionless piston of cross-section area A which is in equilibrium under the action of an external force F acting to the left and a force due to the pressure p of the gas acting to the right, Fig. 9.21. We have

$$F = pA.$$

Let the gas expand moving the piston outwards through a distance δx which is so small that p remains practically constant during the expansion. The external work done δW by the gas against F will be

$$\delta W = F \delta x$$

$$= pA \cdot \delta x$$

$$= p \cdot \delta V$$

where $\delta V (= A\delta x)$ is the increase in volume of the gas. The work done by the gas in this small expansion equals its pressure multiplied by the increase of volume.

The total work done W by the gas in a finite expansion form V_1 to V_2 is given, in calculus notation, by

$$W = \int dW = \int_{V_1}^{V_2} p \, dV.$$

To evaluate this integral and obtain a value for W requires the relation between p and V to be known. If p is constant during the expansion from V_1 to V_2, then

$$W = p(V_2 - V_1).$$

When a gas is compressed, work is done on it and is also given by $\int_{V_1}^{V_2} p \, dV.$

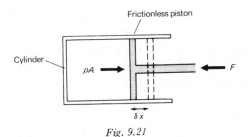

Fig. 9.21

(b) *Reversible changes.* To calculate the work done from the above expressions the pressure of the gas must have a determinable value. This will only be so if the changes of volume occur infinitely slowly, as we can now show. Suppose the external force F in Fig. 9.21 is reduced rapidly during an expansion, there is then not enough time for the gas pressure to be uniform at any stage of the expansion. It will have different values throughout the cylinder, being least near the piston and no proper value can be assigned to it.

Hence if the quantities (p, V, T) defining the state of the system are to be expressed at every stage of the change the system should always be in equilibrium. In other words, we have to regard it as passing through an infinite series of states of equilibrium. Such a change is called a *reversible* one because an infinitesimal change of the controlling conditions will reverse the direction of the change at every stage. In Fig. 9.21 this may be done by increasing F very slightly. The energy transformations must also be reversed exactly in a reversible change and so there must be no energy ' losses '. This means that as well as proceeding infinitely slowly, friction and heat losses must not occur.

In practice the perfect reversible change does not exist but the idea is a useful theoretical standard for judging real changes. A change which is nearly reversible is the very small alternating volume changes in a sound wave, the departures from equilibrium being very small. The reactions in some electrical cells can also be approximately reversible. In other cases, whilst the initial and final states are equilibrium ones, which can be expressed in terms of p, V and T, the intermediate ones are not, especially if rapid changes occur.

(c) *Indicator diagram.* This is a graph showing how the pressure p of a gas varies with its volume V during a change. The work done in any particular case can be derived from it.

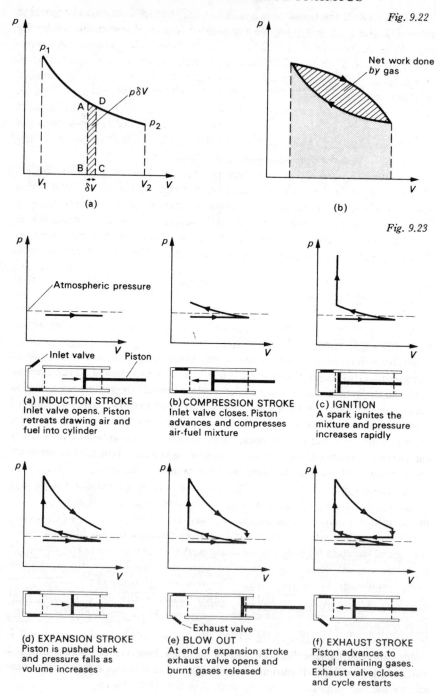

Fig. 9.22

(a)

(b)

Fig. 9.23

Atmospheric pressure

Inlet valve   Piston

(a) INDUCTION STROKE
Inlet valve opens. Piston
retreats drawing air and
fuel into cylinder

(b) COMPRESSION STROKE
Inlet valve closes. Piston
advances and compresses
air-fuel mixture

(c) IGNITION
A spark ignites the
mixture and pressure
increases rapidly

Exhaust valve

(d) EXPANSION STROKE
Piston is pushed back
and pressure falls as
volume increases

(e) BLOW OUT
At end of expansion stroke
exhaust valve opens and
burnt gases released

(f) EXHAUST STROKE
Piston advances to
expel remaining gases.
Exhaust valve closes
and cycle restarts

In Fig. 9.22a if the pressure is p = AB at the start of a very small expansion δV = BC, the work done, $p\delta V$, is represented to a good approximation by the area of the shaded strip ABCD. The total work done *by* the gas in a large expansion from V_1 to V_2 is therefore the sum of the areas of all such strips, that is, area $p_1 p_2 V_2 V_1$ and clearly depends on the shape of the p-V graph. If the graph had represented a compression of the gas from V_2 to V_1, the work done on it would be represented by the same area. When the gas suffers changes which eventually return it to its initial state, it is said to have undergone a *cycle* of operations and the indicator diagram is a closed loop like that in Fig. 9.22b. The net work done by the gas in this case is represented by the shaded area.

Indicator diagrams play a very important part in the theory of heat engines. Those in Fig. 9.23 show the theoretical relations between p and V for different positions of the piston during one complete cycle of operations (known as an *Otto* cycle after the inventor of the internal combustion engine) of a four-stroke petrol engine. A practical p-V diagram for a complete cycle is similar to that in Fig. 9.23f but the sharp corners tend to be rounded, because neither the combustion of the fuel nor the opening and closing of the valves is instantaneous. They are produced while the engine is working by an electronic or mechanical device called an *engine indicator*.

Principal heat capacities of a gas

(*a*) *Definitions.* Heat supplied to a solid, liquid or gas increases its internal energy and, especially in the case of a gas, may enable it to expand and do external work. If the kinetic energy component of the internal energy increases, so also will the temperature but the temperature rise produced by a given amount of heat in a particular case will depend on just how much external work the gas is allowed to do. A gas therefore has an infinite number of heat capacities but only the two simplest are important. These relate to constant volume and constant pressure conditions and are called the *principal* heat capacities.

The molar heat capacity at constant volume (C_v) is the heat required to produce unit rise of temperature in one mole of the gas when the volume is kept constant.

The molar heat capacity at constant pressure (C_p) is the heat required to produce unit rise of temperature in one mole of gas when its pressure remains constant.

It follows that C_p is greater than C_v for a given gas, as Fig. 9.24 helps to show. The cylinders X and Y each contain one mole of gas at the same temperature T and pressure (and so volume). The piston in X is fixed and that in Y can move but has a constant force applied to it. If heat is supplied to each until the temperature has risen by one kelvin, the increase of internal energy must be the same in each case (since the temperature rise is the same). All the heat supplied to X is used to increase the internal energy of the gas. In Y, however, the gas expands and work is done by it on the piston; the heat supplied in this case equals the increase of internal energy plus the (heat equivalent of the) work done.

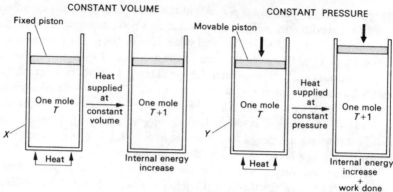

Fig. 9.24

Note that the definitions refer to *reversible* changes, i.e. those that occur infinitely slowly, otherwise pressure, volume and temperature are not determinable properties (see p. 355).

If unit mass of gas is considered instead of one mole we use the term *specific heat capacity*, denoted by c_p or c_v according to whether the pressure or the volume is kept constant.

(*b*) *Relation between C_p and C_v for an ideal gas.* Consider one mole of an *ideal gas* in a cylinder with a frictionless piston, Fig. 9.25. Let a quantity of heat δQ be

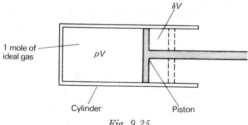

Fig. 9.25

given to the gas which is allowed to expand *reversibly* at constant pressure p, i.e. the external force on the piston does not change during the expansion. Suppose the volume of the gas increases from V to $V + \delta V$ and the temperature from T to $T + \delta T$.

From the first law of thermodynamics we have,

$$\delta Q = \delta U + \delta W \qquad (1)$$

where δQ is the heat to raise the temperature of one mole of gas by δT at constant pressure. Hence, from the definition of molar heat capacity C_p we have

$$\delta Q = \text{number of moles} \times C_p \times \text{temperature rise}$$

$$= C_p \delta T.$$

358

Also, the increase of internal energy δU is, for an ideal gas, the heat needed to raise the temperature by δT at constant volume. Therefore

$$\delta U = C_v \delta V.$$

Further, the external work δW done by the gas is, because the pressure is constant and the change reversible,

$$\delta W = p\delta V.$$

Substituting in (1) for δQ, δU and δW, we get

$$C_p \delta T = C_v \delta T + p\delta V. \tag{2}$$

Applying the ideal gas equation to the initial and final states, we have for 1 mole,

$$pV = RT$$

$$p(V + \delta V) = R(T + \delta T).$$

Subtracting $\qquad\qquad p\delta V = R\delta T.$

Hence from (2) $\qquad\qquad C_p \delta T = C_v \delta T + R\delta T$

$$\therefore \quad C_p - C_v = R.$$

This equation is approximately true for real gases and (as with solids and liquids) the values of C_p and C_v may vary with temperature.

(c) *Calculation of C_p and C_v for an ideal monatomic gas.* The increase of internal energy δU of a mole of an ideal gas for a temperature rise δT is given by

$$\delta U = C_v \delta T \tag{3}$$

where C_v is the molar heat capacity at constant volume. The internal energy of an ideal *monatomic* gas consists entirely of translational kinetic energy, i.e. $U = \frac{1}{2}N_A m\overline{c^2}$ for a mole. From the kinetic theory we have that this also equals $3RT/2$ (p. 339), hence

$$U = \frac{3}{2}RT. \tag{4}$$

If the temperature increases by δT and the internal energy by δU then

$$U + \delta U = \frac{3}{2}R(T + \delta T). \tag{5}$$

Subtracting (4) from (5)

$$\delta U = \frac{3}{2}R\delta T. \tag{6}$$

From (3) and (6)

$$C_v = \frac{3}{2}R.$$

Taking $R = 8.31$ J mol^{-1} K^{-1}

$$C_v = 12.5 \text{ J mol}^{-1} \text{ K}^{-1}.$$

But

$$C_p = C_v + R$$

$$\therefore \quad C_p = \frac{3}{2}R + R = \frac{5}{2}R$$

$$= 20.8 \text{ J mol}^{-1} \text{ K}^{-1}.$$

These values agree well with those for real monatomic gases.

(d) *Importance of* γ. The ratio of the two principal heat capacities of a gas is denoted by γ, that is

$$\gamma = \frac{c_p}{c_v} = \frac{C_p}{C_v}.$$

It is a ratio which appears in the equation for the speed of sound in a gas (p. 250) and in that for a reversible adiabatic change, to be considered in the next section. It can also provide information about the atomicity of gases. Experiments give the following approximate values which do, however, generally decrease with increasing temperature.

Atomicity	γ
monatomic	1.67
diatomic	1.40
polyatomic	1.30

Isothermal, adiabatic and other processes

The processes used to change the state (i.e. the values of p, V and T) of a gas in a heat engine affect its performance in converting heat into mechanical work. In general they are not simple. Here we will consider in terms of the first law of thermodynamics four special cases performed *reversibly* on a mole of an ideal gas. (*Note: isos* is the Greek for ' same'.)

(a) *Isovolumetric process.* This is a constant volume change and one is represented by the continuous line on the p-V graphs of Fig. 9.26a. All the heat entering the system becomes internal energy. No work is done, the temperature rises from T_1 to T_2 and the pressure from p_1 to p_2. Thus $\delta W = 0$ and

$$\delta Q = \delta U = C_v(T_2 - T_1).$$

(b) *Isobaric process.* Pressure remains constant and, of the heat received by the gas, some becomes internal energy as the temperature rises from T_1 to T_2: the rest is used to do work. Hence for the expansion in Fig. 9.26*b*.

$$\delta Q = \delta U + \delta W$$

$$\therefore \quad C_p(T_2 - T_1) = C_v(T_2 - T_1) + p_1(V_2 - V_1).$$

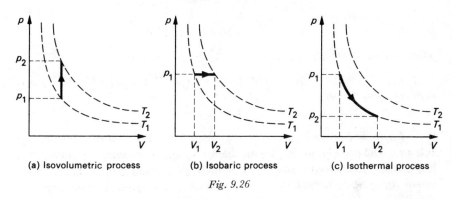

(a) Isovolumetric process   (b) Isobaric process   (c) Isothermal process

Fig. 9.26

(*c*) *Isothermal process.* The change occurs at constant temperature. Work is done at the same rate as heat is supplied so there is no increase of internal energy (for an ideal gas). Hence $\delta U = 0$, $\delta Q = \delta W$. The expansion curve follows a Boyle's law *p*-*V* graph, Fig. 9.26*c*, and the equation relating *p* and *V* is

$$pV = \text{constant}.$$

A reversible isothermal process is an ideal one which requires the gas to be contained, for example, in a cylinder with thin, good-conducting walls having a frictionless piston, surrounded by a constant temperature reservoir. The changes must also occur infinitely slowly. These conditions are necessary to keep the temperature constant at all times, otherwise say in an expansion, external work is done by drawing on the internal energy of the gas and its temperature falls.

(*d*) *Adiabatic process.* This case is different from those discussed so far in that the expansion (or contraction) of the gas takes place without it receiving (or rejecting) heat, i.e. no heat enters or leaves the gas and so $\delta Q = 0$. Hence

$$0 = \delta U + \delta W \qquad \therefore \delta W = -\delta U.$$

In an adiabatic expansion all the work is done at the expense of the internal energy of the gas which therefore cools. Conversely, in an adiabatic compression the work done on the gas by an external agent increases the internal energy and the temperature of the gas rises.

In Fig. 9.27 two *p*-*V* isothermals are given for a mass of ideal gas. If the gas has initially a temperature T_1 and volume V_1, its state (i.e. $p_1 \ V_1 \ T_1$) is represented by point A on the T_1 isothermal. If it then expands adiabatically to volume V_2 so that its temperature falls to T_2, its state is now represented by point B (i.e.

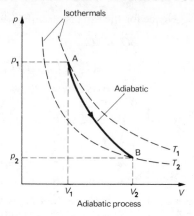

Fig. 9.27

$p_2 V_2 T_2$) on isothermal T_2. The curve through A and B relates the pressure and volume of the mass of gas for this adiabatic change and is called an 'adiabatic'. We see that it is steeper than the 'isothermals' through either A or B. Its equation can be shown to be

$$pV^\gamma = \text{constant}$$

where γ is the ratio of the two principal heat capacities of the gas. It applies to a *reversible adiabatic* change for an ideal gas having a constant value of γ.

A reversible adiabatic process is also an ideal one and requires the gas to be in a thick-walled, poorly-conducting cylinder and piston and to undergo a very rapid, but very small, change of volume to minimize the escape of heat. Very few actual processes can be regarded as adiabatic; those which are approximately so occur (*i*) in insulated vessels, (*ii*) very rapidly, e.g. as in a sound wave, or (*iii*) in a large mass of material. Practical processes fall somewhere between isothermal and adiabatic.

Expressions for the temperature change during a reversible adiabatic process can also be obtained. We have, considering a mole of ideal gas,

$$p_1 V_1^\gamma = p_2 V_2^\gamma. \qquad \text{(1) (from } pV^\gamma = \text{constant)}$$

For *any* type of change
$$\frac{p_1 V_1}{T_1} = \frac{p_2 V_2}{T_2}. \qquad \text{(2) (from } pV = RT)$$

Dividing (1) by (2)

$$T_1 V_1^{\gamma-1} = T_2 V_2^{\gamma-1} \qquad (3)$$

$$\therefore TV^{\gamma-1} = \text{constant}.$$

This gives a relation between T and V. To obtain one between T and P we raise equation (2) to the power γ,

$$\therefore \left(\frac{p_1 V_1}{T_1}\right)^\gamma = \left(\frac{p_2 V_2}{T_2}\right)^\gamma. \qquad (4)$$

Dividing (4) by (1), we get

$$\frac{p_1^{\gamma-1}}{T_1^{\gamma}} = \frac{p_2^{\gamma-1}}{T_2^{\gamma}}$$

$$\therefore \quad \frac{p^{\gamma-1}}{T^{\gamma}} = \text{constant.}$$

It can be shown that a gas does more work in a reversible isothermal expansion than in a reversible adiabatic one.

Carnot's ideal heat engine

In 1824 the French scientist Carnot took the first steps towards developing a scientific theory of heat engines. He imagined an ideal engine, not of any particular type, but free from all imperfections such as friction, in which the working substance was taken reversibly through a cycle (called the *Carnot* cycle) consisting of two isothermal and two adiabatic processes, Fig. 9.28.

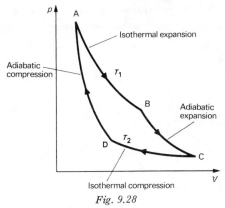

Fig. 9.28

Along AB the substance expands isothermally absorbing heat Q_1 from a source at temperature T_1 and doing external work. Along BC there is an adiabatic expansion, more work is done by it and the temperature ultimately falls to T_2. CD represents an isothermal compression during which work is done on the substance and heat is rejected to a sink at a temperature T_2. Finally along DA, adiabatic compression occurs and more work is done on the substance. The net external work W done by the substance during the cycle equals $(Q_1 - Q_2)$ and we can say

$$\text{efficiency of engine} = \frac{\text{external work done in one cycle}}{\text{heat received from source}}$$

$$= \frac{W}{Q_1} = \frac{Q_1 - Q_2}{Q_1}.$$

363

KINETIC THEORY: THERMODYNAMICS

It can be shown in an advanced treatment that using an ideal gas as the working substance, a heat engine working in a Carnot cycle obeys the relation

$$\frac{Q_1 - Q_2}{Q_1} = \frac{T_1 - T_2}{T_1}$$

where T_1 and T_2 are the temperatures of source and sink respectively. Hence,

$$\text{efficiency} = \frac{T_1 - T_2}{T_1}.$$

Reversible processes are ideal and the most efficient imaginable and so there can be no more efficient engine than a Carnot one. An upper *theoretical* limit thus exists for the efficiency, which practical factors further reduce.

The theoretical efficiency of a steam turbine driven by steam at 830 K (560 °C) and exhausting it directly into the air at 373 K (100 °C) is (830 − 373)/830, i.e. 55 per cent which compares with the actual efficiency of just over 30 per cent. The efficiency of a petrol engine is about 20 per cent and of a large Diesel engine about 40 per cent.

There are similar expressions involving temperature for the coefficients of performance (p. 354) of heat pumps and refrigerators working in Carnot cycles.

Some calculations

1. Calculate the root mean square speed of the molecules of hydrogen at (a) 273 K, and (b) 373 K. Density of hydrogen at s.t.p. = 9.00 × 10⁻² kg m⁻³ and one standard atmosphere = 1.01 × 10⁵ Pa.

(a) *At 273 K*

For an ideal gas of density ρ and pressure p we have from the kinetic theory

$$p = \frac{1}{3}\rho\overline{c^2}$$

$$\therefore \quad \sqrt{\overline{c^2}} = \sqrt{\frac{3p}{\rho}}$$

where $\sqrt{\overline{c^2}}$ is the root mean square molecular speed. Assuming this expression can be applied to hydrogen at s.t.p., for which $\rho = 9.00 \times 10^{-2}$ kg m⁻³ and $p = 1.01 \times 10^5$ Pa, we have

$$\sqrt{\overline{c^2}}_{273} = \sqrt{\frac{3 \times 1.01 \times 10^5}{9.00 \times 10^{-2}}}$$

$$= 1.84 \times 10^3 \text{ m s}^{-1} \quad (\simeq 1.2 \text{ miles per second}).$$

(b) *At 373 K*

For an ideal gas the kinetic theory suggests that $\overline{c^2} \propto T$, hence

$$\frac{\sqrt{\overline{c^2}}_{373}}{\sqrt{\overline{c^2}}_{273}} = \sqrt{\frac{373}{273}}$$

$$\therefore \quad \sqrt{\overline{c^2}}_{373} = \sqrt{\frac{373}{273}} \times 1.84 \times 10^3 \text{ m s}^{-1}$$

$$= 2.15 \times 10^3 \text{ m s}^{-1}.$$

2. *Calculate the two principal molar heat capacities of oxygen if their ratio is 1.40. Density of oxygen at s.t.p. = 1.43 kg m⁻³, one standard atmosphere = 1.01 × 10⁵ Pa and molar mass of oxygen = 32.0 × 10⁻³ kg mol⁻¹.*

For 1 mole of an ideal gas

$$pV = RT$$

where $V = V_m$ = molar volume = molar mass/density = M_m/ρ

$$\therefore \quad R = \frac{pV_m}{T} = \frac{pM_m}{T\rho}.$$

Assuming we can apply this equation to oxygen

$$R = \frac{1.01 \times 10^5 \times 32.0 \times 10^{-3}}{273 \times 1.43} \quad \frac{\text{N m}^{-2} \times \text{kg mol}^{-1}}{\text{K} \times \text{kg m}^{-3}}$$

$$= 8.31 \text{ J mol}^{-1} \text{ K}^{-1}.$$

Also for an ideal gas

$$C_p - C_v = R$$

$$\therefore \quad C_p - C_v = 8.31. \tag{1}$$

But $\qquad \dfrac{C_p}{C_v} = 1.40 \quad \therefore \quad C_p = 1.40 \, C_v. \tag{2}$

Substituting (2) in (1)

$$1.40 \, C_v - C_v = 8.31$$

$$\therefore \quad C_v = \frac{8.31}{0.40} = 20.8 \text{ J mol}^{-1} \text{ K}^{-1}.$$

Hence $\qquad C_p = 20.8 + R = 29.1 \text{ J mol}^{-1} \text{ K}^{-1}.$

3. *A mass of an ideal gas of volume 400 cm³ at 288 K expands adiabatically and its temperature falls to 273 K. (a) What is the new volume if γ = 1.40? (b) If it is then compressed isothermally until the pressure returns to its original value, calculate the final volume of the gas.*

(a) For the reversible adiabatic expansion of an ideal gas we have (p. 362),

$$T_1 V_1^{\gamma-1} = T_2 V_2^{\gamma-1}$$

where $V_1 = 400\text{ cm}^3$, $T_1 = 288$ K, $T_2 = 273$ K and $\gamma = 1.40$

$$\therefore \quad V_2^{0.40} = \frac{288}{273} \times 400^{0.40}.$$

Taking logs,

$$0.40 \ \log \ V_2 = \log 288 + 0.40 \log 400 - \log 273$$

$$= 1.0640$$

$$\therefore \quad \log \ V_2 = 1.0640/0.40 = 2.66$$

$$\therefore \quad V_2 = 457\text{ cm}^3.$$

This is the new volume.

(b) If p_1 and p_2 are the pressures of the gas before and after the adiabatic expansion,

$$p_1 V_1^{\gamma} = p_2 V_2^{\gamma}$$

$$\therefore \quad p_2 = \left(\frac{V_1}{V_2}\right)^{\gamma} p_1 = \left(\frac{400}{457}\right)^{1.40} p_1.$$

After the isothermal compression, the pressure is p_1 and let the final volume be V_3; then from Boyle's law

$$p_2 V_2 = p_1 V_3$$

$$\therefore \quad V_3 = \left(\frac{p_2}{p_1}\right) V_2 = \left(\frac{400}{457}\right)^{1.40} \times 457$$

$$= \frac{400^{1.40}}{457^{0.40}}.$$

Taking logs,

$$\log \ V_3 = 1.40 \log 400 - 0.40 \log 457$$

$$= 2.58$$

$$\therefore \quad V_3 = 380\text{ cm}^3.$$

Thermodynamics and chance

Why does heat flow from a hot to a cold body and never spontaneously in the reverse direction? The highly mathematical subject of statistical mechanics gives insight into the average, but not the individual, behaviour of molecules in such one-way processes. Our treatment will be much simpler and aided by analogies. As a first step we will consider the number of ways in which a situation can arise when things are left to chance, as they are in random changes.

(a) *How many ways?* Suppose we wish to know the number of ways in which six numbered objects can be distributed between two compartments of a box.

We see from Fig. 9.29a that one object can be placed in 2^1 ways; from Fig. 9.29b, two objects can be arranged in $4 = 2^2$ ways; from Fig. 9.29c, three objects can be placed in $8 = 2^3$ ways and so on. In general n objects can be distributed in $W = 2^n$ ways.

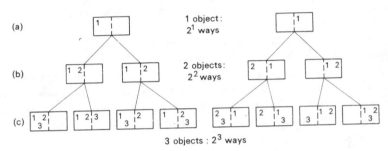

Fig. 9.29

We would expect that any given arrangement would occur once in every W arrangements if these were produced *randomly*. It is worthwhile testing this by starting with all the objects in one compartment and throwing a die to obtain the number of the object to be moved. On average *all* the objects might be in, say, the left-hand compartment once in every $2^6 = 64$ throws but it would not be too common an occurrence even with only six objects.

With a very large number of objects the likelihood of them all ever being in the same compartment together becomes very remote. Thus if we have a gas jar of bromine vapour containing, say, 10^{22} molecules and we place another jar over the first, the chance of seeing all the bromine molecules back in the first jar at some future instant is so unlikely as to be impossible. Diffusion is evidently a one-way process, the result of chance when large numbers of particles are involved. By spreading, more arrangements become possible, which is what chance seems to favour.

Many mixing processes such as diffusion, the addition of milk to a cup of tea etc. occur in one direction only; they appear ridiculous if seen in a film run backwards.[1] Can you think of any *unmixing* processes which occur spontaneously?

(b) *Thermal equilibrium.*[2] This is attained when there is no net flow of energy from one body to another. A picture, at the atomic level, of how it might occur in solids and of the difference between hot and cold bodies can be obtained using a model proposed by Einstein and which we will develop by means of a game that simulates the model.

---

[1] Three such 8-mm film loops made for the Nuffield Advanced Physics course are worth viewing. Penguin XX1668, XX1669, XX1670.

[2] The material of this section is covered by the computer-made film *Change and chance: a model of thermal equilibrium in a solid* made for the Nuffield Advanced Physics course. Penguin XX1673.

In Einstein's solid the atoms vibrate more or less independently about their mean positions in the crystal lattice with internal energies (k.e. + p.e.) which, as we shall see later (p. 438), can only have certain values. If we assume the oscillation is simple harmonic, the values are simple integral multiples of a certain minimum called a *quantum* of energy. It is helpful to think of a ladder whose *equally-spaced* rungs represent the definite energy values or levels, any one of which an atom can occupy. When energy is gained the atom 'jumps' from a lower to a higher level and vice versa when energy is lost. A simple experiment supporting this idea is described in Appendix 9, p. 537.

If we assume quanta can move from atom to atom and do so randomly, the next step is to find out what numbers of atoms in the solid have any particular energy, i.e. what are the relative numbers of atoms in each energy level. An idea of whether there is any pattern in this *distribution* can be obtained from the following game.

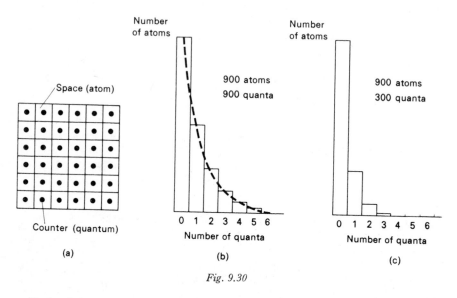

Fig. 9.30

Each of the 36 spaces in the grid of Fig. 9.30a represents an atom whose energy will be indicated by the number of counters in the space. An atom in its lowest energy level has no counters; one in its first level has one counter; in its second level two counters, and so on. We will start with each atom having one counter, as shown. Two dice are then thrown to give a random choice of the atom which is to lose a quantum. If the numbers obtained are 2 and 3 then the atom indicated may be taken as the second from the left and third up from the bottom. The counter on that space (atom) is then moved to the space (atom) given by the next throw of the two dice. This is repeated at least 100 times.

A distribution curve (in the form of a histogram) can be drawn but when chance is involved the results are most reliable if large numbers are concerned. Fig: 9.30b shows the curve that would be obtained by programming a computer to make 10 000 moves with 900 spaces and 900 quanta. A further 10 000 moves make little difference and the result is similar if we start with a different initial distribution or if 400 atoms and 400 quanta are used. The distribution thus depends on the ratio of the number of atoms to the number of quanta and not on their actual numbers.

We see from Fig. 9.30b that the number of atoms with no quanta is twice that with one quantum and the number with one quantum is twice that with two quanta and so on. The curve is an exponential one (dotted) and decreases by a constant fraction as the number of quanta increases by one. It represents a *Boltzmann* distribution.

The curve for an Einstein-type solid with 300 quanta shared among 900 atoms is shown in Fig. 9.30c. The shape is still exponential but steeper, and the ratio of the numbers of atoms with a certain number of quanta to those with one more quantum is still constant, but greater (four compared with two). If we regard temperature as the quantity which determines the likelihood of energy flowing from one body to another, the 900-atom solid with 900 quanta is hotter than the 900-atom solid with 300 quanta (it has more energy per atom), and so it is at a higher temperature.

Temperature can thus be associated with the distribution of atoms among energy levels, i.e. with the steepness of the distribution curve. Theory shows that in general if q quanta are shared among N atoms, the distribution curve ratio is $(1 + N/q)$. (Check this for the cases considered.) The *larger* the value of $(1 + N/q)$ the greater the steepness and the *lower* the temperature of the solid and vice versa. In other words, the temperature is high if the proportion of atoms in higher energy levels is high.

On this interpretation it would be reasonable to suppose that heat flows from an Einstein-type solid with a small distribution curve slope to one with a large slope until both slopes are the same. When this has happened thermal equilibrium will have been reached, although quanta will still be exchanged. The computer, suitably programmed, does in fact show that the ratio of the resultant distribution curve is greater than that of the ' hot ' solid but less than that of the ' cold ' one.

Attaining thermal equilibrium is thus a one-way process, like diffusion, and happens inevitably if left to chance. Heat flows due to the random sharing out of internal energy among the atoms of the bodies involved.

It must be remembered that Einstein's model is a simplification. In fact the atoms in a solid do not vibrate independently and consequently the energy levels of one atom affect those of others so that to talk of ' the ' energy levels of ' an' atom is not quite realistic. Neither are the atoms generally harmonic oscillators. Nonetheless many of the results derived hold for more complex and ' life-like ' models which require advanced mathematical treatment. The use of a model to represent a real system is a common technique in science.

Entropy

The fundamental, difficult, but very useful statistical concept of entropy can be obtained by combining ideas about ' numbers of ways ' and the ' distribution curve steepness ratio '.

(a) *Effect of more quanta on* W. When energy is added to a substance the number of ways W of arranging the quanta among the atoms increases. Thus in Fig. 9.31 we see that there are 3 ways of sharing 1 quantum among 3 atoms and 6 ways

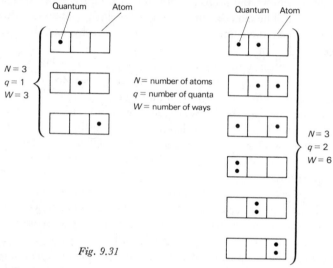

$N = 3$
$q = 1$
$W = 3$

N = number of atoms
q = number of quanta
W = number of ways

$N = 3$
$q = 2$
$W = 6$

Fig. 9.31

of sharing 2 quanta between them. It can be shown for an Einstein solid that, in general, if there are W ways in which q quanta can be distributed among N atoms, the addition of one more quantum increases the number of ways to W_1 where

$$\frac{W_1}{W} = \frac{N + q}{q + 1}.$$

If N and q are both large (as they are in practice) then

$$\frac{W_1}{W} = \frac{N + q}{q} = 1 + \frac{N}{q}.$$

We met the factor $(1 + N/q)$ in the previous section as a measure of the steepness of the Boltzmann distribution curve and we linked it with temperature. We now see that it, and therefore temperature also, are related to the *change* in the number of ways of distributing quanta when energy is added (or removed). Presently we will derive a relation between temperature and change in the number of ways.

Suppose that in a certain solid A, $q = N$, then $W_1/W = 2$ so that the loss (or gain) of one quantum would *halve* (or double) the number of ways of distributing the remaining quanta. On the other hand, in solid B having $q = N/3$, $W_1/W = 4$ and in this case the gain (or loss) of one quantum *quadruples* (or quarters) the number of ways of arranging the quanta.

The transfer of a quantum has a greater effect on W in solid B than in solid A. Now, as we saw with diffusion, chance favours those events which lead to an increase in W and so we would expect that bringing A and B together would lead to the net transfer of quanta from A (the hotter, since q/N is greater) to B. Thermal equilibrium would be established when exchange of quanta between A and B has the same effect on W for each and the process can be regarded as one which must happen inevitably since chance and large numbers are involved.

(*b*) *Temperature and change in number of ways.* From what has been said it follows that every quantum added to an Einstein solid increases W by the same factor, say a, where $a = (1 + N/q)$ and depends only on the temperature of the solid. Then if W_n is the number of ways after adding n quanta

$$\frac{W_n}{W} = a \times a \times a \times \ldots (n \text{ times}).$$

This is conveniently and usually expressed in natural logarithms (i.e. to base e) and if the extra quanta do not materially affect the ratio (i.e. the temperature) we have, writing $\log_e = \ln$,

$$\ln (W_n/W) = \ln W_n - \ln W = \ln a + \ln a + \ln a + (n \text{ times}).$$

The number of terms on the right-hand side equals the number of quanta added and the right-hand side is therefore proportional to the change δU in the internal energy of the solid. Writing $\delta \ln W$ for $\ln W_n - \ln W$,

$$\delta \ln W \propto \delta U$$

the volume being constant and the temperature nearly so.

Earlier we had associated a *high* temperature T with a *small* change in W and so if T is to be incorporated in the above expression we must write

$$\delta \ln W \propto \frac{\delta U}{T}.$$

Introducing a constant k, the relation between W and T is, under the conditions given above,

$$\delta \ln W = \frac{\delta U}{kT}$$

$$\therefore k \, \delta \ln W = \frac{\delta U}{T}.$$

This expression, which is applicable to solids, liquids and gases as well as Einstein solids, defines temperature in terms of energy and number of ways and could, if desired, be used as the basis of a scale of temperature. However, as we have seen, T is defined on the Kelvin scale and then k is found to have the value 1.38×10^{-23} J K^{-1}—which is *Boltzmann's* constant (p. 340).

(c) *Entropy and the direction of a process.* The quantity $k \, \delta \ln W$ is called the *change of entropy*, δS, hence

$$\delta S = k \, \delta \ln W = \frac{\delta U}{T}.$$

It can be calculated by measuring δU and T and gives us information about the increase in the number of ways resulting from adding energy δU to matter (in any phase) at temperature T and constant volume.

Absolute zero is taken as the arbitrary zero of *entropy* S and we can write

$$S = k \ln W.$$

Knowing S we then have information about the direction processes will take, for we have seen that *chance favours those for which W tends to increase*. If W increases so does the entropy $k \ln W$ and an alternative but more basic statement of the second law of thermodynamics than those given earlier (pp. 353 and 354) is as follows.

The direction of a process is such as to increase the total entropy.

Hence any physical or chemical change only occurs if the number of ways W and the entropy S do not diminish as a result. To take a very simple example, ice and water at s.t.p. have entropies of 41 J K^{-1} mol^{-1} and 63 J K^{-1} mol^{-1} respectively and so when heat is supplied to melt ice, the entropy increases.

(d) *Entropy and heat engines.* The efficiency of a heat engine, previously discussed in terms of the Carnot cycle (p. 363), can also be viewed from the standpoint of entropy. Suppose a steam turbine receives Q_1 joules of energy per second from steam at temperature T_1. The entropy *decrease* δS_1 of the steam for the *removal* of this amount of energy is

$$\delta S_1 = \frac{\delta U}{T_1} = \frac{Q_1}{T_1}.$$

The entropy in the *whole* process must increase and does so as a result of the smaller amount of energy, say Q_2 per second, which is *supplied* to the atmosphere as 'exhausted' steam at temperature T_2. The entropy *increase* δS_2 of the atmosphere for this part must be greater than δS_1 and is

$$\delta S_2 = \frac{Q_2}{T_2}.$$

The *net* change of entropy δS is thus

$$\delta S = \frac{Q_2}{T_2} - \frac{Q_1}{T_1}.$$

If δS is to be positive then

$$\frac{Q_2}{T_2} > \frac{Q_1}{T_1}.$$

That is, for the turbine to operate without a net decrease in entropy, T_1 must be high and T_2 low so that even though $Q_1 > Q_2$, the term Q_1/T_1 is smaller than Q_2/T_2. The difference $(Q_1 - Q_2)$ is used by the turbine to do external work, e.g. to drive a dynamo and we are again led to conclude that an engine must necessarily have a maximum efficiency, i.e. it must be inefficient to some extent.

QUESTIONS

Kinetic theory: gas laws

1. If a mole of oxygen molecules occupies 22.4×10^{-3} m^3 at s.t.p. (i.e. 273 K and 1.00×10^5 Pa), calculate the value of the molar gas constant R in J mol^{-1} K^{-1}.

2. (a) Assuming the equation of state for an ideal gas, show that the number of molecules in a volume V of such a gas at pressure p and temperature T is $pVN_A/(RT)$ where N_A is the Avogadro constant and R is the molar gas constant.
    (b) Hence find the number of molecules in 1.00 m^3 of an ideal gas at s.t.p. (One standard atmosphere = 1.00×10^5 Pa.)

3. Distinguish between a *saturated* and an *unsaturated* vapour. Describe how the variation of the saturation vapour pressure of water vapour with temperature may be investigated over a range from about 30 °C to 100 °C.
    A closed vessel of fixed volume contains air and water. The pressures in the vessel at 20 °C and 75 °C are respectively 737.5 mm and 1144 mm of mercury and some of the water remains liquid at 75 °C. If the saturation vapour pressure of water at 20 °C is 17.5 mm of mercury, find its value at 75 °C. *(L.)*

4. Explain what is meant by *a saturated vapour*. Sketch curves which show the relationship between pressure and temperature for (a) an ideal gas, (b) saturated water vapour, and (c) a mixture of an ideal gas and saturated water vapour. The temperature axis should extend from 0–100 °C and the pressure axis from 0–1000 mmHg.
    A flask contains a mixture of air and unsaturated water vapour at a temperature of 50 °C and a pressure of 8.0×10^2 mmHg. The mixture is cooled and when the temperature reaches 20 °C water begins to condense out of the mixture. Given that the s.v.p. of water vapour at 20 °C is 18 mmHg and at 5.0 °C is 7.0 mmHg, calculate the pressure of the mixture if it is cooled to 5.0 °C. You may assume that the unsaturated vapour obeys the gas laws up to the point of saturation.

(A.E.B. part qn.)

5. If the density of nitrogen at s.t.p. is 1.25 kg m⁻³, calculate the root mean square speed of nitrogen molecules at 227 °C. (One standard atmosphere = 1.00×10^5 Pa.)

6. A closed vessel contains hydrogen which exerts a pressure of 20.0 mmHg at a temperature of 50.0 K. At what temperature will it exert a pressure of 180 mmHg? If the r.m.s. velocity of the hydrogen molecules at 50.0 K was 800 m s⁻¹, what will be their r.m.s. velocity at this new temperature? Assume that there is no change in volume of the vessel. (S.)

7. Define *pressure* and explain in qualitative terms how the pressure exerted by a gas may be interpreted in terms of the momenta of the gas molecules.

List the postulates of the simple kinetic theory which are used in the derivation of the expression $p = \frac{1}{3}\rho\overline{c^2}$ for the pressure p exerted by a gas of density ρ whose molecules have a r.m.s. speed $\sqrt{\overline{c^2}}$.

Establish the relation between the temperature of an ideal gas and its molecular kinetic energy.

Give a descriptive account of the arguments which lead to the introduction of the terms a/V^2 and b in van der Waals' equation $(p + a/V^2)(V - b) = RT$. (J.M.B.)

8. A vessel of volume 50 cm³ contains hydrogen at a pressure of 1.0 Pa and at a temperature of 27 °C. Estimate (a) the number of molecules in the vessel, (b) their distance apart, on the average, and (c) their root-mean-square speed.

($R = 8.3$ J mol⁻¹ K⁻¹; Avogadro constant $= 6.0 \times 10^{23}$ mol⁻¹; mass of 1 mole of hydrogen molecules $= 2.0 \times 10^{-3}$ kg mol⁻¹.)

Thermodynamics

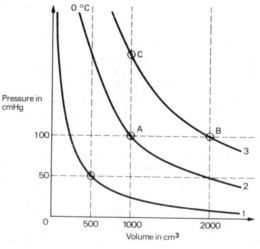

Fig. 9.32

9. The graph in Fig. 9.32 shows three curves relating the pressure and the volume of a fixed mass of a perfect gas at three different temperatures. Curve 2 is for 0 °C.

(*i*) Name and state the law which any one of these three curves represents.

(*ii*) What relationship exists between all the points on the three curves?

(*iii*) Name the law which connects all points on the line through A and B and express it mathematically.

(*iv*) Deduce the pressure at the point C, explaining your reasoning. (*Do not attempt to read it from the graph.*)

(*v*) What temperatures do the curves 1 and 3 refer to?

(*vi*) How much external work (in joules) would have to be done to take the gas from the state represented at A to that at B?

(*vii*) What other energy would have to be supplied during this process?

(*viii*) Under what conditions does an ordinary gas behave as a perfect gas?

(*ix*) What properties must be postulated for the molecules of an ordinary gas to account for its deviations, under other conditions, from perfect gas behaviour?

(Density of mercury = 13.6×10^3 kg m^{-3}; $g = 10$ N kg^{-1}.) (*S.*)

10. Derive an expression for the total translational kinetic energy of the molecules of 1.0 g of helium at T K and calculate the principal specific heats of the gas. You may assume that helium is an ideal monatomic gas of relative molecular mass 4.0 and that the molar (or ' universal ') gas constant R is 8.3 J K^{-1} mol^{-1}.

(*J.M.B. part qn.*)

11. State the first law of thermodynamics. What is the evidence for its validity? When applied to a fixed mass of gas this law can be written in the form

$$dU = dQ - p\ dv.$$

Explain the meaning of each of the three terms.

Explain how the equation is applied when the gas is (*a*) heated at constant volume, and (*b*) compressed adiabatically.

For water at constant pressure of one standard atmosphere the specific latent heat of vaporization is 2.26×10^6 J kg^{-1}. During the transformation to vapour the increase in volume of 1.00 kg of water is 1.67 m^3. Calculate the work done against the external pressure during this process. What has happened to the remainder of the heat supplied during the evaporation?

Density of water = 1.00×10^3 kg m^{-3}; one standard atmosphere = 1.00×10^5 Pa. (*J.M.B.*)

12. Explain why, when quoting the specific heat capacity of a gas, it is necessary to specify the conditions under which the change of temperature occurs. What conditions are normally specified?

A vessel of capacity 10 litres contains 1.3×10^2 grams of a gas at 20 °C and 10 atmospheres pressure. 8.0×10^3 joules of heat energy are suddenly released in the gas and raise the pressure to 14 atmospheres. Assuming no loss of heat to the vessel, and ideal gas behaviour, calculate the specific heat capacity of the gas under these conditions.

In a second experiment the same mass of gas, under the same initial conditions, is heated through the same rise in temperature while it is allowed to expand slowly so that the pressure remains constant. What fraction of the heat energy supplied in this case is used in doing external work? Take 1 atmosphere $= 1.0 \times 10^5$ pascals.

(O. and C.)

13. ' In a reversible adiabatic change in a gas, the pressure p and the volume V obey the relation $pV^\gamma = $ constant.' Explain the terms *adiabatic* and *reversible* and discuss how the value of the index γ depends on the atomicity of the gas.

Air initially at atmospheric pressure is compressed adiabatically and reversibly to a pressure of 4.00 atmospheres and is then allowed to expand isothermally and reversibly to its original volume. Find the final pressure. Sketch on a p-V diagram the curves representing the changes. State, with reasons, which is the greater: the work done on the air during compression or that done by the air in expanding. (γ for air $= 1.40$.)

(L.)

14. Define isothermal and adiabatic changes and give the equation relating the pressure and volume of an ideal gas for each type of change.

Why has it been concluded that the pressure and volume changes accompanying the passage of sound waves through a gas are adiabatic?

Air occupying a volume of 10 litres at 0.0 °C and atmospheric pressure is compressed isothermally to a volume of 2.0 litres and is then allowed to expand adiabatically to a volume of 10 litres. Show the process on a p-V diagram and calculate the final pressure and temperature of the air. At what volume was the pressure momentarily atmospheric? Mark this point on your diagram.

Assume $\gamma = 1.4$ for air (and that all changes are reversible.)

(W.)

15. Sketch on a p-V diagram the theoretical pressure and volume changes which take place in the cylinder of a four-stroke engine working on the Otto cycle. Label each part of the diagram.

Sketch the shape of the indicator diagram which is obtained in practice and account for the differences between the two diagrams. *(J.M.B. Eng. Sc.)*

16. Use the data given below to make *quantitative* predictions, estimates or comparisons concerning man's use of the fuel resources (coal, oil etc.) stored in the Earth.

In particular, you should consider the following problems: (a) The possibility of using up these fuel resources, and any need there may be to find alternative sources of energy. (b) Any differences between groups of people in the amount of energy they use.

Energy reaching Earth from the Sun: 5×10^{24} joules per year.

Maximum energy likely to be available from hydroelectric power if all such sources were used: 5×10^{19} joules per year.

Total energy stored in all known reserves of fuels (coal, oil etc.) 10^{23} joules.

Total energy stored in all known reserves of fuels (coal, oil etc.) that can be extracted economically at present: 10^{22} joules.

Total energy stored in known oil deposits that can be extracted economically at present: 10^{21} joules.

Time taken to produce the Earth's store of coal, oil and other fuels: 10^8 years.
Energy used by one car in a year: 3×10^{10} joules.
Work energy one labourer can produce in one year: 10^{10} joules.

Total energy used in 1964		Population
World	10^{20} joules	3000×10^6
North America	4×10^{19} joules	200×10^6
Western Europe	2×10^{19} joules	300×10^6
Middle East and Africa	0.3×10^{19} joules	300×10^6
Asia	2×10^{19} joules	1700×10^6

(*O. and C. Nuffield*)

10 Atomic physics

Thermionic emission

In a metal each atom has a few loosely-attached outer electrons which move randomly through the material as a whole. The atoms thus exist as positive ions in a 'sea' of free electrons. If one of these electrons near the surface of the metal tries to escape, it experiences an attractive inward force from the resultant positive charge left behind. The surface cannot be penetrated by an electron unless an external source does work against the attractive force and thereby increases the kinetic energy of the electron. If this is done by heating the metal, the process is called *thermionic emission*.[1]

The work function W of a metal is the energy which must be supplied to enable an electron to escape from its surface. It is conveniently expressed in electron-volts (p. 22). (It should, however, be noted that the term is sometimes applied to the potential Φ required, rather than to the energy, in which case $W = \Phi e$ where e is the electronic charge.)

The smaller the work function of a metal the lower the temperature of thermionic emission; in most cases the temperature has to be too near the melting point. Two materials used are (*i*) thoriated tungsten having $W = 2.6$ eV and giving good emission at about 2000 K and, in most cases, (*ii*) a metal coated with barium oxide for which $W = 1$ eV, copious emission occurring at 1200 K.

In many thermionic devices a plate, called the *anode*, is at a positive potential with respect to the heated metal and attracts electrons from it. The latter is then called a *hot cathode*. Hot cathodes are heated electrically either directly or indirectly. In direct heating current passes through the cathode (or filament) itself which is in the form of a wire, Fig. 10.1*a*. An indirectly heated cathode consists of a thin, hollow metal tube with a fine wire, called the *heater*, inside and separated from it by an electrical insulator, Fig. 10.1*b*. Indirect heating is

---

[1] It will be seen in Chapter 11 that modern solid-state theory views conditions in a metal from a slightly different stand-point—that of energy levels; here the above picture is adequate.

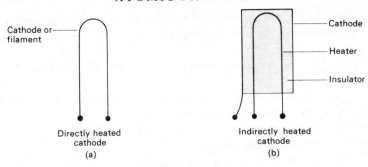

Fig. 10.1

most common since it allows a.c. to be used without the potential of the cathode continually varying. A typical heater supply for many thermionic devices is 6.3 V a.c., 0.3 A.

Cathode rays and the electron

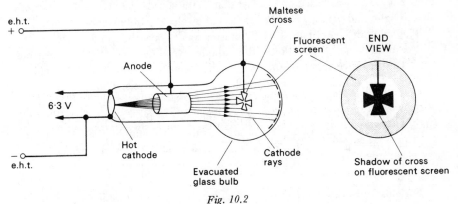

Fig. 10.2

(*a*) *Properties.* Streams of electrons moving at high speed are called *cathode rays.* They exhibit several important and useful effects, some of which can be demonstrated with the Maltese cross tube of Fig. 10.2. It consists of a hot cathode and a hollow cylindrical anode enclosed in an evacuated glass envelope having a coating of fluorescent material on the inside of the bulb. The anode is connected to the positive of an e.h.t. supply of 2–3 kV so that electrons from the cathode are accelerated along the tube in a divergent beam. Most bypass the anode and a dark shadow of the cross appears on the screen against a blue or green fluorescent background. This suggests that the rays are travelling in straight lines from the cathode and those not intercepted by the cross cause the screen to fluoresce.

When a magnet is brought near the side of the tube level with the anode, the beam is deflected vertically and the shadow can be made partially or wholly to disappear. Using Fleming's left-hand rule the direction of the deflection shows that the rays behave like a flow of negative charge, travelling from cathode to anode.

The beam also carries energy since the end of the tube struck by it becomes warm.

These and other properties of cathode rays may be summarized as follows:

(*i*) They travel from the cathode in straight lines.

(*ii*) They cause certain substances to fluoresce.

(*iii*) They possess kinetic energy.

(*iv*) They can be deflected by a magnetic field.

(*v*) They can be deflected by an electric field (p. 382).

(*vi*) They produce X-rays on striking matter (p. 397).

(*b*) *Discovery of the electron.* The first evidence to establish the existence of the electron is usually considered to be provided by J. J. Thomson's experiment in 1897 in which he measured the speeds and the charge to mass ratio (e/m) called the *specific charge*, of cathode rays. The view that cathode rays were not electromagnetic waves was based on the fact that (*i*) the speeds were typically one-tenth that of light, and (*ii*) they could be deflected by electric and magnetic fields.

The value of e/m obtained by Thomson was always the same, whatever the source or method of production of the cathode rays. This suggested that electrons are all alike, universal constituents of matter, and by assuming that the charge carried was equal to that on a monovalent ion in electrolysis, Thomson estimated the mass m of an electron, knowing e/m. Using modern values m is 9.11×10^{-31} kg, i.e. it is 1837 times smaller than the mass of the hydrogen atom. Other interpretations of the value of e/m are possible; it could be that the electron has the same mass as a hydrogen ion but a much greater charge. However, there is now no doubt that an electron carries the fundamental unit of electric charge.

For most purposes the electron can be regarded as a sub-atomic particle having a negative charge of value e, the electronic charge, and a very small mass (since force is required to accelerate it). The value of the mass quoted above is known as the ' rest mass '—it cannot be measured directly. The term has arisen because it has been found, as predicted by Einstein in the theory of relativity, that the mass of a particle accelerated to a speed approaching that of electromagnetic waves in vacuo increases with speed.

Ordinary particles are characterized by size and shape but such properties cannot be stated precisely for sub-atomic particles. To some extent the size of the electron depends on the method of determination and while some measurements indicate that it is a sphere of diameter 10^{-15} m, it is also satisfactory on occasions to regard it as a dimensionless point. By contrast its charge and mass can be uniquely specified. More will be said about the nature of the electron at the end of this chapter.

ATOMIC PHYSICS

Electron dynamics

If cathode rays are assumed to consist of particles (electrons), to which the laws of mechanics apply, we can obtain information about their speed and specific charge from their behaviour in electric and magnetic fields.

(a) *Speed of electrons.* Consider an electron of charge e and mass m which is emitted from a hot cathode and then accelerated by an electric field towards an anode. It experiences a force due to the field and work is done on it. The system (of field and electron) loses electrical potential energy and the electron gains kinetic energy. Let V be the p.d. between anode and cathode responsible for the field. If the electron starts from the cathode with zero speed and moves in a vacuum attaining speed v as it reaches the anode, the energy change W is given by

$$W = eV. \qquad \text{(p. 21)}$$

But
$$W = \tfrac{1}{2}mv^2$$

$$\therefore \quad eV = \tfrac{1}{2}mv^2.$$

From this ' energy equation ' it follows that

$$v = \sqrt{\frac{2eV}{m}}.$$

Substitution of numerical values for e/m and V shows that V is about 1.9×10^7 m s^{-1} (one-sixteenth of the speed of light) when $V = 1000$ V.

(b) *Deflection of electrons by a magnetic field.* It was shown earlier (p. 87) that the force F on a charge Q moving with speed v at right angles to a uniform magnetic field of flux density B is

$$F = BQv.$$

For an electron $Q = e$ and is negative, hence

$$F = Bev.$$

The direction of the force is given by Fleming's left-hand rule (remembering that a negative charge moving one way is equivalent to conventional current in the opposite direction) and is always at right angles to the field and to the direction of motion. Therefore at X in Fig. 10.3 the speed of the electron remains unaltered but it is deflected from its original path to, say, Y. Here the force acting on it still has the same value, Bev, and since the direction of motion and the field continue to be mutually perpendicular, the force is perpendicular to the new direction. The force thus only changes the direction of motion but not the speed

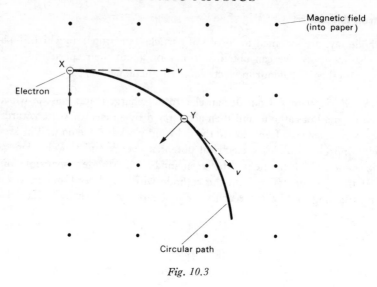

Fig. 10.3

and so the electron of mass m describes a *circular arc* of radius r. The constant radial force Bev is the centripetal force and so

$$Bev = \frac{mv^2}{r}.$$

This equation describes the path of the electron in the magnetic field.

(*c*) *Deflection of electrons by an electric field.* If electrons enter an electric field acting at right angles to their direction of motion, they are deflected from their original path. In Fig. 10.4 a p.d. applied between plates P and Q of length *l*

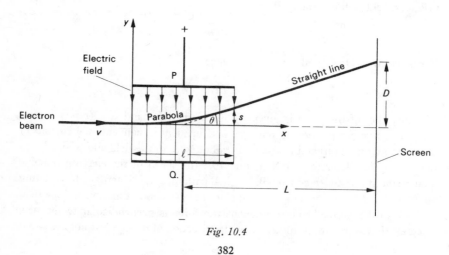

Fig. 10.4

382

creates a uniform electric field of strength E, non-uniformities at the edges of the plates being ignored.

Consider an electron of charge e, mass m and horizontal speed v on entering the field. If the upper plate is positive the electron experiences a force Ee and an acceleration Ee/m (second law of motion) both acting vertically upwards. Since the field is uniform, the acceleration is uniform and combines with the initial horizontal speed v, which the electron retains during the whole of its journey in the vacuum between the plates, to give a path which we shall show is a *parabola*. The behaviour of the electron is similar to that of a projectile fired horizontally; its path (neglecting air resistance) is also a parabola, the resultant of a uniform horizontal velocity and the vertical acceleration due to gravity.

The vertical displacement y of the electron at any time t is given by

$$y = \tfrac{1}{2}at^2$$

But $a = Ee/m$,

$$\therefore \quad y = \tfrac{1}{2} \cdot \frac{Ee}{m} \cdot t^2.$$

The corresponding horizontal displacement x is given by

$$x = vt.$$

Eliminating t,

$$y = \left(\frac{Ee}{2mv^2} \right) \cdot x^2.$$

This equation is of the form $y = kx^2$ (where $k = Ee/(2mv^2)$ = a constant) and so the path of the electron between the plates is a parabola.

The deflection D of the electron (i.e. its displacement from the original direction) on a screen distance L from the centre of the plates, can be obtained using the fact that it continues in a straight line after leaving the field. From Fig. 10.4, $\tan \theta \simeq D/L$, where $\tan \theta$ is the slope of the tangent at the end of the parabolic path. The slope equals, in calculus terms, the differential coefficient ds/dx when $x = l$. Differentiating $s = kx^2$ we get $ds/dx = 2kx$, therefore the slope is $2kl$. Hence $D/L = 2kl$ giving $D = 2klL$. Substituting for k we get $D = EelL/(mv^2)$. If V is the p.d. which has accelerated the electron to speed v then $eV = \tfrac{1}{2}mv^2$ and so

$$D = \frac{ElL}{2V}.$$

Thus D is proportional to E if V is constant and inversely proportional to V if E is constant.

(d) *Worked example.* (i) *An electron emitted from a hot cathode in an evacuated tube is accelerated by a p.d. of 1.0×10^3 V. Calculate the kinetic energy and speed acquired by the electron.* ($e = 1.6 \times 10^{-19}$ C, $m = 9.1 \times 10^{-31}$ kg)

We have

$$\tfrac{1}{2}mv^2 = eV$$

$$= (1.6 \times 10^{-19}\,\text{C})\,(1.0 \times 10^3\,\text{J C}^{-1})$$

$$= 1.6 \times 10^{-16}\,\text{J}$$

$$\therefore\ v = \sqrt{\frac{2 \times 1.6 \times 10^{-16}}{9.1 \times 10^{-31}}}$$

$$= 1.8 \times 10^7\,\text{m s}^{-1}.$$

(ii) *The electron now enters at right angles a uniform magnetic field of flux density 1.0×10^{-3} T. Determine its path.*

The magnetic force Bev, on the electron makes it describe a circular path of radius r given by

$$Bev = \frac{mv^2}{r}.$$

Hence

$$r = \frac{mv}{Be}$$

where $B = 1.0 \times 10^{-3}$ T, $m = 9.1 \times 10^{-31}$ kg, $e = 1.6 \times 10^{-19}$ C and from (i), $v = 1.8 \times 10^7$ m s^{-1}.

$$\therefore\ r = \frac{(9.1 \times 10^{-31}\,\text{kg})\,(1.8 \times 10^7\,\text{m s}^{-1})}{(1.0 \times 10^{-3}\,\text{T})\,(1.6 \times 10^{-19}\,\text{C})}$$

$$= 0.10\,\text{m}.$$

(iii) *Find the intensity of the uniform electric field which, when applied perpendicular to the previous magnetic field so as to be co-terminous with it, compensates for the magnetic deflection. If the electric field plates are 2.0×10^{-2} m apart what is the p.d. between them?*

For the electric and magnetic forces to balance

$$Ee = Bev.$$

Hence

$$E = Bv$$

$$= (1.0 \times 10^{-3}\,\text{T})\,(1.8 \times 10^7\,\text{m s}^{-1})$$

$$= 1.8 \times 10^4\,\text{V m}^{-1}.$$

If V_1 is the p.d. between the plates and d their separation, then $E = V_1/d$ and so

$$V_1 = Ed$$
$$= (1.8 \times 10^4 \text{ V m}^{-1})(2.0 \times 10^{-2} \text{ m})$$
$$= 3.6 \times 10^2 \text{ V.}$$

Specific charge of electron

Two methods that can be performed in a school laboratory will be outlined. A modern value of the specific charge of the electron is

$$\frac{e}{m} = 1.76 \times 10^{11} \text{ C kg}^{-1}.$$

(a) Cathode-ray tube method using crossed fields. This method is similar in principle to that developed by J. J. Thomson in which an electron beam is subjected simultaneously to mutually perpendicular, i.e. crossed, electric and magnetic fields. Here a vacuum-type cathode-ray tube is used, Fig. 10.5, having a hot cathode C and an anode A with a horizontal collimating slit from which the electrons emerge in a flat beam. The beam produces a narrow luminous trace when it hits a vertical fluorescent screen S, marked in squares and set at an angle. S is supported by two parallel deflecting plates Y_1, Y_2, across which a p.d. (about 3 kV) is applied, thereby creating an electric field between them.

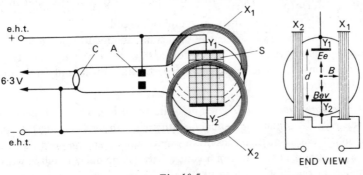

Fig. 10.5

Helmholtz coils X_1, X_2 (p. 83) mounted on opposite sides of the bulb produce a magnetic field between the plates, at right angles to both the direction of travel of the beam and the electric field. The coils are separated by a distance equal to their radius and when connected in series—so that the current has the same direction in each—they give an almost uniform field for a short distance along their common axis midway between them.

Consider an electron of charge e and mass m which emerges from the slit in the anode having been accelerated to speed v. Let E be the electric field strength between Y_1 and Y_2 and B the magnetic flux density along the axis of X_1 and X_2. When E and B are such that the electron suffers no deflection, the electric force Ee on it must be equal and opposite to the magnetic force Bev. Hence

$$Ee = Bev. \tag{1}$$

If the electron is emitted from the cathode with zero speed and moves in a good vacuum, its kinetic energy $\frac{1}{2}mv^2$ is given by

$$\tfrac{1}{2}mv^2 = eV \tag{2}$$

where V is the accelerating p.d. between anode and cathode. Eliminating v from (1) and (2)

$$\frac{e}{m} = \frac{E^2}{2B^2V}.$$

If the p.d. between Y_1 and Y_2 creating the electric field equals the accelerating p.d., then $E = V/d$ (i.e. field strength = potential gradient) where d is the separation of Y_1 and Y_2. Hence

$$\frac{e}{m} = \frac{V}{2B^2d^2}.$$

Thus e/m can be found if V, B and d are known. B may be determined experimentally by removing the tube and investigating the region between the coils with a current balance (p. 79) or it can be calculated from the expression $B \simeq 0.72\,\mu_0 NI/r$ (p. 83) where $\mu_0 = 4\pi \times 10^{-7}$ H m^{-1}, N is the number of turns on one coil, I is the current and r the radius of the coil.

In the above simple treatment the fields are assumed to be uniform and coterminous, i.e. to extend over the same length of the electron beam. In practice such conditions are not achieved and this partly accounts for only an approximate value of e/m being obtained.

(b) *Fine beam tube method.* The fine beam tube, Fig. 10.6, is a special type of cathode-ray tube containing a small quantity of gas (often hydrogen) at a pressure of about 10^{-2} mmHg. Electrons from a hot cathode emerge as a narrow beam from a small hole at the apex of a conical-shaped anode and collide with atoms of the gas in the tube. As a result, the latter may lose electrons (i.e. inelastic collisions occur) and form positive gas ions.

The electrons created by ionization are easily scattered but the relatively heavy gas ions form a line of positive charge along the path of the beam which attracts the fast electrons from the anode, focusing them into a ' fine ' beam. It will be seen later (p. 417) that a gas atom which has lost an electron can emit light when it recaptures an electron. The gas, therefore, not only focuses the beam but also reveals its path.

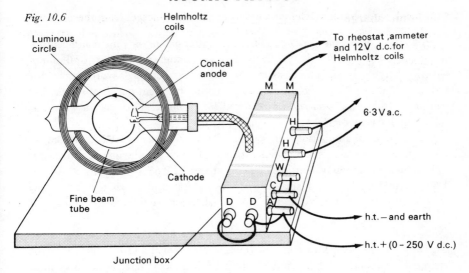

Fig. 10.6

Helmholtz coils are arranged one on each side of the tube and produce a fairly uniform magnetic field at right angles to the beam. If the field is sufficiently strong the electrons are deflected into a circular orbit and a luminous circle of low intensity appears when the tube is viewed in the dark. The diameter of the circle may be measured by placing a large mirror with a scale behind the tube so that the observer sees the circle, its image and the corresponding marks on the scale all in line. The circle diameter is altered by varying either the anode voltage or the current in the coils.

If B is the magnetic flux density between the coils in the region of the tube, r the radius of the luminous circle and e, m and v are the charge, mass and orbital speed of the electron then the circular motion equation gives

$$Bev = \frac{mv^2}{r}. \tag{3}$$

We shall assume electrons are emitted from the cathode with zero speed and that their orbital speed v is constant and equal to that with which they emerge from the anode after being accelerated by the p.d. V between anode and cathode. The energy equation then gives

$$\tfrac{1}{2}mv^2 = eV. \tag{4}$$

Eliminating v from (3) and (4),

$$\frac{e}{m} = \frac{2V}{B^2r^2}.$$

As in the cathode-ray tube method, B is obtained experimentally or by calculation.

387

Electronic charge

(*a*) *Electrolysis and the Faraday.* During the latter half of the nineteenth century it was suggested that electricity, like matter, was atomic and that a natural unit of electric charge existed. The basis for this belief was Faraday's work on electrolysis (1831–34) which may be summarized by the statement *a mole of monovalent ions of any substance is liberated by 9.65 × 10⁴ coulombs.* This electric charge is called the *Faraday constant* and is denoted by F.

The number of atoms (ions) in a mole of any substance is 6.02×10^{23} and if we assume every atom is associated with the same charge during electrolysis, it follows that each monovalent ion carries a charge of $9.65 \times 10^4/(6.02 \times 10^{23}) = 1.60 \times 10^{-19}$ C. It would appear that the natural unit of charge, called the *electronic charge*, has this value. For a monovalent ion we therefore have

$$F = N_A e$$

where N_A is the Avogadro constant $(6.02 \times 10^{23}\ \text{mol}^{-1})$ and e is the electronic charge. Nowadays, the charge indicated by the Faraday is referred to as a ' mole of electrons '. A mole of divalent ions is liberated by two ' moles of electrons ' (i.e. $2 \times 9.65 \times 10^4$ C) and of trivalent ions by three ' moles of electrons ' (i.e. $3 \times 9.65 \times 10^4$ C).

The above expression provides one of the most accurate ways of measuring e. F is found by electrolysis and N_A from X-ray crystallography measurements.

(*b*) *Millikan's oil drop experiment.* In 1909 Millikan started a series of experiments lasting many years which supplied evidence for the atomic nature of electricity and provided a value for the magnitude of the electronic charge. The principle of his method is to observe very small oil drops, charged either positively or negatively, falling in air under gravity and then either rising or being held stationary by an electric field.

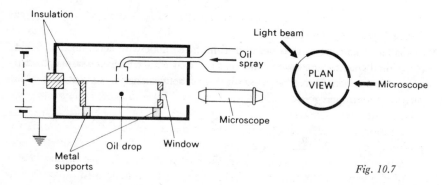

Fig. 10.7

The essential features of the apparatus are shown in Fig. 10.7. A spray of oil drops is formed above a tiny hole in the upper of two parallel metal plates and some find their way into the space between them. The drops are strongly illuminated and appear as bright specks on a dark background when viewed through a microscope.

With no electric field between the plates, one drop is selected and its velocity of fall found by timing it over a convenient number of divisions on a scale in the eyepiece of the microscope. (To find the actual distance fallen, the eyepiece scale is calibrated by viewing a millimetre scale through the microsope and comparing it with the eyepiece scale.) For a spherical drop of radius r, moving with uniform velocity v through a homogeneous medium having coefficient of viscosity η, Stokes' law states that the viscous force retarding its motion is $6\pi\eta\, rv$. In falling, the drop attains its terminal velocity almost at once because it is so small. The retarding force acting up then equals its weight, given by $\frac{4}{3}\pi r^3 \rho g$, where ρ is the density of the oil and g the acceleration due to gravity. If v is the terminal velocity and the small upthrust of the air is neglected

$$6\pi\eta rv = \frac{4}{3}\pi r^3 \rho g. \tag{1}$$

From this the radius r of the drop can be found.

Some of the drops become charged either by friction in the process of spraying or from ions in the air. Suppose the drop under observation has a negative charge Q. When a p.d. is applied to the plates so that the top one is positive, an electric field is created which exerts an upward force on the drop. If V is the p.d. and E is the intensity of the field required to keep the drop at rest, then the electric force experienced by it is EQ (since E is the force per unit charge). The electric force on the drop then equals its weight and so

$$EQ = \frac{4}{3}\pi r^3 \rho g. \tag{2}$$

$E = V/d$ where d is the distance between the plates, r is known from (1) and hence Q can be calculated.

Certain measures were adopted by Millikan to improve accuracy.

(*i*) He used non-volatile oil to prevent evaporation altering the mass of the drop.

(*ii*) Convection currents between the plates and variation of the viscosity of air due to temperature change were eliminated by enclosing the apparatus in a constant-temperature oil bath.

(*iii*) Stokes' law assumes fall in a homogeneous medium. The air consists of molecules and, as Millikan put it, very small drops ' fall freely through the holes in the medium '. He investigated this effect and corrected the law to allow for it.

(*iv*) To simplify the theory we have considered the drop held at rest by the electric field but Millikan reversed the motion and found the upward velocity of the drop.

Millikan found that the charge on an oil drop, whether positive or negative, was always an integral multiple of a basic charge. He studied drops having charges many times the basic charge and by using X-rays he was able to change the charge on a drop. The same minimum charge, equal to that on a monovalent ion, was always involved. The value of the ' atom ' of electric charge, i.e. the electronic charge e, is

$$e = 1.60 \times 10^{-19} \text{ C.}$$

Photoelectric emission

In photoelectric emission electrons are ejected from metal surfaces when electro-magnetic radiation of high enough frequency falls on them. The effect is given by zinc when exposed to X-rays or ultraviolet. Sodium gives emission with X-rays, ultraviolet and all colours of light except orange and red, while preparations containing caesium respond to infrared as well as to high frequency radiation.

(a) *Simple demonstration of photoelectric effect.* Ultraviolet from a mercury vapour lamp is allowed to fall on a small sheet of zinc, freshly cleaned with emery cloth and connected to an electroscope as in Fig. 10.8a. If the electroscope is given a positive charge the pointer is unaffected by the ultraviolet. When negatively charged, however, the electroscope discharges quite rapidly when the zinc is illuminated with ultraviolet. A sheet of glass between the lamp and the zinc plate halts the discharge.

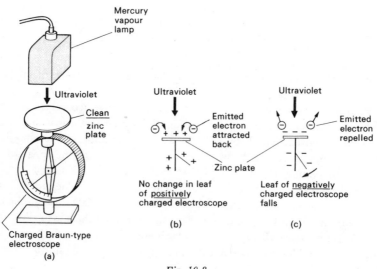

Fig. 10.8

The photoelectric effect was discovered in 1887 but was not explained until the electron had been ' discovered ' by J. J. Thomson. When the zinc plate is positively charged the electrons ejected from it by the ultraviolet fail to escape, being attracted back to the plate, Fig. 10.8*b*. A negative charge on the zinc repels the emitted electrons and both the zinc and the electroscope lose negative charge, Fig. 10.8*c*. The insertion of the sheet of glass cuts off much of the ultraviolet but allows the passage of violet light, also emitted by the lamp, and shows that violet light does not produce the effect with zinc.

Experiment shows that ' photoelectrons ' have the same specific charge as any other electrons.

(*b*) *Laws of photoelectric emission.* An experimental study of the photoelectric effect yields some surprising results which may be summarized as follows:

Law I. The *number* of photoelectrons emitted per second is proportional to the *intensity* of the incident radiation.

Law II. The photoelectrons are emitted with a range of kinetic energies from zero up to a *maximum* which increases as the *frequency* of the radiation increases and is independent of the intensity of the radiation. (Thus a faint blue light produces electrons with a greater maximum kinetic energy than those produced by a bright red light, but the latter releases a greater number.)

Law III. For a given metal there is a certain minimum frequency of radiation, called the *threshold frequency*, below which no emission occurs irrespective of the intensity of the radiation. For zinc, the threshold frequency is in the ultraviolet.

Most of these facts appear to be inexplicable on a wave theory of electromagnetic radiation. Law I can be vindicated in terms of waves because if the radiation has greater intensity, more energy is absorbed by the metal and it is possible for more electrons to escape. Also, it is reasonable to suppose that the range of emission speeds (and kinetic energies) from zero to a maximum is due to electrons having a range of possible kinetic energies inside the metal. Those with the highest kinetic energy are emitted with the maximum speeds. However, we would expect a certain number of photoelectrons to be ejected with greater speeds when the radiation intensity increases but this is not so according to Law II.

The increase of maximum kinetic energy with frequency and the existence of the threshold frequency are even more enigmatic. Furthermore, according to the wave theory, radiation energy is spread over the wavefront and since the amount incident on any one electron would be extremely small, some time would elapse before an electron gathered enough energy to escape. No such time lag between the start of radiation and the start of emission is observed, even when the radiation is weak.

Quantum theory

By the end of the nineteenth century the wave theory, despite its earlier notable successes, was unable to account for most of the known facts concerning the interaction of electromagnetic radiation with matter. One of these, as we have just seen, was photoelectric emission and another was black body radiation, which is considered in Chapter 8.

(*a*) *Planck's theory.* In 1900 Planck tackled the problem of finding a theory which would fit the facts of black body radiation. Whereas others had considered the radiation to be emitted continuously, Planck supposed this to occur intermittently in integral multiples of an ' atom ' or *quantum* of energy, the size of which depended on the frequency of the oscillator producing the radiation. A body would thus emit one, two, three etc. quanta of energy but no fractional amounts.

According to Planck the quantum E of energy for radiation of frequency f is given by

$$E = hf$$

where h is a constant, now called *Planck's constant.* For electromagnetic radiation of wavelength λ, $c = f\lambda$, where c is its speed in vacuo and so we also have $E = hc/\lambda$. The energy of a quantum is thus inversely proportional to the wavelength of the radiation but directly proportional to the frequency. It is convenient to express many quantum energies in electron-volts; the quantum for red light has energy of about 2 eV and for blue light of about 4 eV.

Using the equation $E = hf$, Planck derived an expression for the variation of energy with wavelength for a black body which agreed with the experimental curves (Fig. 8.63, p. 316), at all wavelengths and temperatures. At the time the quantum theory was too revolutionary for most scientists and little attention was paid to it. Nevertheless the interpretation of black body radiation was the first of its many successes.

(*b*) *Einstein's photoelectric equation.* Einstein extended Planck's ideas in 1905 by deriving an equation which explained in a completely satisfactory way the laws of photoelectric emission. He assumed that not only were light and other forms of electromagnetic radiation emitted in whole numbers of quanta but that they were also absorbed as quanta, called *photons.* This implied that electromagnetic radiation could exhibit particle-like behaviour when being emitted and absorbed and led to the idea that light etc. has a dual nature; under some circumstances it behaves as waves and under others as particles. Wave-particle duality will be considered later in the chapter.

When dealing with thermionic emission it was explained that to liberate an electron from the surface of a metal a quantity of energy, called the *work function*, W, which is characteristic of the metal, has to be supplied. In photoelectric emission, Einstein proposed that a photon of energy hf causes an electron to be emitted if $hf \geqslant W$. The excess energy $(hf - W)$ appears as kinetic energy of the emitted electron which escapes with a speed having any value up to a maximum v_{max}. The actual value depends on how much energy the electron has inside the metal. Hence

$$hf - W = \tfrac{1}{2}mv_{max}^2.$$

This is Einstein's photoelectric equation.

If the photon has only just enough energy to liberate an electron, the electron gains no more kinetic energy. It follows that since W is constant for a given metal, there is a minimum frequency, the threshold frequency f_0, below which no photoelectric emission is possible. It is given by

$$hf_0 = W.$$

Einstein's equation can also be written as

$$h(f - f_0) = \tfrac{1}{2}mv_{max}^2.$$

The increase of maximum emission speed with higher frequency radiation can now be seen to be due to the greater photon energy of such radiation. It should also be noted that it is assumed that a photon imparts all its energy to one electron and then no longer exists.

(c) *Millikan's experiment on photoelectric emission.* In 1916 Millikan verified Einstein's photoelectric equation experimentally and provided irrefutable evidence for the photon model of light. He was the first to obtain photoelectrically a value for Planck's constant which agreed with values from other methods. His apparatus is shown in simplified form in Fig. 10.9a.

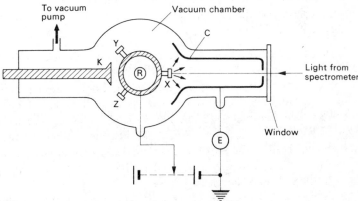

Fig. 10.9a

Monochromatic light from a spectrometer entered the window of a vacuum chamber and fell on a metal X mounted on a turntable R controlled from outside the chamber. The photoelectrons emitted were collected by an electrode C and detected by a sensitive current measuring device E. The minimum positive potential, called the *stopping potential*, which had to be applied to X to prevent the most energetic photoelectrons reaching C and causing current flow, was found for different frequencies of the incident radiation. The procedure was repeated with Y and then Z opposite C. X, Y and Z were made from the alkali metals lithium, sodium and potassium, since these emit photoelectrons with light and each one can therefore be studied over a wide range of frequencies. Immediately before taking a set of readings R was rotated and the knife K adjusted so that a fresh surface was cut on the metal, thus eliminating the effects of surface oxidation.

In such experiments the p.d. between the electrodes, as measured by any form of voltmeter, is not the same as the p.d. in the space between them unless their work functions are equal. The error is called the *contact p.d.* In Millikan's experiment this was a few volts, that is, the correction was as large as the effect measured. One feature of his experiment was an ingenious device (not shown in Fig. 10.9a) by which an accurate correction was made.

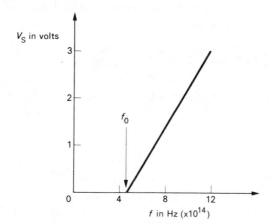

Fig. 10.9b

The relation between the stopping potential V_s and the maximum kinetic energy of the photoelectrons is given by the energy equation (p. 381), $eV_s = \frac{1}{2}mv_{max}^2$. Einstein's equation may then be written

$$\tfrac{1}{2}mv_{max}^2 = eV_s = hf - W$$

$$\therefore \quad V_s = \frac{h}{e} \cdot f - \frac{W}{e}.$$

The graph of V_s against f should be a straight line, a fact which Millikan's results confirmed. Einstein's relation was thus verified. Furthermore, the stopping potential for a given frequency of light was independent of the intensity of the light. One of Millikan's graphs is shown in Fig. 10.9b. He found that whatever the metal all graphs had the same slope and from the above equation this is seen to be h/e. Knowing e, h can be calculated and the value obtained agrees with that found from black body radiation experiments. The threshold frequency f_0 and the work function W are characteristic for each metal and may also be deduced from the graph. The values of W are in good agreement with those found from thermionic emission.

Planck's constant h is a fundamental physical constant and occurs in many formulae in atomic physics. Its value to three figures is

$$h = 6.63 \times 10^{-34} \text{ J s.}$$

It is because of this smallness that quantum effects are not normally apparent.

Because of the need for very clean surfaces, quantitative photoelectric experiments are difficult with simple apparatus in a school laboratory but one which gives fair results is outlined in Appendix 10, p. 538.

(*d*) *Worked example. If a photoemissive surface has a threshold wavelength of 0.65 μm, calculate (i) its threshold frequency, (ii) its work function in electron volts, and (iii) the maximum speed of the electrons emitted by violet light of wavelength 0.40 μm. (Speed of light $c = 3.0 \times 10^8$ m s⁻¹, $h = 6.6 \times 10^{-34}$ J s, $e = 1.6 \times 10^{-19}$ C and mass of electron $m = 9.1 \times 10^{-31}$ kg.)*

(*i*) $\qquad\qquad \lambda_0 = 0.65 \text{ μm} = 6.5 \times 10^{-7} \text{ m}$

$$\therefore \ f_0 = c/\lambda_0$$

$$= \frac{3.0 \times 10^8 \text{ m s}^{-1}}{6.5 \times 10^{-7} \text{ m}}$$

$$= 4.6 \times 10^{14} \text{ Hz (s}^{-1}).$$

(*ii*) We have $\qquad\qquad W = hf_0$

$$= (6.6 \times 10^{-34} \text{ J s}) (4.6 \times 10^{14} \text{ s}^{-1})$$

$$= 6.6 \times 10^{-34} \times 4.6 \times 10^{14} \text{ J.}$$

But $\qquad\qquad 1 \text{ eV} = 1.6 \times 10^{-19} \text{ J}$

$$\therefore \ W = \frac{6.6 \times 4.6 \times 10^{-20} \text{ eV}}{1.6 \times 10^{-19}}$$

$$= 1.9 \text{ eV.}$$

(*iii*) For violet light $\quad\quad\quad \lambda = 0.40\,\mu m = 4.0 \times 10^{-7}\,m$

$$f = c/\lambda$$

$$= \frac{3.0 \times 10^8\,m\,s^{-1}}{4.0 \times 10^{-7}\,m}$$

$$= 7.5 \times 10^{14}\,Hz.$$

From the photoelectric equation $\frac{1}{2}mv_{max}^2 = hf - W,$

$$\frac{1}{2}mv_{max}^2 = (6.6 \times 10^{-34} \times 7.5 \times 10^{14} - 1.9 \times 1.6 \times 10^{-19})\,J$$

$$= 1.9 \times 10^{-19}\,J$$

$$\therefore\;\; v_{max} = \sqrt{\frac{2 \times 1.9 \times 10^{-19}}{9.1 \times 10^{-31}}}\,m\,s^{-1}$$

$$= 6.5 \times 10^5\,m\,s^{-1}.$$

Photocells and their uses

There are several kinds of photocell, only the photoemissive type will be considered here; those based on semiconductor materials are discussed later (pp. 463–4).

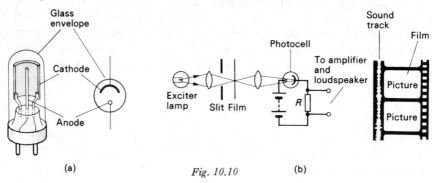

Fig. 10.10

(*a*) *Photoemissive cell.* The construction of a typical cell and its symbol are shown in Fig. 10.10a. Two electrodes are enclosed in a glass bulb which may be evacuated or contain an inert gas at low pressure. The cathode, often called the *photocathode*, is a curved metal plate having an emissive surface facing the anode, here shown as a single metal rod. When electromagnetic radiation falls on the cathode, photoelectrons are emitted and are attracted to the anode if it is at a suitable positive potential. A current of a few microamperes flows and increases with the intensity of the incident radiation. An inert gas in the cell gives greater current but causes a time lag in the response of the cell to very rapid changes of radiation which may make it unsuitable for some purposes.

The choice of material for the cathode surface depends on the frequency range over which the cell is to operate, and should be such that a good proportion of the incident photons yield electrons. Ideally, every photon should release one electron but in practice the yield is much lower. Pure metals are rarely used because of their high reflecting power; most photocathode surfaces are composite. Caesium on oxidized silver has a peak response near the red end of the spectrum and a threshold wavelength of 12 μm which makes it suitable for use with infrared.

(b) *Some uses.* Objects on a conveyor belt in a factory can be counted if the beam of light or other radiation falling on the photocell is cut off by the passing object.

Something in the wrong place, e.g. a safe-breaker in a bank vault or the hand of a worker too close to a piece of machinery, can be detected if it interrupts a beam—often of infrared—falling on a photocell connected to an ' intruder ' alarm.

In the reproduction of sound from a film, light from an exciter lamp, Fig. 10.10b, is focused on the ' sound track ' at the side of the moving film and then falls on a photocell. The sound track varies the intensity of the light passing through it so that the photocell creates a varying current which is a replica of that obtained in the recording microphone when the film was made. The fluctuating p.d. developed across a load R is amplified and the output converted to sound by a loudspeaker. In the variable area sound track the width of the ' white ' part varies in the same way as the amplitude of the original sound wave. The track is produced by allowing the electrical variations from the recording microphone to control the width of a narrow slit through which light passes on its way to the side of the moving film.

X-rays

X-rays, so named because their nature was at first unknown, were discovered in 1895 by Röntgen. They are produced whenever cathode rays (electrons) are brought to rest by matter.

(a) *Production.* A modern X-ray tube is highly evacuated and contains an anode and a tungsten filament connected to a cathode, Fig. 10.11. Electrons are obtained from the filament by thermionic emission and are accelerated to the anode by a p.d., typically up to 100 kV. The anode is a copper block inclined to the electron stream and having a small target of tungsten, or another high-melting-point metal, on which electrons are focused by the concave cathode. The tube has a lead shield with a small window to allow the passage of the X-ray beam.

Less than $\frac{1}{2}$ per cent of the kinetic energy of the electrons is converted into X-rays—the type of electromagnetic radiation produced in cases where electrons are rapidly brought to rest by striking matter (see p. 310). The rest of the kinetic

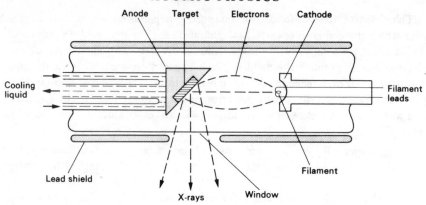

Fig. 10.11

energy becomes internal energy of the anode which has to be kept cool by circulating oil or water through channels in it or by the use of cooling fins.

The *intensity* of the X-ray beam increases when the number of electrons hitting the target increases and this is controlled by the filament current. The *quality* or penetrating power of the X-rays is determined by the speed attained by the electrons and increases with the p.d. across the tube. ' Soft ' X-rays only penetrate such objects such as flesh, ' hard ' X-rays can penetrate much more solid matter.

The p.d. required to operate an X-ray tube may be obtained from a half-wave rectifying circuit containing a step-up transformer and in which the X-ray tube itself acts as the rectifier.

(*b*) *Properties.* These may be summarized as follows:

(*i*) They travel in straight lines.

(*ii*) They readily penetrate matter; penetration is least in materials containing elements of high density and high atomic number. Thus, while sheets of cardboard, wood and some metals fail to stop them, all but the most penetrating are absorbed by a sheet of lead 1 mm thick. Lead glass is a much better absorber than ordinary soda glass.

(*iii*) They are not deflected by electric or magnetic fields.

(*iv*) They eject electrons from matter by the photoelectric effect and other mechanisms. The ejected electrons are responsible for the next three effects.

(*v*) They ionize a gas, permitting it to conduct. An electrified body such as an electroscope charged positively or negatively is discharged when the surrounding air is irradiated by X-rays.

(*vi*) They cause certain substances to fluoresce, e.g. barium platinocyanide.

(*vii*) They affect a photographic emulsion in a similar manner to light.

(*c*) *Nature: von Laue's experiment.* X-rays cannot be charged particles since they are not deflected by electric and magnetic fields. The vital experiment which

398

established their electromagnetic nature was initiated by von Laue in 1912. After unsuccessful attempts had been made to obtain X-ray diffraction patterns with apparatus similar to that used for light, it was realized that the failure might be due to X-rays having much smaller wavelengths.

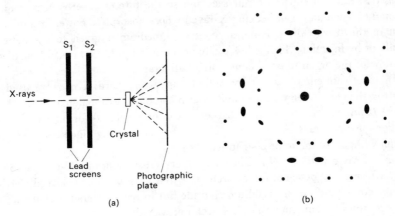

Fig. 10.12

Von Laue's suggestion was that if the regular spacing of atoms in a crystal is of the same order as the wavelength of the X-rays, the crystal should act as a three-dimensional diffraction grating. This idea was put to the test with conspicuous success by Friedrich and Knipping, two of von Laue's colleagues. The arrangement of their apparatus is shown in Fig. 10.12a. A narrow beam of X-rays from two slits S_1 and S_2 fell on a thin crystal in front of a photographic plate. After a long exposure the plate (on developing) revealed that most of the radiation passed straight through to give a large central spot, but some of it produced a distinct pattern of fainter spots around the central one, Fig. 10.12b.

The electromagnetic wave behaviour of X-rays was thus demonstrated and analysis of diffraction patterns confirmed they had wavelengths of about 10^{-10} m (0.1 nm). They are considered to be electromagnetic waves because (i) their method of production involves accelerated charges, (ii) they eject electrons from matter (implying strong electric fields), (iii) they give line spectra similar in general character to the optical spectrum of hydrogen (p. 299).

(d) *Uses of X-rays.* The usefulness of X-rays is largely due to their penetrating power.

(i) *Medicine.* Here, radiographs or X-ray photographs are used for a variety of purposes. Since X-rays can damage healthy cells of the human body, great care is taken to avoid unnecessary exposure. In radiography the X-ray film is sandwiched between two screens which fluoresce when subjected to a small amount of X-radiation. The film is affected by the fluorescent light rather than by the X-rays; formerly X-rays were used to sensitize the film directly and much longer exposures were required.

Suspected bone fractures can be investigated since X-rays of a certain hardness can penetrate flesh but not bone. In the detection of lung tuberculosis by mass radiography use is made of the fact that diseased tissue is denser than healthy lung tissue which consists of air sacs and so the former absorbs X-rays more strongly. When an organ is being X-rayed whose absorptive power is similar to that of the surrounding tissue, a 'contrast' medium is given to the patient, orally or by injection. This is less easily penetrated by the X-rays and enables a shadow of the organ to be obtained on a radiograph.

In the treatment of cancer by radiotherapy, very hard X-rays are used to destroy the cancer cells whose rapid multiplication causes malignant growth.

(*ii*) *Industry.* Castings and welded joints can be inspected for internal imperfections using X-rays. A complete machine may also be examined from a radiograph without having to be dismantled.

(*iii*) *X-ray crystallography.* The study of crystal structure by X-rays is now a powerful method of scientific research. The first crystals to be analysed were of simple compounds such as sodium chloride but in recent years the structure of very complex organic molecules has been unravelled.

X-ray diffraction

(*a*) *Bragg's law.* In von Laue's method of producing X-ray diffraction, the crystal is used as a transmission grating and interpretation of the patterns is not easy. Sir William Bragg and his son Sir Lawrence developed a simpler technique in which the crystal acts as a reflection diffraction grating.

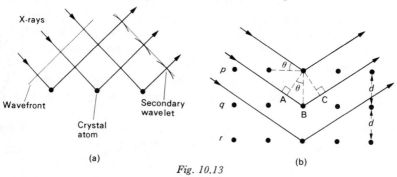

Fig. 10.13

When X-rays fall on a single plane of atoms in a crystal, each atom scatters a small fraction of the incident beam and may be regarded as the source of a weak secondary wavelet of X-rays. In most directions destructive interference of the wavelets occurs but in the direction for which the angle of incidence equals the angle of 'reflection', there is reinforcement and a weak reflected beam is obtained. The X-rays thus behave as if they were weakly ' reflected ' by the layer of atoms, Fig. 10.13a.

Other planes of atoms, such as p, q and r in Fig. 10.13b, to which the X-rays penetrate, behave similarly. The reflected beams from all the planes involved interfere and the resultant reflected beam is only strong if the path difference between successive planes is a whole number of wavelengths of the incident X-radiation. Thus reinforcement only occurs for planes p and q when

$$AB + BC = n\lambda$$

where n is an integer and λ is the wavelength of the X-rays. If d is the distance between planes of atoms and θ is the angle between the X-ray beam and the crystal surface, called the *glancing angle*, then $AB + BC = 2d \sin \theta$ and the reflected beam has maximum intensity when

$$2d \sin \theta = n\lambda.$$

This equation is Bragg's law; note that it refers to the glancing angle and not to the angle of incidence. Intensity maxima occur for several glancing angles. The smallest angle is given by $n = 1$ and is the first-order reflection; $n = 2$ gives the second order, and so on.

This process is sometimes termed X-ray ' reflection ' and although it may appear that reflection occurs it is in fact a diffraction effect, since interference takes place between X-rays from secondary sources, i.e. crystal atoms, on the *same* wavefront.

The atoms in a crystal can be considered to be arranged in several different sets of parallel planes, from all of which strong reflections may be obtained to give a pattern of spots characteristic of the particular structure.

A microwave analogue of X-ray ' reflection ' from the planes of ' atoms ' in a polystyrene ball crystal can be demonstrated. (See *Advanced Physics: Materials and Mechanics*, pp. 22–24.)

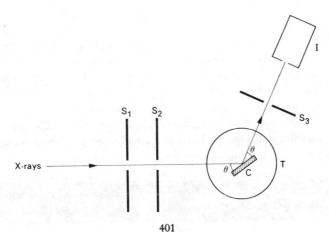

Fig. 10.14

(b) *X-ray spectrometer.* The Bragg X-ray spectrometer was developed to measure (i) X-ray wavelengths, and (ii) the spacing of atoms in crystals. The principle of the instrument is shown in Fig. 10.14. X-rays from the target of an X-ray tube are collimated by two slits S_1 and S_2 (made in lead sheets) and the narrow beam so formed falls on a crystal C set on the table T of the spectrometer. The reflected beam passes through a third slit S_3 into an ionization chamber I (p. 480) where it creates an ionization current which is a measure of the intensity of the reflected radiation.

As the crystal and the ionization chamber are rotated, the angle of reflection always being kept equal to the angle of incidence, the ionization current is found. Strong reflection occurs for glancing angles satisfying Bragg's law $2d \sin \theta = n\lambda$. Knowing either d or λ, the other can be calculated.

X-ray wavelengths are now measured directly using mechanically ruled diffraction gratings (similar to optical gratings) provided the X-rays strike the grating at a glancing angle much less than $1°$.

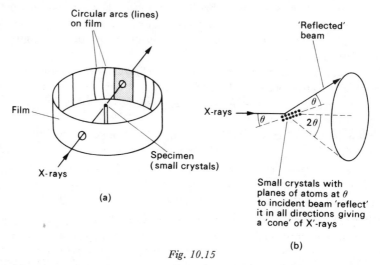

Fig. 10.15

(a)

(b)

Small crystals with planes of atoms at θ to incident beam 'reflect' it in all directions giving a 'cone' of X'-rays

(c) *X-ray powder photography.* If instead of a single crystal, a polycrystalline specimen or a crystalline powder is used, many planes are involved at once in X-ray ' reflection ' and thousands of spots are produced from ' reflection ' at all possible angles. As a result, circles or circular arcs are obtained on a suitably placed X-ray film. Fig. 10.15a shows the arrangement in the X-ray powder (or polycrystalline) technique using a strip of film and Fig. 10.15b indicates how the ' lines ' are formed. Figs. 10.16a and b are photographs for sodium chloride powder and a polycrystalline copper wire.

Fig. 10.16a

Fig. 10.16b

X-ray spectra and the quantum theory

(a) *Continuous and line spectra.* The radiation from an X-ray tube can be analysed with a spectrometer and an intensity wavelength graph obtained to show the spectral distribution. A typical X-ray spectrum is given in Fig. 10.17. It has two parts.

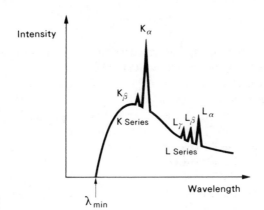

Fig. 10.17

(*i*) A *continuous* spectrum which has a definite lower wavelength limit, increases to a maximum and then decreases gradually in the longer wavelengths. All targets emit this type of radiation.

(*ii*) A *line* spectrum consisting of groups or series of two or three peaks of high-intensity radiation superimposed on the continuous spectrum. The series are denoted by the letters K, L, M etc., in order of increasing wavelength, and the peaks by α, β, γ. The wavelengths of the peaks are characteristic of the target element; all the series are not normally given by one element.

(b) *Quantum theory and the continuous spectrum.* Certain features of the continuous spectrum are readily explained by the quantum theory. Thus the existence of a definite *minimum* wavelength can be justified if we assume that this radiation consists of X-ray photons produced by electrons which have given up all their kinetic energy in a single encounter with a target atom. If such an electron has mass m and speed v on striking the target, the energy hf of the photon is given by

$$hf = \tfrac{1}{2}mv^2 \tag{1}$$

h being Planck's constant and f the frequency of the radiation. With a p.d. V

403

across the X-ray tube, an electron of charge e has work eV done on it by the electric field and so

$$eV = \tfrac{1}{2}mv^2. \tag{2}$$

From (1) and (2)

$$hf = eV. \tag{3}$$

The value of f given by (3) is the maximum frequency of the X-rays emitted at p.d. V, since all the energy of the electron is converted to the photon. The corresponding wavelength will have a minimum value, and if this is λ_{min} then $c = f\lambda_{min}$ where c is the speed of travel of X-rays. It follows that

$$\lambda_{min} = \frac{hc}{eV}. \tag{4}$$

As V increases we see from (4) that λ_{min} decreases, i.e. X-rays of higher frequency and greater penetrating power are emitted. The values of λ_{min} calculated from this equation agree with those found experimentally.

Most of the electrons responsible for the X-radiation usually have more than one encounter before losing all their energy. Several photons are produced with smaller frequencies than f and therefore with greater wavelengths than λ_{min}. Different electrons lose different amounts of energy and so a continuous spectrum covering a range of wavelengths is obtained. The great majority of electrons however lose their kinetic energy too gradually for X-rays to be emitted and merely increase the internal energy of the target.

(c) *Explanation of line spectra.* This will be considered later (p. 420).

(d) *Worked example. An X-ray tube operates at 30 kV and the current through it is 2.0 mA. Calculate (i) the electrical power input, (ii) the number of electrons striking the target per second, (iii) the speed of the electrons when they hit the target, and (iv) the lower wavelength limit of the X-rays emitted.*

(i) If V is the p.d. across the tube and I the tube current then

$$\text{power input} = VI$$

$$= (30 \times 10^3 \,\text{V})\,(2.0 \times 10^{-3}\,\text{A})$$

$$= 60 \,\text{W}.$$

(ii) Current through the tube is given by $I = ne$, where n is the number of electrons striking the target per second and e is the electronic charge (i.e. 1.6×10^{-19} C).

$$\therefore\ n = \frac{I}{e} = \frac{2.0 \times 10^{-3}\,\text{A}}{1.6 \times 10^{-19}\,\text{C}}$$

$$= 1.3 \times 10^{16}.$$

(*iii*) If m is the mass of an electron (i.e. 9.0×10^{-31} kg) and v its speed at the target, then from equation (2) above

$$\tfrac{1}{2}mv^2 = eV$$

$$\therefore \quad v = \sqrt{\frac{2eV}{m}}$$

$$= \sqrt{\frac{2 \times 1.60 \times 10^{-19} \times 30 \times 10^3}{9.0 \times 10^{-31}}}$$

$$= 1.0 \times 10^8 \text{ m s}^{-1}.$$

(*iv*) From equation (4) above, the lowest X-ray wavelength emitted is given by

$$\lambda_{\min} = \frac{hc}{eV}$$

where $h = 6.6 \times 10^{-34}$ J s and $c = 3.0 \times 10^8$ m s^{-1}

$$\therefore \quad \lambda_{\min} = \frac{(6.6 \times 10^{-34} \text{ J s}) (3.0 \times 10^8 \text{ m s}^{-1})}{(1.6 \times 10^{-19} \text{ C}) (30 \times 10^3 \text{ V})}$$

$$= 0.41 \times 10^{-10} \text{ m}.$$

Electrical conduction in gases

(*a*) *Ionized gases.* Gases at s.t.p. are very good electrical insulators and about 30 kV is required to produce a spark discharge between two rounded electrodes 1 cm apart in air; the value for pointed electrodes 1 cm apart is 12 kV. To conduct, a gas must be ionized.

Ionization occurs when an electron is removed from an atom (or molecule) by supplying a certain amount of energy to overcome the attractive force securing the electron to the atom. The two resulting charged particles form an *ion-pair*, the electron being a negative ion and the atom, now deficient of an electron, a positive ion. When a p.d. is applied between two electrodes in the ionized gas, positive ions move towards the cathode and electrons towards the anode. The ions thus act as charge carriers and current flows in the gas. If an atom gains an electron it becomes a heavy negative ion but these are generally few in number.

Various agents can ionize a gas and these include a flame, sufficiently energetic electrons, ultraviolet, X-rays and the radiation from radioactive substances, i.e. α-, β- and γ-rays. The few ions always present in the atmosphere are caused by cosmic rays from outer space and radioactive minerals in the earth. They are responsible for a charged, insulated body such as an electroscope, gradually discharging.

(*b*) *Current-p.d. relationship.* The gas between two parallel plates P and Q, Fig. 10.18*a*, is ionized by a beam of ionizing radiation. When a p.d. is applied the resulting ionization current is recorded by a sensitive current detector, e.g. a d.c. amplifier as described in Chapter 12. If the p.d. is varied while the intensity of the radiation remains constant, the current variation is shown by the curve OABCD in Fig. 10.18*b*.

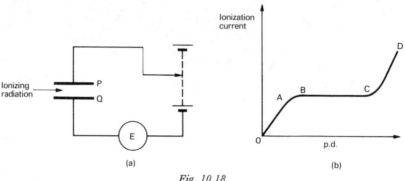

(a)

(b)

Fig. 10.18

The shape of the curve is explained as follows. Small p.d.s cause the electrons and positive ions to move slowly to the anode P and cathode Q respectively. On the way some ions recombine to form neutral atoms. As the p.d. increases, the ions travel more quickly and there is less opportunity for recombination; the ionization current increases. At B all the ions produced by the radiation reach the electrodes, no recombination occurs and further increase of p.d. between B and C does not affect the current. Along BC the current has its *saturation value* and is independent of the applied p.d.

Beyond C the current rises rapidly with p.d. and indicates that a new source of ion-pairs has become operative. The original ions due to the radiation are now accelerated sufficiently by the large p.d. to form new electrons and new positive ions by collision with neutral gas molecules between the plates. Thus along CD each original ion-pair creates several other ion-pairs, the process being known as *ionization by collision*. It can be shown that electrons are mainly responsible for this.

Two points should be noted:

(*i*) Ohm's law is obeyed between O and A

(*ii*) the saturation current is proportional to the rate of production of ions by the radiation and so measures the intensity of the radiation.

(*c*) *Gas discharge effects.* The conductivity of a gas increases as its pressure is reduced. Whereas a large p.d. is required at atmospheric pressure to produce a brief current in the form of a spark, by contrast at low pressure a smaller p.d. causes a steady current. The conduction of electricity by gases, called *gas discharge*, is accompanied by certain luminous effects as the pressure is reduced.

Fig. 10.19

(a)

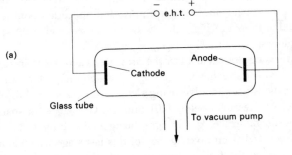

(b)   20 mm

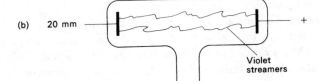

(c)   5 mm

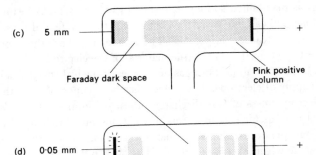

(d)   0·05 mm

(e)   0·01 mm
       (or less)

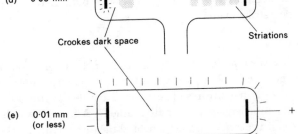

The passage of electricity through air can be studied at different pressures using a glass tube which is connected to a vacuum pump by a side tube, Fig. 10.19*a*. A p.d. of several thousand volts d.c. is applied between two electrodes at opposite ends of the tube, from an e.h.t. supply. The appearance of the discharge as air is removed is shown in Figs. 10.19*b* to *e*; other gases give the same stages but different colours.

Luminous effects do not occur until the pressure is about 20 mmHg and take the form of wavy violet streamers. The stage at which the pink positive column almost fills the tube is used in advertising signs; neon gives a red discharge and argon pale blue. At pressures below 0.01 mm the Crookes dark space fills the tube but the walls fluoresce, green light being emitted by soda glass. The electron was discovered from investigation of this last stage which alone will be considered here, since the full explanation of all the phenomena of the gas-discharge tube is complex.

A gas normally contains a few ions and in a discharge tube at low pressure these are accelerated by the applied p.d. towards the anode or cathode, depending on their signs. The positive ions bombard the atoms of the cathode and cause the emission of electrons. In their journey towards the anode some of the ejected electrons produce more positive ions by collision with neutral gas molecules they encounter. In this way fresh ions are made available for carrying the current and the discharge is maintained so long as residual gas is present. A beam of high-speed electrons, i.e. cathode rays, travels towards the anode and those striking the walls of the tube cause fluorescence. The cathode rays have the same properties whatever the gas in the tube or the material of the electrodes.

Although historically important, gas discharge is now little used as a means of obtaining a supply of electrons. The process has been superseded by thermionic and photoelectric emission. Gas-discharge tubes, sometimes called *cold cathode tubes*, are not only inefficient producers of electrons but they may also present (at cathode ray pressures) an X-ray hazard. In many cases they require an operating p.d. much in excess of 5 kV and at such p.d.s X-rays are produced when cathode rays strike matter.

Positive rays

(*a*) *Production and properties.* The explanation of the production of cathode rays in a discharge tube requires positive ions of the residual gas to travel towards the cathode. The resulting streams of such ions are called *positive rays* and their path is visible in a tube having a perforated cathode. A faint glow, whose colour depends on the gas, appears in the space behind the cathode, Fig. 10.20.

In addition to the sign of their charge, positive rays differ markedly from cathode rays.

(*i*) They are deflected to a much smaller extent than cathode rays by electric and magnetic fields. This is not surprising since, being ionized atoms and molecules, they are more massive.

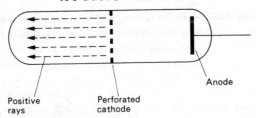

Fig. 10.20

Positive rays

Perforated cathode

Anode

(*ii*) Whereas most of the electrons in a beam of cathode rays have the same speed, positive rays exhibit a range of speeds. This arises from the fact that the positive ions comprising the rays are formed at various points between the anode and cathode and so are accelerated to different speeds before reaching the cathode.

(*iii*) They are related to the gas in the tube.

(*iv*) They cause fluorescence and affect a photographic film.

(*b*) *Thomson's experiment.* J. J. Thomson (in 1912) measured the specific charge of positive rays from their deflections in electric and magnetic fields. He concluded from his results that the value for a hydrogen ion was roughly two thousand times smaller than for an electron and, assuming both carry the same charge, the mass of the hydrogen ion must be about two thousand times greater than that of the electron.

By having other gases in the discharge tube he was able to compare the masses of different positive ions and this led to the discovery of *isotopes* (see below).

Terms and definitions

(*a*) *Atomic number* (proton number). After it had been established that the negatively charged electron was one of the basic constituents of all atoms, the search began for a positively charged counterpart. Positive rays were thoroughly investigated but the lightest particle to be detected was obtained with hydrogen in the discharge tube and its specific charge clearly indicated that it was a positive hydrogen ion. The latter, being the lightest and simplest positively charged particle then known, was called a *proton* (from the Greek *protos* meaning first) symbol p.

Subsequently, experiments on the bombardment of atoms by suitable high-speed particles revealed that in certain cases a proton is ejected from the nucleus of an atom (p. 499). This provided further evidence for the belief that protons are one of the fundamental particles from which atoms are made. The proton has a charge equal in magnitude but opposite in sign to that on the electron and its rest mass is 1836 times the rest mass of the electron.

The number allotted to an element in the periodic table was called its *atomic number*. The significance of this term was not apparent until the nuclear theory of the atom had been proposed and evidence obtained (p. 416) that the *atomic number, Z, of an element equals the number of protons in the nucleus.*

An atom is normally electrically neutral and so the atomic number is also the number of electrons in a neutral atom of the element. Hydrogen with $Z = 1$ has 1 proton and 1 electron while uranium with $Z = 92$ has 92 protons and 92 electrons.

(b) *Mass number* (nucleon number). Since the atomic number of an element is about half its relative atomic mass, it follows that if (as the nuclear theory supposes, p. 414) the mass of an atom is due to its nucleus, then there must be other constituents besides protons in the nucleus. The possibility of the existence of an electrically neutral particle, having a similar mass to that of the proton, was suggested in 1920. This particle, named the *neutron*, symbol n, remained undiscovered until 1932 (p. 501), partly on account of the difficulty of detecting a particle which, being uncharged, is not deflected by electric or magnetic fields and produces no appreciable ionization in its path.

Neutrons can also be expelled from certain nuclei by bombardment and there is now no doubt that they are basic constituents of matter. Each does not consist of a proton in close association with an electron but is an entity in itself. The rest mass of the neutron is 1839 times that of the electron. Protons and neutrons are called *nucleons*.

The mass number A of an atom is the number of nucleons in the nucleus.

The simplest nucleus is that of hydrogen which consists of 1 p; in symbolic notation it is written $_1^1\text{H}$, where the superscript gives the mass number and the subscript the atomic number. Helium has 2 p and 2 n, giving $A = 4$ and $Z = 2$ and symbol $_2^4\text{He}$. Lithium $_3^7\text{Li}$ has $A = 7$ and $Z = 3$ and its atom has 3 p and 4 n. In general, atom X is represented by $_Z^A\text{X}$. The neutron is written $_0^1\text{n}$ since it has $A = 1$ and zero charge, i.e. $Z = 0$; the proton can be written $_1^1\text{p}$ and the electron $_{-1}^0\text{e}$. Sometimes the atomic number is omitted and the name or symbol of the element is given followed by the mass number, e.g., lithium 7 or Li 7.

(c) *Isotopes.* If two atoms have the same number of protons but different numbers of neutrons, their atomic numbers are equal but not their mass numbers. Each atom is said to be an *isotope* of the other. They are chemically indistinguishable (since they have the same number of electrons) and occupy the same place in the periodic table.

Few elements consist of identical atoms: most are isotopic mixtures. Hydrogen has three forms: $_1^1\text{H}$ with 1 p, heavy hydrogen or deuterium $_1^2\text{D}$ with 1 p and 1 n and tritium $_1^3\text{T}$ with 1 p and 2 n. Ordinary hydrogen contains 99.99 per cent of $_1^1\text{H}$ atoms. Isotopes are not as a rule given separate names and symbols; exception is made in the case of hydrogen because there is an appreciable difference in the physical properties of the three forms. Water made from deuterium is called *heavy water* and is denoted by D_2O; it has density 1.108 g cm^{-3}, a freezing point of 3.82 °C and a boiling point of 101.42 °C. The nucleus of the deuterium atom is called a *deuteron*.

Isotopes account for fractional atomic masses. For example, chlorine with atomic mass 35.5 has two forms: $^{35}_{17}$Cl and $^{37}_{17}$Cl and these are present in ordinary chlorine in the approximate ratio of three atoms of the former to one of the latter. Chemically they are identical but one is slightly denser than the other.

Isotopes were discovered among the radioactive elements in 1906 but their nature was not understood. The first hint of their existence among non-radioactive elements was obtained in 1912 by Thomson with neon in his positive rays apparatus, when isotopes of atomic masses 20 and 22 were obtained. Although there are only 104 elements (89 in nature and 15 man-made), each one has isotopes and the total number known at present is about 1500. Of these about 300 occur naturally and the rest are artificial; all of the latter and some of the former are radioactive.

The term *nuclide* is used to specify an atom with a particular proton-neutron combination. $^{6}_{3}$Li and $^{7}_{3}$Li are isotopes and nuclides, $^{9}_{4}$Be and $^{10}_{5}$B are nuclides (they have the same number of neutrons, i.e. 5, but different numbers of protons).

Atomic mass: mass spectrograph

(*a*) *Atomic mass* (more correctly, *relative atomic mass*). This is denoted by A_r and was previously called the *atomic weight*. It is defined as the ratio

$$\frac{mass\ of\ an\ atom}{1/12\ mass\ of\ {}^{12}_{6}C\ atom}.$$

Formerly atomic masses were referred to the hydrogen atom as 1 and then to the oxygen isotope $^{16}_{8}$O as 16, but in 1960 physicists and chemists agreed to adopt the carbon 12 scale. This was chosen because in measuring atomic masses by mass spectroscopy (to be explained shortly), carbon 12 is a convenient standard for comparison since it forms many compounds. Furthermore, there are only two isotopes of carbon and their proportions vary very little in naturally-occurring carbon.

Atomic masses were first measured to a high degree of accuracy by Aston who established between 1919 and 1927 that (*i*) most elements exhibit isotopy, and (*ii*) most isotopic masses are very nearly but not quite whole numbers.

In Aston's apparatus the deflecting electric and magnetic fields were arranged so that all particles of the same mass, irrespective of their speed, were brought to a line focus. When ions of different masses were present, a series of lines, i.e. a mass spectrum, Fig. 10.21, was obtained on a photographic film. The relative

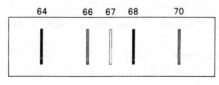

Fig. 10.21

intensities of the lines enabled an estimate to be made of the relative amounts of isotopes. Aston called his instrument a *mass spectrograph*.

(*b*) *Mass spectrograph.* Many types have been constructed for the accurate determination of atomic mass; the essential features of that due to Bainbridge are shown in Fig. 10.22.

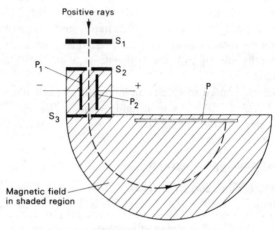

Fig. 10.22

Positive rays are produced by passing electrons from a hot cathode into the gas or vapour to be investigated. The positive ions so formed are directed through slits S_1 and S_2 and emerge as a narrow beam with a range of speeds and specific charges. In the region between S_2 and S_3 crossed uniform electric and magnetic fields are applied. The electric field between P_1 and P_2 exerts a force acting to the left on the positive ions. The magnetic field, which acts normal to and into the plane of the diagram, tends to deflect the ions to the right. If Q is the charge of an ion and E the electric field strength, the electric force is EQ. The magnetic force is BQv where v is the velocity of the ion and B the flux density. When the forces are equal

$$BQv = EQ$$

$$\therefore \quad v = \frac{E}{B}. \tag{1}$$

The ion will be undeflected and will emerge from the *selector system*, as S_2-S_3 is called, if its velocity equals the ratio E/B. All ions leaving S_3 thus have the same velocity v whatever their specific charge and *velocity selection* is said to have occurred.

Beyond S_3 only the magnetic field acts, the ions describe circular arcs and strike the photographic plate P. For particles of mass M, the radius r of the path is given by

$$BQv = \frac{Mv^2}{r}$$

$$\therefore \quad r = \frac{Mv}{BQ}. \tag{2}$$

From (1) and (2)

$$r = \frac{M}{Q} \cdot \frac{E}{B^2}.$$

If B and E are constant, r is directly proportional to M (assuming Q is the same for all ions). When ions with different masses are present each set produces a definite line and from their positions the masses can be found. In a mass spectrograph the masses of individual atoms are being measured; by contrast chemical methods give the average atomic mass for a large number of atoms.

The mass spectrograph, which uses photographic detection, is employed primarily for precise mass determinations and can measure the very small differences of mass occurring in nuclear reactions when one nuclide changes to another. The mass spectrometer uses an electrometer as a detector and gives the exact relative abundances, as in gas analysis. Such an instrument is shown in Fig. 10.23. The focusing magnet (here an electromagnet) is on the right surrounding the analyser tube from which cables can be seen going to the various control and recording instruments on the left. In this case the results of the analysis are displayed on a meter or a chart recorder.

Fig. 10.23

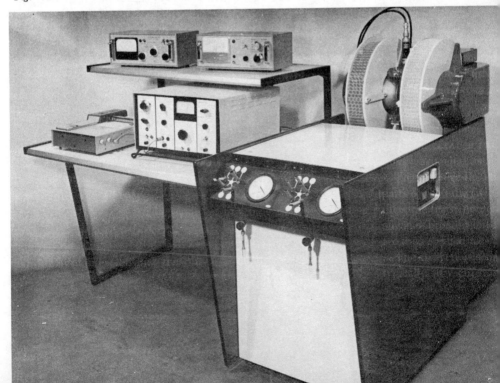

Nuclear model of the atom

(a) *Rutherford's nuclear atom.* The nuclear atom is the basis of the modern theory of atomic structure and was proposed by Rutherford in 1911. He had observed that the passage of alpha particles through a very thin metal foil was accompanied by some scattering of the particles from their original direction. Two of his assistants, Geiger and Marsden made a more detailed study of this effect. They directed a narrow beam of alpha particles on to gold foil about 1 μm thick and found that while most of the particles passed straight through, some were scattered appreciably and a very few—about 1 in 8000—suffered deflections of more than 90°. In effect they were reflected back towards the radioactive source.

To account for this very surprising result Rutherford suggested that *all the positive charge and nearly all the mass were concentrated in a very small volume or nucleus at the centre of the atom.* The large-angle scattering of alpha particles would then be explained by the strong electrostatic repulsion to which the alpha particles (also positively charged) are subjected on approaching closely enough to the tiny nucleus; the closer the approach the greater the scattering. We now believe that protons are responsible for the positive charge on the nucleus and protons and neutrons together for the nuclear mass.

Rutherford considered the electrons to be outside the nucleus and at relatively large distances from it so that their negative charge did not act as a shield to the positive nuclear charge when an alpha particle penetrated the atom. The electrons were supposed to move in circular orbits round the nucleus (like planets round the sun), the electrostatic attraction between the two opposite charges being the required centripetal force for such motion.

On this planetary model it would be reasonable to expect that many physical and chemical properties of atoms could be explained in terms of the number and arrangement of the electrons on account of their greater accessibility.

Fig. 10.24

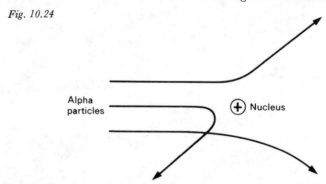

Alpha particles — Nucleus

(b) *Geiger-Marsden scattering experiment.* To test his theory Rutherford derived an expression for the number of alpha particles deflected through various angles. The derivation was complex but involved, among other factors, the charge on the nucleus of the scattering atom, the thickness of the foil, the charge, mass and speed of the bombarding alpha particles and was based on the assumption that the repulsive force between the two positive charges obeys an inverse square law. The path predicted for the scattered alpha particle was a hyperbola, Fig. 10.24.

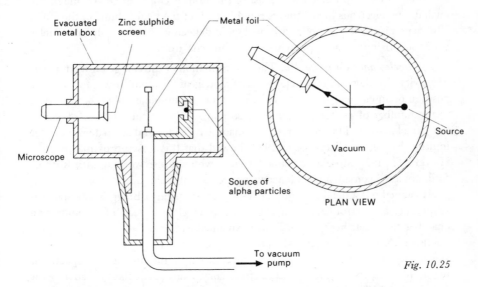

Fig. 10.25

The test was performed by Geiger and Marsden using the apparatus shown in Fig. 10.25. A fine beam of alpha particles from a radioactive source fell on a thin foil of gold, platinum or other metal in an evacuated box. The angular deflection of the particles was measured by using a microscope to observe the scintillations (flashes of light) on a glass screen coated with zinc sulphide. The screen and microscope could be rotated together relative to the foil and source, which were fixed.

Geiger and Marsden spent many hours in a darkened room counting the scintillations for a wide range of angles. Their results completely confirmed Rutherford's deductions and vindicated the use of the inverse square law. (We will see later however (p. 501), that in certain cases where the alpha particle approaches extremely closely to the nucleus the inverse square law no longer holds.)

This experiment represents one of the great landmarks in physics. As well as putting the nuclear model on a sound footing, it inaugurated the technique of using high-speed particles as atomic probes. The subsequent exploitation of the technique was responsible for profound discoveries in nuclear physics.

(c) *Nuclear size*. The maximum angle of scattering will occur when the distance between the centres of an alpha particle and the atomic nucleus involved in the encounter is a minimum. This distance gives an upper limit for the sum of the radii of an alpha particle and the nucleus. It is about 10^{-14} m, therefore the radius of the nucleus must be of the order of 10^{-15} m. With the reservation mentioned earlier about the *size* of atomic particles (p. 380), if the radius of the nucleus is compared with that of the atom (about 10^{-10} m), it is evident, since electrons are similar in size to nuclei, that most of the atom is empty. The volume of the nucleus and electrons in an atom is roughly 10^{-12} of the total volume of the atom. The penetration of thin foil, with negligible deflection, by most alpha particles is not surprising; close approaches to the nucleus are rare.

(*d*) *Moseley and X-ray spectra*. The determination of the charge on the nucleus was a vital problem in the development of models of the atom. It was felt that the position of an element in the periodic table, i.e. its atomic number Z, was equal to the number of protons in the nucleus and so also to the number of extra-nuclear electrons. The measurements made by Geiger and Marsden and later improved upon by Chadwick in 1920 showed that this was approximately true. However, in 1913 Moseley, a young physicist who was also working with Ruther-ford, obtained convincing evidence using a different technique.

He carried out a detailed study of X-ray spectra using a Bragg X-ray spectro-graph (i.e. a spectrometer with a photographic plate instead of an ionization chamber to detect the X-rays). It was explained earlier (p. 403) that in the pro-duction of X-rays, a characteristic line spectrum is superimposed on a continuous spectrum. Moseley measured the frequencies of the lines in the line spectra of nearly 40 elements and found that the frequency of any particular line, such as the K_α of Fig. 10.17 (p. 403), increased progressively from one element to the next in the periodic table. If f is the frequency of this line for a certain element, it is related to a number Z which always works out to be the atomic number, by the expression

$$\sqrt{f} = a(Z - b)$$

where a and b are constants for this particular line.

Moseley wrote, ' We have here a proof that there is in the atom a fundamental quantity which increases by regular steps from one element to the next.' He supported his belief by theoretical arguments that this fundamental quantity was the positive charge on the nucleus.

The Bohr atom

Rutherford's model of the atom, although strongly supported by evidence for the nucleus, is inconsistent with classical physics. An electron moving in a circular orbit round a nucleus is accelerating and according to electromagnetic theory it should therefore emit radiation continuously and thereby lose energy. If this happened the radius of the orbit would decrease and the electron would spiral into the nucleus. Evidently either this model of the atom or the classical theory of radiation requires modification.

In 1913, in an effort to overcome this paradox, Bohr, drawing inspiration from the success of the quantum theory in solving other problems involving radiation and atoms, made two revolutionary suggestions:

(*i*) Electrons can revolve round the nucleus only in certain *allowed orbits* and while they are in these orbits they do not emit radiation. An electron in an orbit has a definite amount of energy. It possesses kinetic energy because of its motion and potential energy on account of the attraction of the nucleus. Each allowed orbit is therefore associated with a certain quantity of energy, called the *energy of the orbit*, which equals the total energy of an electron in it.

(*ii*) An electron can 'jump' from one orbit of energy E_2 to another of lower energy E_1 and the energy difference is emitted as one quantum of radiation of frequency f given by Planck's equation $E_2 - E_1 = hf$.

By choosing the allowed orbits correctly Bohr was able to explain quantitatively why particular wavelengths appeared in the line spectrum of atomic hydrogen and this provided evidence for his ideas concerning how electromagnetic radiation originates in an atom.

Despite its considerable achievements the Bohr atom had certain shortcomings. First, it could not interpret the details of the optical spectra of atoms containing more than one electron. Second, the very arbitrary method of selecting allowed orbits had no theoretical basis and third, it involved quantities, such as the radius of an orbit, which could not be checked experimentally. Nevertheless great credit is due to Bohr for linking spectroscopy and atomic structure and for introducing quantum ideas into atomic theory.

Bohr's model of the atom has been superseded by a new theory based on *wave mechanics* in which there is no need to make assumptions because they give correct results. But whereas the Bohr atom was easily visualized and involved fairly simple mathematics, its successor is abstract and the mathematics more difficult.

For the present it is sufficient to say that while wave mechanics preserves the general idea of a hollow, nuclear atom it discards the Bohr picture of electrons moving in allowed orbits. However, the essential characteristic of Bohr's orbits, i.e. definite energy values, is retained.

Energy levels in atoms

Wave mechanics permits the electrons in an atom to have only certain energy values. These values are called the *energy levels* of the atom. They are not something we can observe in the usual sense but, as we will see shortly, there is fairly direct experimental evidence to support our belief in their existence. Any theory of atomic structure must be able to explain how they arise; the wave mechanical justification will be given later.

The levels can be represented by horizontal lines, arranged one above the other to form an energy level diagram (or a ladder, of unequally-spaced rungs),

417

each line indicating by its position a particular energy value. Every *atom* has a characteristic set of energy levels whose values can be found experimentally or calculated using wave mechanics. Whilst an electron is permitted to pass from one level to another by gaining or losing energy, it is not allowed to have an amount of energy that would put it between two levels. An atom can thus only accept ' parcels ' of energy of certain definite sizes, i.e. its energy is *quantized*.

All levels have negative energy values because the energy of an electron at rest outside an atom is taken as zero and when the electron ' falls ' into the atom, energy is lost as electromagnetic radiation (compare the loss of p.e. when a body falls in the earth's attractive gravitational field). In effect the electron is passing from a higher to a lower energy level and the lower the level the larger the negative energy value. The most stable or *ground state* of the atom is the condition in which every electron is in the lowest energy state available. An atom is in an *excited state* when an electron is in a state of energy above that of a state which is unoccupied.

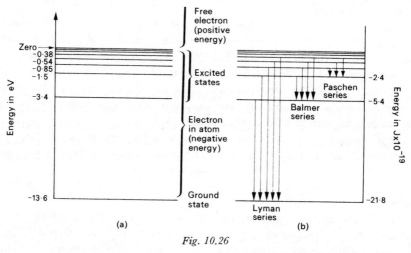

Fig. 10.26

Energy level diagrams can be drawn for every atom—that for atomic hydrogen is shown in Fig. 10.26a. In this case the lowest level has energy – 13.6 eV and is the one normally occupied by the single electron of a hydrogen atom. Above this state are the excited states to which the hydrogen atom may be raised by absorbing the correct amount of energy. If the energy absorbed is sufficient to allow the electron to escape from the atom, the latter becomes ionized; for hydrogen the ionization energy is 13.6 eV (21.8×10^{-19} J).

Electrons in the lower energy levels are held strongly by the nucleus. Those in higher levels need less energy to escape and are more loosely held in the atom. We may think of them as being near the outside of the electron cloud which surrounds the nucleus and partly screened from the attraction of the positive nuclear charge by inner, lower energy electrons.

Detailed study of the periodic variation of the chemical properties and the ionization energies (Fig. 10.29, p. 422) of the elements and of their optical and X-ray spectra suggests that the electrons in an atom fall into groups—called *shells*. All the energy states which the electrons in one group can occupy have *about* the same value. Each shell is given a *principal quantum number n* and they are labelled in ascending order of energy by the letters K, L, M etc.

Wave mechanics indicates that the number of electrons that can be accommodated in a shell is $2(n)^2$. Thus the K-shell ($n = 1$) can take at most $2(1)^2 = 2$ electrons, the L-shell ($n = 2$) can take a maximum of $2(2)^2 = 8$ electrons and so on. For the first five shells the numbers are 2, 8, 18, 32, 50. Normally the electrons occupy the lowest energy shells first, i.e. starting with K, but this is not always so. Thus whilst argon with 18 electrons has its full complement of 2 and 8 electrons in the K- and L-shells respectively and 8 electrons in the incomplete M-shell, in potassium (atomic number 19) the electronic configuration is 2–8–8–1, i.e. the extra electron is in the N-shell although the M-shell can take 10 more electrons.

Evidence of energy levels

(a) *Optical line spectra.* A line in an optical emission spectrum indicates the presence of a particular frequency (and wavelength) of light and is considered to arise from the loss of energy which occurs in an excited atom when an electron ' jumps ', directly or in stages, from a higher to a lower level. The frequency f of the quantum of electromagnetic radiation emitted in a transition between levels of energies E_1 and E_2 $(E_2 > E_1)$ is given by

$$E_2 - E_1 = hf$$

where h is Planck's constant. The transitions for some of the lines in the visible spectrum of atomic hydrogen, called the *Balmer series*, are shown in Fig. 10.26b. An electronic transition from the -2.4 J level to the -5.4 J level represents an energy loss to the atom of $[-2.4 - (-5.4)] \times 10^{-19} = 3.0 \times 10^{-19}$ J. The wavelength λ of the radiation emitted is given by

$$3.0 \times 10^{-19} = hf = \frac{hc}{\lambda}$$

where c is the speed of light $= 3.0 \times 10^8$ m s^{-1}. Hence

$$\lambda = \frac{hc}{3.0 \times 10^{-19}} = \frac{6.6 \times 10^{-34} \times 3.0 \times 10^8}{3.0 \times 10^{-19}} \text{ m}$$

$$= 6.6 \times 10^{-7} \text{ m}$$

$$= 0.66 \ \mu m.$$

The spectrum of atomic hydrogen contains a line of this wavelength (Fig. 8.36, p. 299). The lines of the *Lyman series* are in the ultraviolet and of the *Paschen* in the infrared. Thus, assuming the energy levels for hydrogen have been calculated from wave mechanics, the optical spectrum of hydrogen can be explained.

(*b*) *X-ray line spectra.* The energy of a light photon is a few electron-volts, that of an X-ray photon is several kiloelectron-volts. This suggests that whereas optical line spectra are due to transitions of loosely held electrons in higher energy levels, X-ray line spectra originate from electronic transitions in the lowest energy levels which have their full complement of electrons. They are only produced when materials are bombarded by high-energy electron beams (of the order of thousands of electron-volts) which are able to penetrate deep into atoms and displace electrons from very ' deep ' energy levels. The subsequent fall of an electron from a higher level into one of these gaps in an otherwise complete, but appreciably lower energy level, causes the emission of a high-energy X-ray photon.

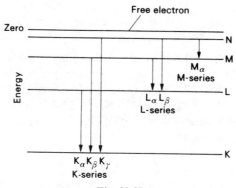

Fig. 10.27

The K-series of X-ray lines is produced when an electron is knocked out of the lowest or K-shell. The return of the same or another electron to the gap in the K-shell causes the emission of a line of the K-series. If the electron falls from the L-shell the K_α line occurs, when from the M-shell we have the K_β line and so on, Fig. 10.27. Similarly the L-series arises when electrons return to vacancies in the L-shell. This series is excited by smaller energies than the K-series because the L-electrons are less strongly held; the X-rays emitted have lower frequencies and longer wavelengths and are less penetrating.

The progressive increase of frequency with increasing atomic number, in say the K_α line, is due to small energy differences in the K- and L-shells of different atoms. This arises from the increasing positive charge on the nucleus and it makes possible the determination of atomic numbers from a study of X-ray line spectra (p. 416; Moseley).

(c) *Electron collision experiments.* Direct evidence for the existence of energy levels is provided by experiments, first successfully performed in 1914 by Franck and G. Hertz (nephew of H. Hertz), in which electrons have collisions with the atoms of a gas at low pressure. Whilst a free electron completely detached from an atom can be accelerated to any energy, these experiments lead to the conclusion that the electrons *in an atom* can have only *certain* values, i.e. those permitted by its energy levels.

When an electron has an encounter with a gas atom one of three things can happen:

(*i*) an elastic collision occurs in which the bombarding electron, being much less massive than the gas atom, suffers only a slight loss of kinetic energy (see p. 516, Qn. 10),

(*ii*) an inelastic collision occurs in which an electron in a gas atom gains *exactly* the amount of energy it requires to reach a higher energy level. Excitation has then occurred and the gas atom is in an excited state. The collision is inelastic since the kinetic energy lost by the bombarding electron does not reappear as kinetic energy of the gas atom but is emitted as electromagnetic radiation when the gas atom returns to a lower state or the ground state, usually after about 10^{-9} s. The p.d. through which the bombarding electron has to be accelerated from rest to cause excitation is called the *excitation potential* and since every atom has many energy levels, there are numerous excitation potentials characteristic of a particular atom,

(*iii*) an inelastic collision occurs in which an electron in a gas atom gains enough energy to escape from the atom, so causing ionization. The accelerating p.d. is then the *ionization potential* of the atom.

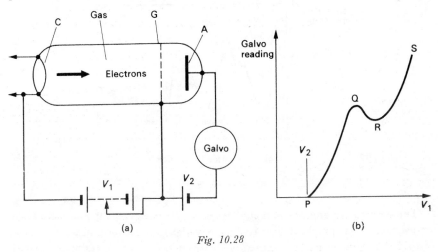

Fig. 10.28

The principle of a Franck-Hertz type of experiment is illustrated in Fig. 10.28a. Electrons emitted by the hot cathode C in a tube containing gas at a low pressure are accelerated by a positive potential V_1 on the wire mesh G, called the grid. When V_1 just exceeds the small negative potential V_2 which the anode A has with respect to G, electrons reach A and a small current is indicated on the galvanometer. As V_1 is increased the current increases. During this phase, PQ in Fig. 10.28b, the bombarding electron energies are small and all collisions between electrons and gas atoms are elastic. At a certain value of V_1, called the *first excitation potential*, some electrons have inelastic collisions with gas atoms near G and raise them to their first excited energy level. The negative potential V_2 ensures that the electrons, having lost their kinetic energy, are carried back to G. The anode current therefore falls abruptly, shown by QR. Further increase of V_1 allows even those electrons which have inelastic collisions to overcome V_2 and reach A. The galvanometer reading increases again, RS.

Other effects may be obtained at higher values of V_1 depending on the tube and circuit conditions. Thus it is possible to find the ionization potential of the gas, i.e. the potential at which the bombarding electrons have enough energy to cause gas molecules to be ionized.

Collision experiments, as just outlined, indicate that an atom cannot absorb any amount of energy but only certain definite amounts which are determined by its energy levels.

Franck-Hertz type tubes that give reliable results are difficult to make and so are expensive. An experiment with a commercial tube containing the inert gas xenon is described in Appendix 11 (p. 540) and gives rough values of the ionization potential and one excitation potential.

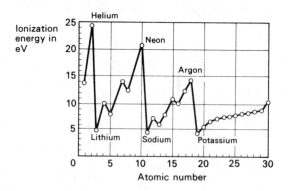

Fig. 10.29

The ionization potentials and energies of many gases and vapours have been investigated. The graph of Fig. 10.29 shows the variation of ionization energy with atomic number for the first thirty elements in the periodic table. A periodicity, like that displayed by other properties, is evident. The small ionization energies of the alkali metals suggests they have a loosely-held electron in a higher energy level. By contrast the inert gases must have very stable electronic structures since they require most energy for ionization.

(d) *Spectrum of mercury vapour.* Electron collision experiments with mercury vapour show that there is an energy difference of 4.9 eV (7.8 × 10⁻¹⁹ J) between two of the energy levels in an isolated mercury atom. We might therefore expect photons with this energy to be emitted by a mercury vapour lamp. The wavelength of such radiation is 2.5 × 10⁻⁷ m, i.e. 0.25 μm (check this); it is in the ultraviolet region as may be shown using the arrangement of Fig. 10.30a (*in which the lamp should be well-screened from observers*). The ultraviolet lines A and B in Fig. 10.30b appear only on the fluorescent paper (the visible lines are also on the white paper) and are cut off if a piece of glass is held in front of the lamp. Assuming the

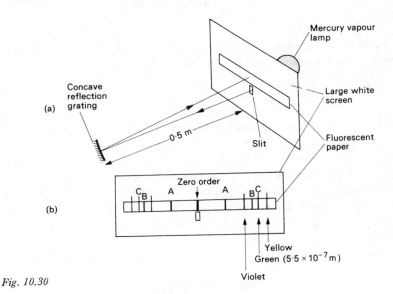

Fig. 10.30

wavelength of the green line C is 5.5 × 10⁻⁷ m (it can be measured readily using a transmission diffraction grating of known spacing), the wavelength λ of the ultraviolet line A is given approximately by

$$\frac{\lambda}{5.5 \times 10^{-7}} = \frac{AA}{CC}.$$

If *cold* mercury vapour is puffed *gently* into the air from a polythene bottle held in front of the reflection grating, both A and B fade or disappear. What does this suggest about B? *N.B. Mercury vapour is highly poisonous and on no account must the mercury be warmed.*

Wave-particle duality of matter

The wave-particle nature of electromagnetic radiation discussed previously (p. 392), led de Broglie (pronounced ' de Broy ') to suggest in 1923 that matter might also exhibit this duality and have wave properties. His ideas can be expressed quantitatively by first considering radiation. A photon of frequency f and wavelength λ has, according to the quantum theory, energy $E = hf = hc/\lambda$ where c is the speed of light and h is Planck's constant. By Einstein's energy-mass relation (p. 503) the equivalent mass m of the photon is given by $E = mc^2$. Hence

$$\frac{hc}{\lambda} = mc^2$$

$$\therefore \quad \lambda = \frac{h}{mc}.$$

By analogy de Broglie suggested that a particle of mass m moving with speed v behaves in some ways like waves of wavelength λ given by

$$\lambda = \frac{h}{mv}.$$

Calculation shows that electrons accelerated through a p.d. of about 100 V should be associated with *de Broglie* or *matter waves*, as they are called, having a wavelength of the order of 10^{-10} m. This is about the same as for X-rays and it was suggested that the conditions required to reveal the wave nature of X-rays might also lead to the detection of electron waves. At first de Broglie's proposal was no more than speculation, but within a few years a variety of experiments proved beyond dispute that moving particles of matter had wave-like properties associated with them.

Interference and diffraction patterns can be obtained with electrons. Fig.

Fig. 10.31 a and b

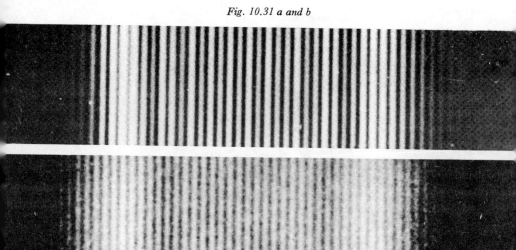

10.31 shows interference fringes produced by Young's double-slit type experiments (a) with light and (b) with a stream of electrons. An arrangement for producing electron interference is given in Fig. 10.32a. The stream of electrons is split into two by a very thin wire between two metal plates. When the wire is made a few volts positive with respect to the plates, the electron streams passing on each side of the wire are brought closer together and overlap on the photographic film as if they had come from two sources, Fig. 10.32b. The interference pattern is so small that it has to be magnified by an electron microscope and then by an optical enlarger to give fringes of the width shown in Fig. 10.31b.

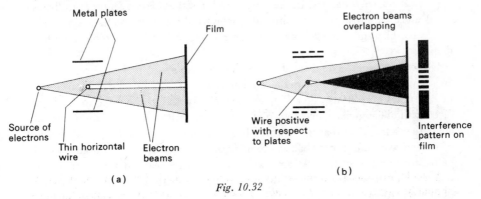

Fig. 10.32

Electron diffraction may be shown in a school laboratory using the arrangement of Fig. 10.33 in which a beam of electrons strikes a thin film of graphite on a metal grid just beyond a hole in the anode. Diffraction effects have been obtained with streams of protons, neutrons and alpha particles, but it is evident from de Broglie's equation that the greater the mass of the moving particle, the smaller is the associated wavelength and so the more difficult does detection become. Ordinary objects have extremely small wavelengths.

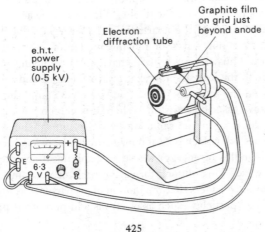

Fig. 10.33

The idea of matter waves is useful when we consider the various kinds of microscopes. The resolving power of any microscope increases as the wavelength of whatever is used to 'illuminate' the object decreases. Thus electron waves have a smaller wavelength than light waves and so an electron microscope reveals much more detail. The field ion microscope gives even greater resolution because the waves associated with the helium ions used have an even shorter wavelength.

Wave mechanics

(a) *Wave-particle duality*. In the early decades of the present century physicists faced an apparently puzzling situation. How could matter and radiation have *both* wave-like and particle-like properties? It seemed to be a complete contradiction.

However, to some extent the dilemma was of man's own making for originally an attempt was made to explain this dual behaviour in terms of the real waves and real particles of everyday life. This was a perfectly reasonable thing to do when faced with phenomena which appeared to have similarities. But perhaps it was not altogether surprising that ideas which helped to deal with the large-scale world of water waves and cricket balls were not completely suited to the atomic world of photons and electrons. At any rate it came to be realized that part of the dilemma, at least, arose from the use of models based on our experience of the macroscopic and that such analogies might not be valid when pushed too far.

It is now appreciated that a scientific model is an aid to understanding and not necessarily a true description. In the same way a map showing only towns and roads is no more a complete representation of a country than is one showing only physical features; each describes one aspect and neither is wrong. Both models are necessary for an adequate description of the behaviour of matter and radiation: they are complementary and not contradictory. The sensible thing to do is to use one or other model when it is appropriate and helpful.

Such arguments help to make the duality dilemma more acceptable but they do not explain the nature of the 'particles' and 'waves'. The present view is that the 'waves' are not waves of moving matter or varying fields but are probability waves. Thus the wave-like behaviour of a moving electron is considered to be due to the fact that the chance of it being found in an element of volume depends on the intensity (i.e. the square of the amplitude) at the point of the electron wave associated with it. Where this is high we are likely to have a high electron density. In the same way an electromagnetic wave, although undoubtedly connected with electric and magnetic fields, is regarded as consisting of photons (i.e. bundles of energy) whose probable locations are given by the intensity of the wave. This outlook forms the basis of wave mechanics which, as the name implies, considers that the motion of a 'particle' is determined by a wave equation. For 'particles' such as electrons and photons it predicts quite a different behaviour from that given by the particle mechanics of Newton but agrees with the latter for macroscopic bodies. It deals in probabilities and not certainties.

(b) *Wave-mechanical model of the atom.* The real objection to the Bohr model of the atom is that it pin-points an electron in a definite orbit and takes no account of its wave-like aspect. The wave-mechanical model requires advanced mathematics but we can attempt to explain it by making analogies with real waves.

In a vibrating string fixed at both ends, waves are reflected to and fro and a stationary wave system is set up having nodes and antinodes. In its simplest mode of vibration the string has a node at each end and an antinode in the centre but it can vibrate with two, three or any integral number of loops. The vibration is restricted to those modes having a complete number of loops and each mode produces a note of different frequency.

Similarly we can imagine an electron, confined to an atom by the attractive force exerted on it by the nucleus, being associated with electron waves which are reflected when they reach the boundary of the atom. A stationary wave system is established inside the atom and can vibrate in various modes. Each mode corresponds to a particular frequency and, therefore, to a particular energy level of the atom. When an electron wave changes its mode of vibration, the energy difference between the levels is emitted (or absorbed) as radiation. This argument, although much too realistic, enables us to see how wave mechanics can account for the existence of energy levels.

QUESTIONS

Electrons

1. What is an *electron-volt*? Assuming that the charge on an electron is -1.60×10^{-19} C express one electron-volt in terms of another unit.

Calculate the kinetic energy and velocity of protons after being accelerated from rest through a potential difference of 2.00×10^5 V. (Assume that the mass of a proton = 1.67×10^{-27} kg.) (*J.M.B.*)

2. How may cathode rays be produced? What are their chief properties?

An electron starts from rest and moves freely in an electric field whose intensity is 2.4×10^3 V m^{-1}. Find (a) the force on the electron, (b) its acceleration and (c) the velocity acquired in moving through a potential difference of 90 V. (The charge on an electron = 1.6×10^{-19} C and the mass of an electron = 9.1×10^{-31} kg.)

(*W.*)

3. What do you understand by an *electron*?

Electrons in a certain cathode-ray tube are accelerated through a potential difference of 2.0 kV between the cathode and the screen. Calculate the velocity with which they strike the screen. Assuming they lose all their energy on impact and given that 10^{12} electrons pass per second, calculate the power dissipation. (Charge on the electron $= 1.6 \times 10^{-19}$ C; mass of electron $= 9.1 \times 10^{-31}$ kg.)

(O. and C. part qn.)

4. Describe a method for measuring the ratio of the charge to mass (e/m) for an electron.

Calculate (a) the speed achieved by an electron accelerated in a vacuum through a p.d. of 2.00×10^3 V and (b) the magnetic flux density required to make an electron travelling with speed 8.00×10^6 m s^{-1} traverse a circular path of diameter 10.0×10^{-2} m. Take e/m for an electron as 1.76×10^{11} C kg^{-1}.

5. Give an account of a method by which the charge associated with an electron has been measured.

Taking this electronic charge to be -1.60×10^{-19} C, calculate the potential difference in volts necessary to be maintained between two horizontal conducting plates, one 5.00×10^{-3} m above the other, so that a small oil drop, of mass 1.31×10^{-14} kg with two electrons attached to it, remains in equilibrium between them. Which plate would be at the positive potential? ($g = 9.81$ N kg^{-1}) (L.)

6. Two plane parallel conducting plates 1.50×10^{-2} m apart are held horizontally, one above the other, in air. The upper plate is maintained at a positive potential of 1.50×10^3 V while the lower plate is earthed. Calculate the number of electrons which must be attached to a small oil drop of mass 4.90×10^{-15} kg if it remains stationary in the air between the plates. (Assume that the density of air is negligible in comparison with that of oil.)

If the potential of the upper plate is suddenly changed to -1.50×10^3 V what is the initial acceleration of the charged drop? Indicate, giving reasons, how the acceleration will change. (The charge of an electron is 1.60×10^{-19} C and g is 9.81 N kg^{-1}.)

(J.M.B.)

Photoelectric effect

Speed of light $= 3.0 \times 10^8$ m s^{-1}
Planck's constant $= 6.6 \times 10^{-34}$ J s
Mass of electron $= 9.1 \times 10^{-31}$ kg
Electronic charge $= 1.6 \times 10^{-19}$ C

7. (a) If a surface has a work function of 3.0 eV, find the longest wavelength light which will cause the emission of photoelectrons from it.

(b) What is the maximum velocity of the photoelectrons liberated from a surface having a work function of 4.0 eV by ultraviolet radiation of wavelength 0.20 μm?

8. When light of wavelength 0.50 μm falls on a surface it ejects photoelectrons with a maximum velocity of 6.0×10^5 m s^{-1}. Calculate (a) the work function in electron-volts, and (b) the threshold frequency for the surface.

9. What do you understand by the *quantum theory?* Describe the evidence provided by experiments on the photoelectric effect in favour of this theory.

Calculate the energy of the photons associated with light of wavelength 3.0×10^{-7} m. Give your answer in joules. (Planck's constant $= 6.6 \times 10^{-34}$ J s; speed of light $= 3.0 \times 10^8$ m s^{-1}.) (*C.*)

10. When light is incident in a metal plate electrons are emitted only when the frequency of the light exceeds a certain value. How has this been explained?

The maximum kinetic energy of the electrons emitted from a metallic surface is 1.6×10^{-19} J when the frequency of the incident radiation is 7.5×10^{14} Hz. Calculate the minimum frequency of radiation for which electrons will be emitted. Assume that Planck's constant $= 6.6 \times 10^{-34}$ J s. (*J.M.B.*)

X-rays

11. If the p.d. applied to an X-ray tube is 20 kV, what is the speed with which electrons strike the target? Take the specific charge of the electron as 1.8×10^{11} C kg^{-1}.

12. Describe the modern hot cathode X-ray tube, and give a diagram of a circuit suitable for its operation.

Discuss briefly the energy conversions that take place in the tube.

If the p.d. across the tube is 1.5×10^3 V, and the current 1.0×10^{-3} A, find (*a*) the number of electrons crossing the tube per second, and (*b*) the kinetic energy gained by an electron traversing the tube without collisions. (Take the electronic charge *e* to be 1.6×10^{-19} C.) (*O.*)

13. What is the minimum wavelength of the X-rays produced when electrons are accelerated through a potential difference of 1.0×10^5 V in an X-ray tube? Why is there a minimum wavelength? ($e = 1.6 \times 10^{-19}$ C; $h = 6.6 \times 10^{-34}$ J s ; $c = 3.0 \times 10^8$ m s^{-1}.)

Structure of atom

14. The atomic nucleus may be considered to be a sphere of positive charge with a diameter very much less than that of the atom. Discuss the experimental evidence which supports this view. Describe briefly how this experimental evidence was obtained. (*J.M.B.*)

15. How are X-rays produced, and how are their wavelengths determined?

Discuss briefly the origin of the lines in the spectrum produced by an X-ray tube that are characteristic of the target metal.

Give a brief account of Moseley's work and the part it played in establishing the idea of atomic number. (*O.*)

16. What are the chief characteristics of a line spectrum? Explain how line spectra are used in analysis for the identification of elements.

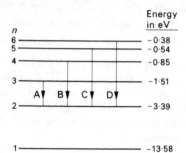

Fig. 10.34

Fig. 10.34, which represents the lowest energy levels of the electron in the hydrogen atom, specifies the value of the principal quantum number n associated with each state and the corresponding value of the energy of the level, measured in electron-volts. Work out the wavelengths of the lines associated with the transitions A, B, C and D marked in the figure. Show that the other transitions that can occur give rise to lines that are either in the ultraviolet or the infrared regions of the spectrum. (Take 1 eV to be 1.6×10^{-19} J; Planck's constant h to be 6.5×10^{-34} J s; and c, the velocity of light in vacuo, to be 3.0×10^8 m s^{-1}.) (O.)

17. Some of the energy levels of the mercury atom are shown in Fig. 10.35.

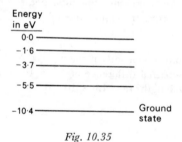

Fig. 10.35

(a) How much energy in electron-volts is required to raise an electron from the ground state to each of the levels shown?

(b) What is the ionization energy of mercury?

(c) If a mercury vapour atom in the ground state has a collision with an electron of energy (i) 5 eV and (ii) 10 eV, how much energy might be retained by the electron in each case?

(d) What would happen to a photon of energy (i) 4.9 eV and (ii) 8 eV, which has a collision with a mercury atom?

11 Electronics

Introduction

The striking advances in electronics during recent years have been due to the development of semiconductor devices such as junction diodes, transistors and integrated circuits. On account of their small size, low power requirements and very long life they have largely replaced thermionic valves in modern electronic equipment. They are, however, damaged by too high temperatures and p.d.s that are too great or of the wrong polarity.

In this chapter the main emphasis will be on semiconductors but two obsolescent thermionic devices—diode and triode valves—will be discussed briefly since there is still equipment in use containing them. First, however, we shall consider the cathode-ray oscilloscope—an important tool in electronics and a thermionic device.

Cathode-ray oscilloscope

The cathode-ray oscilloscope (CRO) consists of a cathode-ray tube and associated circuits. It has four main parts, as described in (a) to (d) below.

(a) *Electron gun*. This is an electrode assembly for producing a narrow beam of cathode rays. A typical CRO tube is shown in Fig. 11.1 connected to a potential divider which supplies appropriate voltages from an e.h.t. power supply.

The gun comprises an indirectly heated cathode C, a cylinder G called the *grid* and two anodes A_1 and A_2. G has a negative potential (or bias) with respect to C and controls the number of electrons reaching A_1 from C. R_1 determines the potential of G and thus acts as a *brightness* control.

A_1 and A_2 are metal discs or cylinders with central holes and both have positive potentials relative to C, A_2 more so than A_1. They accelerate the electrons to a high speed down the tube (which is highly evacuated) and their shapes and potentials are such that the electric fields between them focus the beam to a small

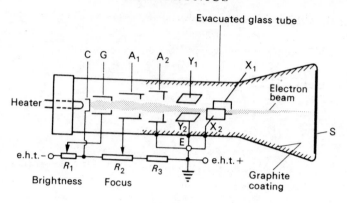

Fig. 11.1

spot on the fluorescent screen S at the end of the tube. Adjustment of the potential of A_1 by R_2 provides a *focus* control. A_1 and A_2 are an ' electron lens ' system. Typical potentials for a small tube are 1000 V for A_2, 200 to 300 V for A_1 and -50 V to zero for G.

There must be a return path to A_2 (and hence e.h.t.+) for electrons reaching S. Its provision depends on an effect known as *secondary emission* in which electrons, called *secondary electrons*, are emitted from a surface struck by high-speed electrons. The secondary electrons from S are collected by a coating of graphite on the inside of the tube which is connected to A_2. S gains bombarding electrons but loses about the same number of secondary ones and has a potential roughly equal to that of A_2. The electron beam therefore travels with *constant speed* between A_2 and S. In addition, the coating shields the beam from external electric fields. A mumetal screen round the tube protects it from stray magnetic fields.

To prevent earthed objects (e.g. the experimenter) near the screen affecting the beam, A_2 (and the coating) are usually earthed. This means e.h.t.+ is also at earth potential, Fig. 11.1, and the other electrodes become negative with respect to A_2. The electrons are still accelerated through the same p.d. between A_2 and C.

(*b*) *Deflecting system.* The beam from A_2 passes first between a pair of horizontal metal plates, the Y-plates whose electric field causes vertical deflection and then between two vertical plates, the X-plates that cause horizontal deflections when a p.d. is applied to them.

To minimize the effect of electric fields between the deflector plates and other parts of the tube (which might cause defocusing), one of each pair of plates is at the same (earth) potential as A_2. In practice X_2, Y_2 and A_2 are connected internally and brought out to a single terminal marked E (for earth). Deflecting p.d.s are then applied to Y_1 (marked Y or input) and E or to X_1 (marked X) and E.

The deflection sensitivity for p.d.s applied directly to the deflector plates is typically 50 V cm^{-1}, which is not high. X and Y deflection amplifiers are usually built into the CRO to amplify p.d.s that are too small to give measurable deflections before they are connected to the plates.

X and Y shift controls are used to move the spot ' manually ' in the X or Y directions respectively. They apply a positive or negative voltage to one of the deflecting plates according to the shift required.

(c) *Fluorescent screen.* The inside of the wide end of the tube is coated with a *phosphor* which emits light when struck by fast-moving particles. This may occur as fluorescence, i.e. the emission stops with the bombardment, and phosphorescence, when there is ' afterglow ', possibly for a few seconds. Zinc sulphide is commonly used in cathode-ray tubes; it emits blue light and has no afterglow.

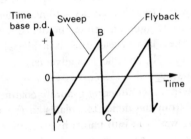

Fig. 11.2

(d) *Time base.* When the variation of a quantity with time is to be studied, an alternating p.d. representing the quantity is applied to the Y-plates via the vertical or Y-amplifier and the X-plates are connected via the horizontal or X-amplifier to a circuit in the CRO, called the *time base*, which generates a sawtooth p.d. like that in Fig. 11.2. The ' sweep ' AB must be linear so that the deflection of the beam is proportional to time and causes the spot to travel across the screen from left to right at a speed which is steady but which can be varied by the time base control(s) to suit the frequency of the alternating input p.d.

To maintain a stable trace on the screen each horizontal sweep must start at the same point on the waveform being displayed. This is done by feeding part of the input signal to a trigger circuit that gives a pulse (at a chosen point on the input signal selected by the TRIG LEVEL control) which is used to start the sweep of the time base sawtooth, i.e. it ' triggers ' the time base and initiates the horizontal motion of the spot at the left-hand side of the screen. However, automatic triggering by the input can be obtained (and is used for most applications) by selecting the AUTO position on the TRIG LEVEL control.

A trace appears only during the sweep part of the sawtooth p.d. since when the time base is not triggered by an input signal, the electron beam, in one arrangement, is deflected so that it does not reach the screen. The deflection is achieved by a set of plates (not shown in Fig.11.1) at suitable potentials immediately beyond A_1. When the time base is triggered a positive ' unblanking ' pulse, derived from the time base, is applied to the set of plates and allows the electrons to travel along the tube to the screen.

(e) *Mono and PDA tubes.* Two desirable properties of a CRO are high light output from the phosphor (especially for viewing high frequency signals) and good deflection sensitivity. The former requires the electrons to be accelerated through a large p.d. so that they strike the screen at high speed. The latter is attained when the electrons travel slowly between the deflecting plates (see p. 383).

The cathode-ray tube described above is a monoaccelerator type (*mono* for short) in which the electrons are accelerated to their final high speed in the electron gun, i.e. *before* they reach the deflection plates; this is not conducive to high deflection sensitivity. In the post-deflection acceleration tube (PDA for short), the electrons leave the gun and enter the deflection region at a comparatively low speed and the main acceleration is achieved *after* deflection. The deflection factor is, as a result, affected less adversely than in a mono tube with the same overall accelerating p.d.

A mono tube usually has a conducting graphite coating (as stated earlier) on the inside of the tube from the deflection plates to the screen. In some early PDA tubes this coating was split into bands by insulating gaps, and each band was at a different potential. In many present-day PDA tubes a continuous helically-wound resistive material is used instead of the graphite bands as the final accelerating ' anode '.

Fig. 11.3*a* is a block diagram of a CRO; Figs. 11.3*b* and *c* show single and double beam CROs respectively.

Some uses of the CRO

(a) *As a voltmeter.* The CRO can be used as a d.c./a.c. voltmeter by connecting the p.d. to be measured across the Y-plates, with the time base off and the X-plates earthed. The spot is deflected by d.c.; a.c. causes it to move up and down and if the motion is fast enough (e.g. at 50 Hz) a vertical line is observed.

In the CRO of Fig. 11.3*b* the ' gain ' control (marked VOLTS/CM) of the Y-deflection amplifier is calibrated. If it is on the ' 50 ' setting a deflection of 1 cm would be given by a 50-V d.c. input; a line 1 cm long would be produced by an a.c. input of 50 V peak-to-peak (i.e. peak voltage $V_0 = 25$ V and r.m.s. voltage $V_{\text{r.m.s.}} = V_0/\sqrt{2} = 18$ V). On the ' 0.1 ' setting the gain of the amplifier is greatest and small p.d.s can be measured, 0.1 V causing a deflection of 1 cm. (The calibration can be checked by applying known input p.d.s.) The use of the CRO as a voltmeter is shown in Fig. 4.6. (p. 112) and Fig. 4.29 (p. 131).

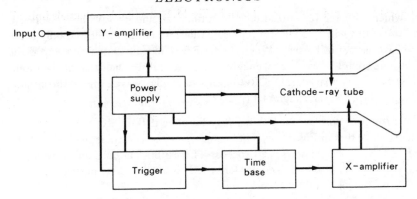

Fig. 11.3a

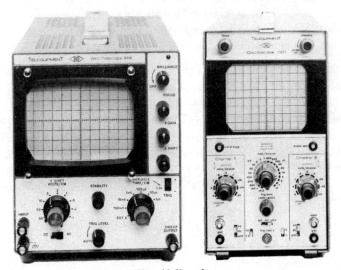

Fig. 11.3b and c

Among the advantages of the CRO as a voltmeter are:

(*i*) the electron beam behaves as a pointer of negligible inertia, responding instantaneously and having a perfect ' dead-beat ' action,

(*ii*) direct and alternating p.d.s can be measured, the latter at frequencies of several megahertz,

(*iii*) it has an almost infinite resistance to d.c. and a very high impedance to a.c. and so the circuit to which connection is made is little affected, and

(*iv*) it is not damaged by overloading.

(*b*) *Displaying waveforms.* The signal to be examined is connected to the Y-plates and the time base to the X-plates. As the spot is drawn horizontally across the screen by the time base, it is also deflected vertically by the alternating signal p.d. which is therefore ' spread out ' on a time axis and its waveform displayed. The waveform will be a faithful representation, free from distortion, only if the time base has a linear sweep. When the time base has the same frequency as the input, one complete wave is formed on the screen; if it is half that of the input, two waves are displayed. The CRO is used to display waveforms in the circuits of Fig. 5.34 (p. 184) and Fig. A4.1 (p. 532).

(c) *Measuring time intervals.* This may be done if the CRO has a calibrated time base like that in Fig. 11.3*b*. The coarse time-base control is marked TIME/ CM and gives six pre-set sweep speeds. The calibration only holds when the fine time-base control (marked VARIABLE) is in the CAL position and the X-GAIN control is fully anticlockwise.

If the TIME/CM control is on 10 ms, the spot takes 10 ms to move 1 cm and so travels 10 cm on the screen graticule in 100 ms (0.1 s). How could the calibration of the time base be checked using a 50–Hz a.c. mains signal?

A CRO was used to estimate a very small time interval in the measurement of the speed of sound in air and in a metal rod (pp. 252 and 255).

(d) *Measurement of phase relationships.* If two sinusoidal p.d.s of the same frequency and amplitude are applied simultaneously to the X- and Y-plates (time base off), the electron beam is subjected to two mutually perpendicular simple harmonic motions and the trace produced on the screen by the spot depends on the phase difference between the p.d.s. Fig. 11.4 shows the traces obtained using 50-Hz supplies (of a few volts) from, say, a signal generator and a step-down transformer. In general, the resultant motion of the spot is an ellipse except when the phase difference is 0°, 90° or 180°.

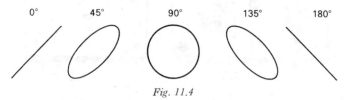

$$0° \qquad 45° \qquad 90° \qquad 135° \qquad 180°$$

Fig. 11.4

The method can be used to find the phase relationships between p.d.s and currents in a.c. circuits if a double beam CRO is not available.

(e) *Comparison of frequencies.* When two p.d.s of different frequencies f_x and f_y are applied to the X- and Y-plates (time base off), more complex figures than those in Fig. 11.4 (known as *Lissajous' figures*) are obtained. Some are shown in Fig. 11.5. In any particular case the frequency ratio can be found from inspection by imagining a horizontal and a vertical line to be drawn at the top and side of the trace. It is given by

$$\frac{f_y}{f_x} = \frac{\text{number of loops touching horizontal line}}{\text{number of loops touching vertical line}}.$$

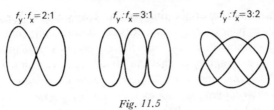

$f_y:f_x=2:1$ $f_y:f_x=3:1$ $f_y:f_x=3:2$

Fig. 11.5

The pattern is stationary when f_y/f_x is a ratio of whole numbers and the phase difference is constant.

When f_y/f_x is large, comparison by Lissajous' figures is difficult and in such cases a *circular time base* method can be used in which one alternating p.d. of known frequency (say 50 Hz) applied to a ' phase-splitting ' *R-C* series circuit, produces a circular trace on the CRO. (For a 2-V, 50-Hz input and a CRO gain setting of 1 V, convenient values of R and C are 10 kΩ and 0.47 μF respectively.) Three connections from the *R-C* circuit to the Y-, E- and X-terminals are necessary, Fig. 11.6*a*. If an unknown, much higher frequency (from a signal generator), is applied to the Z- and E-terminals on the CRO the brightness of the trace varies and a broken circle is obtained, Fig. 11.6*b*. Here the frequency ratio is 8:1. Why? (The Z-input on the CRO is joined to the grid of the tube via an isolating capacitor, and an external p.d. applied to it brightens or blacks-out the trace, depending on whether it is positive or negative. The beam is thus ' intensity modulated '.)

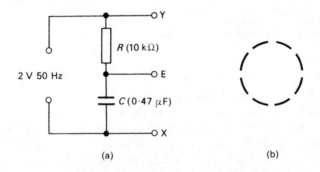

2 V 50 Hz R (10 kΩ)   O Y   O E C (0·47 μF)   O X

Fig. 11.6   (a)   (b)

This method can be used to check the calibration of a signal generator at frequencies up to 2 kHz against the 50-Hz mains supply. If the generator has a square wave output it should be used in preference to the sine wave output. Why?

Energy bands, conductors and insulators

(*a*) *Band theory.* Charge carriers are necessary for electrical conduction. The ' free ' electron theory, described earlier for metals (p. 378), suggests that solids possessing ' free ' electrons are conductors and those which do not are insulators. This theory was developed at the start of the present century and whilst it deals adequately with certain aspects of conduction, it cannot satisfactorily account for others. One of its assumptions is that the ' free ' electrons can have any energy. The present theory of conduction, which is based on wave mechanics and is called the *band theory*, permits them to have only certain ranges of energies.

We saw previously that an *isolated* atom, as in a gas, has a characteristic set of well-defined energy levels that electrons can occupy. When two similar atoms are brought together, the electric fields due to their charges overlap and mutual interaction occurs. As a result, according to the band theory, each of the higher energy levels is slightly altered in value, one becoming a little higher and the other a little lower than it was in the isolated atoms. (Why are the lower levels less likely to be affected?) The closer the atoms are together, the greater is the energy difference between the two levels, Fig. 11.7*a*.

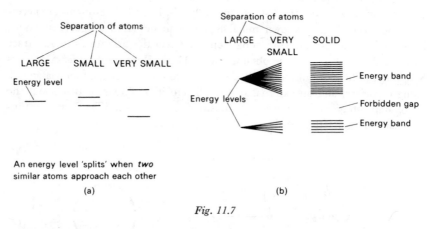

An energy level 'splits' when *two* similar atoms approach each other

(a)

(b)

Fig. 11.7

In a solid, many atoms are closely packed and for each original higher level in an isolated atom, there will be many levels. Since the spacing between atoms varies, the levels will be altered by different amounts and will, in effect, form a band of extremely closely spaced levels corresponding to one particular level in the isolated atoms, Fig. 11.7*b*. A solid can thus be considered to have energy bands that are shared by all atoms. The bands, like the levels in an isolated atom, are separated by gaps which are forbidden to the electrons.

(*b*) *Conductors.* The electrical conductivity of a solid depends on the spacing of its energy bands and the extent to which these are occupied by electrons. In a conductor such as copper the lower bands are full but the highest occupied band, called the *conduction band*, is only half-full. An electron in it can accept from an electric field (due to the p.d. applied to it) the very small amount of energy it requires to be raised to an empty, slightly higher energy level in the band where it is so loosely held that it can drift through the solid under the action of the applied field (as a ' free ' electron) thus creating an electric current.

It is the conduction band electrons in copper which form chemical bonds with other atoms and so they are also called ' valence ' electrons and the band the ' *conduction-valence* ' band. Conductors in general are characterized by having either a partly filled conduction-valence band, like copper, Fig. 11.8*a*, or an empty conduction band overlapping a filled valence band, like magnesium, so that there is in effect one continuous partly-filled band.

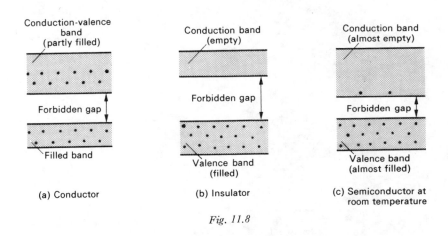

| (a) Conductor | (b) Insulator | (c) Semiconductor at room temperature |

Fig. 11.8

(*c*) *Insulators*. An insulator is considered on this model to have neither partly-filled bands nor overlapping bands. The highest occupied band, the valence band, is full but the band above it, the conduction band, is empty Fig. 11.8*b*. There is, however, a substantial energy gap—the forbidden gap—between the two bands. The energy required to raise a valence electron into the conduction band is about 5 eV, which is more than can be supplied by the electric field of an ordinary source of p.d.

Semiconductors

A semiconductor may be defined as a material whose conductivity lies between that of a good conductor and a good insulator. Silicon and germanium are the best known semiconductors because they are used to make the most common devices, i.e. transistors and diodes, the present trend being towards silicon devices. Other semiconductor materials are in common use such as cadmium sulphide, lead sulphide, gallium arsenide. These tend to be used in specialist devices such as photoelectric cells (p. 463). A selection of semiconductor devices is shown in Fig. 11.9.

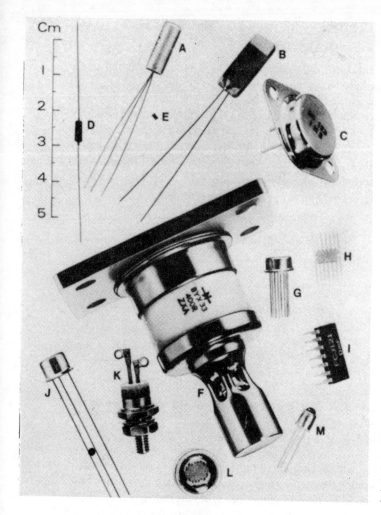

Fig. 11.9
A, J. Low-power transistors
B. Plate thermistor
C. Power transistor
D. Low-power diode
E. Leadless inverted device (LID)
F. 250-ampere rectifier diode
G, H, I, Integrated circuits
K. Thyristor
L. Photoconductive cell
M. Phototransistor

Intrinsic semiconductors

When semiconductors such as germanium and silicon are extremely pure, they resemble insulators but their forbidden gap is narrower, about 1 eV. At absolute zero the valence band is completely filled and the conduction band is empty. At room temperatures a few valence electrons gain enough energy from the vibration of atoms in the crystal lattice of the material (due to their internal energy) to reach the conduction band, Fig. 11.8c. There they behave like the conduction electrons in a conductor, and when a p.d. is applied they move under the action of the electric field to form a current. There is also a second type of charge carrier which arises as follows.

When a valence electron enters the conduction band, it leaves in the valence band an empty energy level. A position in the crystal which was previously electrically neutral is now deficient of an electron and behaves like a positive charge. Such positive vacancies are called ' holes '. Under an applied electric field a nearby electron in the valence band can occupy the hole but in doing so creates another positive hole. Thus, whilst the actual particle that moves is a valence electron, it seems that a hole with a positive electronic charge has moved in the opposite direction. The very small current obtainable in a *very pure* semiconductor therefore consists of electrons in the nearly empty conduction band moving one way and an equal number of positively charged holes in the almost full valence band drifting the opposite way, Fig. 11.10. Such a semi-conductor is said to exhibit *intrinsic* conduction, i.e. the charge carriers have their origin inside the material, arising in this case from the transfer of electrons from the valence to the conduction band.

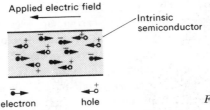

Fig. 11.10

It is possible for a hole to be filled by a conduction electron thereby reducing the number of charge carriers. But at any temperature the rate at which holes are filled equals the rate at which they are created when a valence electron gains enough energy from the lattice to move to the conduction band.

The behaviour of valence electrons and holes in a semiconductor can be likened to a row of chairs, all occupied except one at the end. If everyone moves one place towards the vacant seat, the vacancy appears at the other end. The vacancy (hole) seems to have moved along the row but the motion has really been of the occupants (electrons) of the chairs in the opposite direction.

The conductivity of a pure semiconductor at room temperature is very small but may be increased greatly by raising the temperature so that many more valence electrons can reach the conduction band. This fact is used in thermistors. Above 100 °C for germanium and 150 °C for silicon the conductivity is no longer under control and precautions have to be taken to limit their working temperatures.

Extrinsic semiconductors

The use of semiconducting materials in devices such as the transistor depends on increasing their conductivity by introducing tiny, but controlled, amounts of certain ' impurities ' into very pure (i.e. intrinsic) semiconductors. The process is known as ' doping ' and the material obtained is an *extrinsic* semiconductor because the impurity introduces charge carriers, additional to the intrinsic ones.

The impurity atom must have a similar size to that of the semiconducting atom so that it can occupy a position in the crystal lattice of the semiconductor without distorting it.

The valency of germanium and silicon is four, i.e. each atom has four electrons in the valence band. Each valence electron forms a covalent bond with a valence electron of four neighbouring atoms. The impurity atoms must be either pentavalent or trivalent.

(*i*) *n-type*. Suppose a pentavalent atom, such as phosphorus, arsenic or antimony, is introduced into the lattice of a pure silicon crystal. The impurity atom has five valence electrons but only four are required to form covalent bonds with adjacent tetravalent silicon atoms. One electron from every pentavalent atom added is spare and being loosely held it can take part in conduction. The impurity atom is called a *donor* and the ' impure ' silicon is known as an *n*-type semiconductor because the main charge carriers are *n*egative electrons. However, note that the overall charge in the crystal remains zero because each atom present is electrically neutral. In terms of energy bands, the donor atom provides *filled* energy levels in the forbidden gap of silicon, just below its conduction band, Fig. 11.11. An impurity electron in one of these donor levels can easily acquire from the lattice vibrations, the small amount of energy it needs to reach the empty conduction band in silicon.

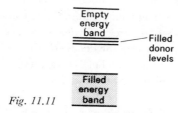

Fig. 11.11

(*ii*) *p-type*. A trivalent impurity such as aluminium, gallium or indium, also increases the conductivity of a pure semiconductor but in a different way. In this case the impurity atom has three valence electrons and can only form covalent bonds with three of the four surrounding silicon atoms in the crystal lattice. One bond is incomplete and, as before, the position of the missing electron behaves like a positive hole. This can accept a valence electron from an adjacent silicon atom (when it receives just a small amount of energy) and thereby form another hole. The impurity atoms are therefore called *acceptors* and since the conduction is now chiefly by *p*ositive charge carriers (i.e. holes), the ' impure ' silicon is said to be a *p*-type semiconductor. Note again that the semiconductor is electrically neutral. In the energy band diagram of Fig. 11.12, the acceptor atom is shown as supplying *empty* energy levels in the forbidden gap just above the filled valence band of silicon. A valence electron can easily occupy one of these slightly higher levels and create a vacancy in the valence band which makes increased conduction possible.

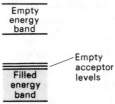

Fig. 11.12

To sum up, the *majority carriers* in an n-type semiconductor are electrons, positive holes being very much *minority carriers*. In p-type material, the reverse is true. The sign and concentration of the majority carriers can be found from the Hall effect (p. 89). In both types of material a temperature rise increases the proportion of minority carriers (due to increased intrinsic conduction) and upsets the semiconductor device whose action depends entirely on its extrinsic charge carriers.

In the manufacture of n- and p-type semiconductors, the semiconducting material is first purified (e.g. by zone refining, p. 117) to 1 part in 10^{10} and then impurity atoms added to produce the required conductivity.

Junction diode

A crystal of silicon or germanium with a junction of p- and n-type materials can act as a rectifier. The junction must be grown as one crystal so that the lattice continues across it; merely having p- and n-type materials is not enough. The action is as follows.

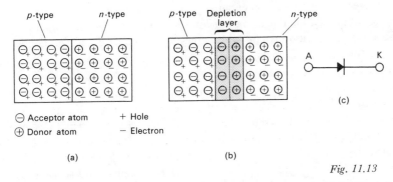

Fig. 11.13

As soon as the p-n junction is formed, electrons from the n-type material diffuse into the p-type to fill its holes and holes from the p-type diffuse into the n-type to be filled by its electrons, Fig. 11.13a. The exchange is only momentary since, due to both materials being initially neutral, a narrow region on either side of the junction loses its charge carriers and static, negative acceptor atoms are left in the p-type and static positive donor atoms remain in the n-type. The p-type region of this so-called *depletion layer*, Fig. 11.13b, thus becomes negative and the n-type region positive. A ' potential barrier ' is thus established; it sets up an electric field which prevents further movement of charge carriers. The p.d.s have values of about 0.1 V for germanium and 0.6 V for silicon, but depend on the doping and the temperature. The rectifying properties of the junction are due to the potential barrier: the circuit symbol for a junction diode is given in Fig. 11.13c, A being the anode and K the cathode.

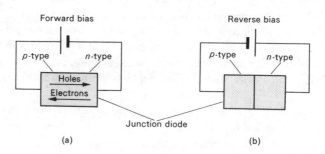

Fig. 11.14

(a)   (b)

If an actual battery whose e.m.f. exceeds the barrier potential is joined across the junction with its positive pole to the p-type material, then the p-type material is now at a higher potential than the n-type. Positive holes again cross the junction from p-type to n-type and electrons pass in the opposite direction, i.e. conduction occurs by the motion of both positive and negative charge carriers, Fig. 11.14a. This method of connecting the battery is called ' forward bias ' and a current is obtained with a battery e.m.f. just greater than the barrier potential. Reversing the polarity of the battery gives ' reverse bias ' connection and increases the barrier potential so that very few charge carriers cross the junction, Fig. 11.14b. The reverse or *leakage current* may be a few microamperes arising from minority carriers: it also flows during forward bias and opposes the flow of the majority carriers.

Fig. 11.15

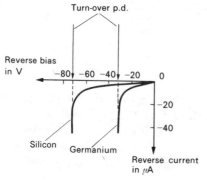

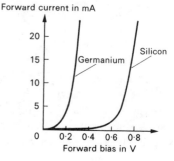

Typical characteristic curves for silicon and germanium junction diodes are given in Fig. 11.15. They are non-linear devices which can be used as rectifiers (p. 185) and detectors (p. 463) on account of their low forward resistance and high reverse resistance. If the reverse p.d. exceeds a certain value, called the *turn-over* p.d., the reverse current increases sharply and rectification is no longer possible. This is called the *Zener effect*. Silicon diodes withstand higher reverse p.d.s than those made from germanium and their reverse currents are also much smaller.

The construction of a junction diode is shown in Fig. 11.16. The *p-n* junction is formed by melting a pellet of a trivalent impurity into a thin slice of *n*-type semiconductor in an inert atmosphere. Connections are made and the diode sealed to exclude light and moisture.

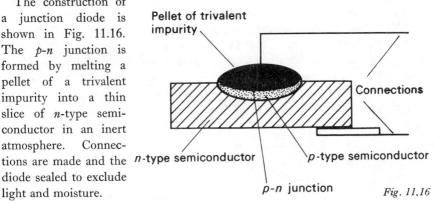

Fig. 11.16

Transistors

Transistors can be classed as *bipolar* or *unipolar*. The first uses both types of charge carriers (electrons and holes) and has two *p-n* junctions. The second uses only one type (either electrons or holes) and has only one *p-n* junction: the field effect transistor (f.e.t.) is a unipolar type.

(*a*) *Description.* A bipolar junction transistor consists of a very thin wafer of *p*- or *n*-type semiconductor called the *base*, between two layers of semiconductor material of the opposite type, one of which is called the *emitter* and the other the *collector*. There are two types, *n-p-n* and *p-n-p*, both shown diagrammatically in Fig. 11.17 with their symbols. In effect a transistor consists of two *p-n* diodes back to back.

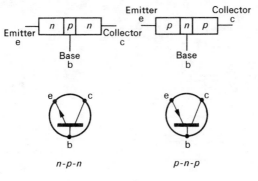

Fig. 11.17

Various manufacturing techniques are used. Most of the earliest transistors were of germanium, since this is easier to work with, and were *p-n-p* types made by an *alloy-junction* process. However, these are unsuitable for use at frequencies above about 1 MHz on account of the difficulty of getting the base thin enough. The silicon planar transistor, which is usually of the *n-p-n* type, is now preferred and is made by a *diffusion* technique. Diffusion processes occur slowly and so can be controlled better thereby enabling much thinner base regions (3 to 5 μm in thickness) to be formed in the transistor. The high frequency performance of planar transistors is therefore much improved, being well above 1000 MHz, i.e. 1 GHz. They are also very reliable. A simplified section of a *n-p-n* silicon planar transistor is shown in Fig. 11.18a, in practice there are no well-defined demarcation lines between regions and the boundaries are not straight; leads are attached before encapsulation, Fig. 11.18b. This technique, which cannot be used with germanium, enables thousands of transistors to be made together on a thin slice of silicon about 3 cm in diameter.

Fig. 11.18

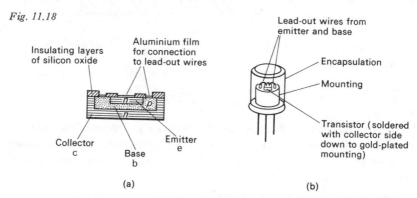

(a)   (b)

(b) *Action.* An *n-p-n* transistor is represented in Fig. 11.19a and its symbol is shown in Fig. 11.19b. An *n-p-n* silicon transistor requires a *positive* base potential of about 0.6 V with respect to the emitter before there is any base current I_b. I_b controls the collector current I_c which flows when the collector is *positive* with respect to the emitter.

Fig. 11.19

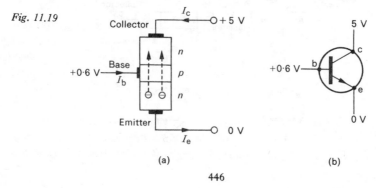

(a)   (b)

The base-emitter junction is forward biased and electrons, the majority charge carriers in the n-type material of the emitter, are able to cross the junction into the base because the applied electric field overcomes the barrier potential. At the same time there is a flow of positive holes from the base (where they are the majority carriers) to the emitter, but since the transistor materials used are such that the concentration of electrons in the n-type regions is much greater than that of holes in the p-type base, only electron flow need be considered.

In the base, which is very thin, only a few electrons (less than 1 per cent) combine with holes of the p-type material; most cross the base-collector junction because the collector is positive with respect to the base. Although the collector base junction is reverse biased and prevents electrons passing from collector to base, it allows them to move in the opposite direction.

If I_e, I_b and I_c are the currents flowing in the emitter, base and collector connections respectively, we can say

$$I_e = I_b + I_c.$$

I_e is only slightly greater than I_c (typically 1 mA and 0.995 mA) on account of the small electron-hole recombination in the base. The little recombination which does occur gives rise to the small base current I_b, in the form of electrons flowing out of the base into the external circuit. The current directions shown for I_e, I_b and I_c in Fig. 11.19a refer to conventional current flow, as does the arrow on the symbol in Fig. 11.19b.

The reason for the terms *emitter* and *collector* is evident from the action of the transistor. In the p-n-p transistor, whose behaviour is similar to that of the n-p-n type, the polarities of the applied p.d.s are reversed, i.e. the collector and base are negative with respect to the emitter, and it is holes, not electrons, that are injected at the emitter and received by the collector.

The word *transistor* is derived from the ' transfer of resistance ' action of the device. The base emitter circuit is forward biased and so has a low resistance, while the collector-base circuit is reverse biased and has a high resistance. The emitter current is transferred from a low-resistance circuit to a high-resistance circuit.

Transistor characteristics

These are curves, found by experiment, which show the relationships between the various currents and p.d.s and enable us to see how transistors can be used. Transistors can be connected in different ways, the commonest is the *common-emitter* circuit in which the emitter is common to both base and collector circuits and is frequently earthed, i.e. taken as zero potential. A circuit for studying an n-p-n transistor is shown in Fig. 11.20. The fixed resistor R_2 is such that when the maximum p.d. is applied from E_1, the base current I_b does not exceed its maximum permissible value for the transistor. Both voltmeters must have a very high resistance and E_1 and E_2 are low voltage batteries. Three curves are of interest.

Fig. 11.20

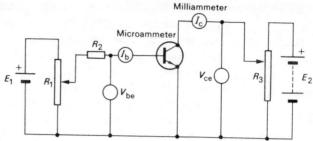

(a) *Base (input) characteristic.* The collector-emitter p.d. V_{ce} is kept constant and V_{be} measured for different values of I_b, obtained by varying R_1. A typical curve for a silicon planar transistor is given in Fig. 11.21; it resembles that for a *p-n* junction diode but the base currents are much smaller and do not flow until V_{be} exceeds $+0.6$ V. The base (input) resistance equals the reciprocal of the slope of the tangent at any point and is comparatively small (due to the forward bias of the emitter-base).

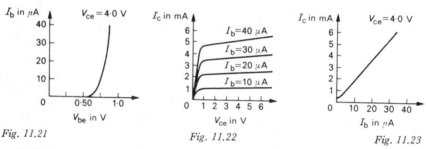

Fig. 11.21   Fig. 11.22   Fig. 11.23

(b) *Collector (output) characteristics.* The base current I_b is fixed at a low value and I_c measured as V_{ce} is increased by steps, small at first and then greater. This is repeated for different base currents to give a series of curves, Fig. 11.22. We see that I_c depends almost entirely on I_b and hardly at all on V_{ce} except when V_{ce} is less than about 0.3 V. As an amplifier a transistor is used well beyond the sharp bend or 'knee' of the characteristic where the collector (output) resistance is high and I_c varies linearly with V_{ce} for a given I_b. Below the knee (i.e. $V_{ce} <$ 0.3 V) I_c falls abruptly to zero and the transistor is said to be 'cut-off'; it is used like this as a switch (p. 453).

(c) *Transfer characteristic.* This gives the relationship between I_c and I_b for a fixed value of V_{ce}, Fig. 11.23. It shows that a small change in I_b, say 10 μA, causes a large change in I_c, about 1.5 mA here, a ratio of $1.5 \times 10^3/10 = 150$. The graph is almost a straight line whose slope is called the *static value of the forward current transfer ratio*, h_{FE}, of the transistor in common emitter connection (formerly the *current amplification factor* α). A value of about 150 is typical for planar silicon transistors and 50 for alloy junction germanium ones. Basically the transistor is a *current amplifier* in the common emitter connection.

Transistor amplifiers

Amplifiers are necessary in many pieces of electronic equipment such as radios, oscilloscopes and record players. Sometimes it is a small alternating p.d. that has to be amplified, as in a CRO when we wish to deflect the electron beam by applying a sufficiently large p.d. to the deflecting plates; on other occasions power has to be magnified to drive, say, a loudspeaker. We shall be concerned with the first kind of task, for which voltage amplifiers are used.

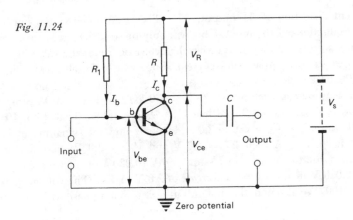

Fig. 11.24

The circuit for a simple transistor amplifier in common-emitter connection is shown in Fig. 11.24; the base-emitter circuit is the input circuit and the collector-emitter circuit is the output circuit. With no input signal, resistor R_1 determines the steady base current I_b (called the *bias* current) and we have $I_b \simeq V_s/R_1$ since the base is only slightly positive (about 0.6 V) with respect to the emitter, i.e. the p.d. across R_1 is nearly the battery p.d. V_s. The steady collector current I_c is then given by $I_c \simeq h_{FE} \times I_b$ where h_{FE} is the static value of the forward current transfer ratio of the transistor. I_c flows through R, called the *collector load* and the steady p.d. across it is $V_R = I_c R$. Hence the steady collector-emitter p.d. $V_{ce} = V_s - V_R$.

When a signal is applied to the input and causes a small alternating current to be superimposed on the steady bias current, the latter changes and corresponding large changes of collector current occur. These produce changes of p.d. across R which may be many times greater than the change in the base-emitter signal p.d. responsible for the base current change. Thus V_{ce} becomes a varying direct p.d. and may be regarded as an alternating p.d. varying like the input, superimposed on a steady direct p.d., (i.e. the steady value of V_{ce} when there is no input), Fig. 11.25. Only the alternating component is required at the output and capacitor C, called the *blocking* or *coupling capacitor*, blocks the direct component but allows the alternating one to pass. The transistor and load *together* act as an amplifier of p.d. *changes*.

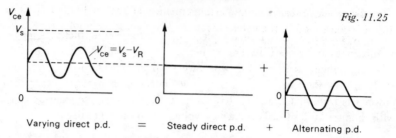

Fig. 11.25

Varying direct p.d.   =   Steady direct p.d.   +   Alternating p.d.

To prevent distortion of the output, the collector current changes must be faithful reproductions of the input. This will only be so if the transistor always operates on a straight part of its transfer characteristic, i.e. if the value of R_1 is chosen to give an appropriate bias current about which the alternating input current to the base varies.

Consider a numerical example. Take $R = 2.0$ kΩ and $V_s = 6.0$ V. Suppose that when $I_b = 10$ μA, $I_c = 1.0$ mA then $V_R = I_c R = 1.0 \times 10^{-3} \times 2.0 \times 10^3 = 2.0$ V and $V_{ce} = V_s - V_R = 6.0 - 2.0 = 4.0$ V. If now the signal makes $I_b = 20$ μA and $I_c = 2.0$ mA then $V_R = 4.0$ V and $V_{ce} = 2.0$ V. Assuming this change in I_b is due to a change in V_{be} of 20 mV (0.020 V) then we see that an *increase* of 0.020 V in V_{be} causes a *decrease* of 2.0 V in V_{ce}. The voltage gain is $2.0/0.020 = 100$. We also note that the output p.d. is in antiphase with the input signal.

The very simple method used in the circuit of Fig. 11.24 to obtain a steady bias current for the base causes difficulties if the temperature varies. Other more complex self-stabilizing circuits have been developed.

Transistor oscillators

An oscillator is a generator of alternating current and in essence consists of an amplifier which feeds back a small part of its output to its input. If the feedback is positive, i.e. in phase with the input and sufficient to compensate for resistive energy losses, then undamped oscillations are obtained.

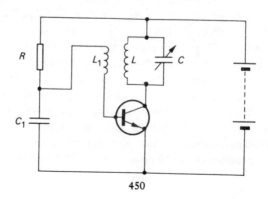

Fig. 11.26

(a) *Tuned oscillator*. A simple arrangement is shown in Fig. 11.26. The *L-C* circuit is connected in the collector (as the load) and oscillations start in it when the supply is switched on. Left to themselves these would decay but changes of current in L are fed back by mutual induction to the base-emitter (input) circuit by coil L_1, arranged close to L. The frequency f of the oscillations is given by $f = 1/(2\pi\sqrt{LC})$, (p. 184), i.e. the natural frequency of the LC circuit: the transistor merely ensures that energy is fed back at the correct instant from the battery.

The current bias for the base of the transistor is obtained through R; C_1 allows the a.c. component of the base-emitter current, at the frequency of the oscillations, to pass whilst not short-circuiting the d.c. bias.

Fig. 11.27

(b) *Astable multivibrator*. There are three types of multivibrator, the astable, the bistable (p. 456) and the monostable (p. 460). The *astable* (not stable) or free-running one will be considered here. It produces a train of almost square waves or pulses, Fig. 11.27 and is also called a *pulse generator* or *square wave oscillator*. A square wave can be analysed into a large number of sinusoidal waveforms with frequencies that are multiples (harmonics) of the fundamental frequency. This accounts for the name *multivibrator*.

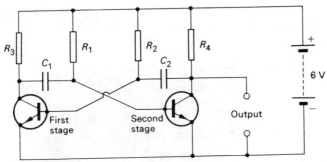

Astable multivibrator

Fig. 11.28

The circuit is shown in Fig. 11.28; its detailed operation need not concern us here but in essence it is a two stage common-emitter amplifier with the output of the second stage being fed back to the input of the first stage. The feedback is positive since the output of each stage is in antiphase with its input. The stages are coupled by capacitors C_1 and C_2. On switching on, a state is reached where one transistor is conducting and the other is not but because of the cross-coupling the situation soon reverses and continues to do so repeatedly. As a result, the output, which can be taken from the collector of either transistor, is alternately high (+6 V) and low (0 V) and a series of square pulses are produced. The change-over rate of the transistors from the conducting to the non-conducting state depends almost entirely on the time constants C_1R_1 and C_2R_2 and this governs the frequency of the pulses. How can the frequency be increased? What will be the effect on the waveform if $C_1R_1 \neq C_2R_2$?

The astable multivibrator has many uses including producing timing pulses for computer circuits, acting as a timing switch in radar equipment and generating musical notes in an electronic organ.

(c) *Demonstrations.* The 'oscillations' of an astable multivibrator may be shown in slow motion by flashing lamps using special modules connected as in Fig. 11.29a (initially with R_1 and R_2 short-circuited). The *transistor modules*,[1]

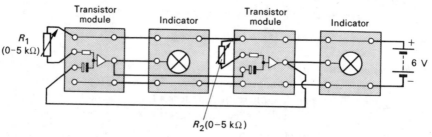

Fig. 11.29a   Astable multivibrator

Fig. 11.29b, are essentially one-stage transistor amplifiers (–▷– is the amplifier symbol) and the *indicators* contain 6-V, 60-mA lamps. The effect of including and altering the variable resistors R_1 and R_2 should be tried.

Fig. 11.29b

[1] Designed for Nuffield Advanced Physics to do many jobs and called 'basic units' in that course.

Higher frequency oscillations that are audible can be obtained as in Fig. 11.29*c*. Why is the frequency increased by connecting C_1 and C_2 in series with the internal coupling capacitors? The waveform can also be viewed by replacing the earpiece with an oscilloscope.

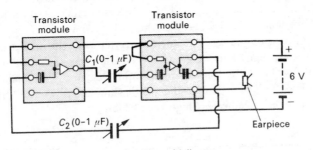

Fig. 11.29c   Astable multivibrator

Switching circuits

Electronic switching circuits perform essential roles in high-speed digital computers. In a calculation with numbers, a particular series of steps has to be taken, each of which may depend on the result of the previous step. Computer switches must therefore be such that their outputs depend on certain input conditions being satisfied and they need to have more than one input. Such switching circuits are called *logic gates*—they only 'open' and give an output if specified input conditions are met. In a transistor, the collector current is 'switched on' and controlled by conditions in the base-emitter circuit and this fact is used in transistorized logic gates.

Fig. 11.30

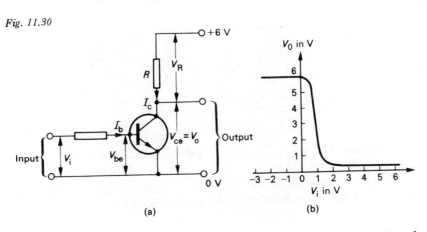

(a)   (b)

(a) *NOT gate* (or *inverter*). The circuit of Fig. 11.30*a* is basically that of a transistor amplifier. Experiment and theory indicate that its output-input characteristic of V_0 against V_i is as shown in Fig. 11.30b. So if $V_i = 0$, then $I_c = 0$, $V_R = 0$ and $V_0 = 6$ V. If $I_c = 2$ mA when $V_i = 1$ V then, taking $R = 2$ kΩ, $V_R = 4$ V and so $V_0 = 6 - 4 = 2$ V. Therefore if V_i is low (less than about 0.5 V or negative), V_0 is high (6 V) and if V_i is high (greater than about 1.5 V), the output is low (0.2 V or less). In the first state the transistor is 'cut-off', i.e. $I_c = 0$, and in the second, 'saturated', i.e. I_c does not increase if the base current increases. In general $V_0 = V_s - I_c R$ (here $V_s = 6$ V).

The circuit is acting as a NOT gate or inverter since the output is *high* only if the input is *not* high. This may be checked using the transistor module and indicator of Fig. 11.31a by connecting the input first to the $+6$-V line and then to the 0-V line. The behaviour of the gate is summarized by the 'truth' table of Fig. 11.31b where 1 indicates 'high' and 0 'low'.

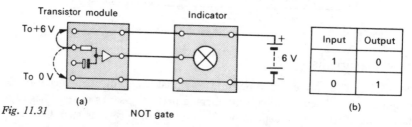

Input	Output
1	0
0	1

Fig. 11.31   (a)   NOT gate   (b)

A NOT gate can be used to do certain useful jobs such as making a lamp come on when it gets dark. In Fig. 11.32 the resistance of the cadmium sulphide photocell (p. 463) increases when it is shielded from light and the lamp comes on. Why? Draw (and test if possible) a circuit containing a thermistor (e.g. TH 3) connected to a NOT gate so that a lamp comes on when there is a sharp fall in temperature. (An aerosol freezer can be used to cool the thermistor.)

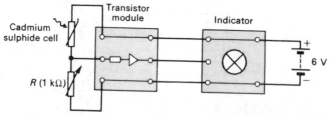

Fig. 11.32

(*b*) *NOR gate.* This has two (or more) resistive inputs and only when both are low is the output high. That is, the output is high if neither one *nor* the other input is high. A circuit and a truth table for a NOR gate are given in Fig. 11.33; *if any of the inputs is high, the output is low.* A NOT gate is a NOR gate with one input.

(*c*) *OR gate.* In this case the output is high if one input *or* the other (or both) is high (i.e. connected to $+6$ V). It is the opposite of the NOR gate due to the

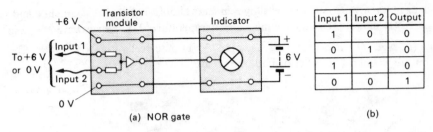

Input 1	Input 2	Output
1	0	0
0	1	0
1	1	0
0	0	1

(a) NOR gate

(b)

Fig. 11.33

addition of the inverting NOT gate and *if any of the inputs is high the output is high*. A truth table may be obtained using the circuit of Fig. 11.34.

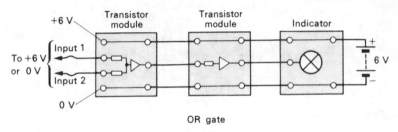

OR gate

Fig. 11.34

(*d*) *AND gate.* Here the output is high only if *both* inputs are high, as may be found from the circuit of Fig. 11.35 by touching both inputs to the + 6 V line The explanation of this gate is as follows. The third transistor module only gives a high output if both its inputs are low, i.e. when the outputs for each of the first two modules are low and this only occurs *if both their inputs are high*.

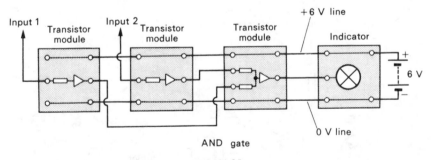

AND gate

Fig. 11.35

An AND gate might be used in a low temperature warning system to combine the outputs from a thermistor–operated NOT gate and an astable multivibrator so that the note from the latter is heard when the temperature falls below the danger level. Such a system is shown in Fig. 11.36 using an *astable* module and an AND module. Design a similar high-temperature warning system.

(*e*) *NAND gate*. This consists of an AND gate feeding a NOT gate. The output is ' high ' if either or both inputs are ' low '.

Fig. 11.36

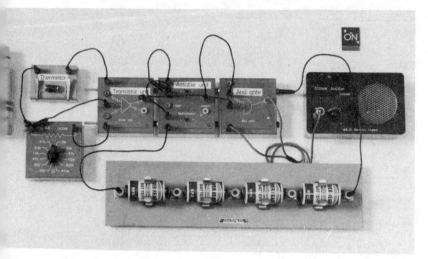

Counting circuits

At the heart of computer counting circuits is the bistable multivibrator.

(*a*) *Bistable multivibrator* (*flip-flop*). As its name suggests this has two stable states and its circuit, Fig. 11.37*a*, resembles that of the astable multivibrator (Fig. 11.28) but the feedback from one stage to the other is through resistors and not capacitors, i.e. the coupling is resistive.

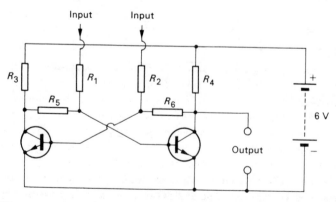

Fig. 11.37a   Bistable multivibrator

456

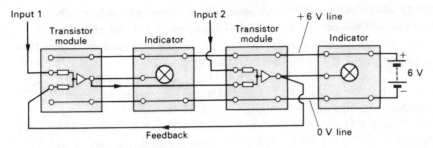

Fig. 11.37b

Bistable multivibrator

The action can be seen using the circuit of Fig. 11.37b. If input 1 is made high (by connecting it to the +6-V line), the output from the first transistor module is low and makes the input low to the second transistor module. The first lamp is therefore off. The output of the second transistor is high (second lamp on) and is fed back to the input of the first and, being high, *keeps* the output of the first transistor low, i.e. the feedback is positive. The state of the first transistor is thus reinforced and is a stable one (input 1 can be removed).

If input 2 is now made high, its output becomes low (second lamp off) and is fed back to the input of the first transistor to make its output go high (first lamp on). This output is fed to the second transistor and maintains the new, second stable state of the system. The process can be repeated indefinitely and always when the output from one transistor goes high, the other goes low but each state persists until there is a change at one input.

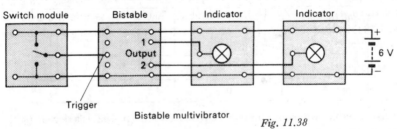

Bistable multivibrator

Fig. 11.38

A bistable multivibrator consists of two NOT gates connected so that the output of each is the input of the other. Both gates can be housed together to give a bistable module. This has another input, called the *trigger*, which saves switching over from one input to the other. Pulses applied to the ' trigger ' input are directed automatically by diodes inside the module to the gate not used by the previous pulse. Switching occurs when the input pulse to trigger *falls from* +6 V *to* 0 V. This may be demonstrated using the circuit of Fig. 11.38 in which successive +6 V to 0 V pulses are applied by pressing and releasing the switch as often as is required. The indicators show that *half* as many pulses appear at one output as are fed to the ' trigger ', i.e. the bistable divides by two.

(b) *Binary counter.* A digital computer performs calculations using binary arithmetic. Only two digits are used, 0 and 1, which are represented by the low (off) and high (on) states respectively of the outputs of a bistable multivibrator.

If one output of a bistable is connected to the input of a second bistable, the latter will change its state only when the input to its trigger *falls* from $+6$ V to 0 V. It therefore counts every second (2^1) pulse applied at the input of the first bistable. A third bistable taking the output of the second as its trigger input, will change state only when its input goes from $+6$ V to 0 V, i.e. every fourth (2^2) pulse and so on. Each bistable represents a *power* of 2 starting with the 2^0 ($= 1$) for the first.

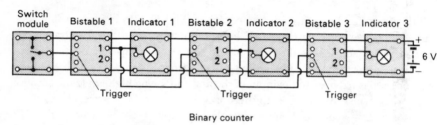

Binary counter

Fig. 11.39

A counter to show addition up to 7 on the binary scale may be made with three bistables in series and three indicators, Fig. 11.39. A lighted lamp indicates the 'on' (digit 1) state and an unlit one the 'off' (digit 0) state of its bistable. If pulses $+6$ V to 0 V are fed in by a switch, the following results are obtained.

Pulse number	1	2	3	4	5	6	7
Lamp 1	1	0	1	0	1	0	1
Lamp 2	0	1	1	0	0	1	1
Lamp 3	0	0	0	1	1	1	1

Twelve bistables will count up to over 4000. A scaler (p. 482) is a decade (scale of ten) counter.

As well as their role in computers, counters can be used—if suitable inputs are provided—to do jobs like counting the number of people entering a room or the number of particles from a radioactive source entering a Geiger-Müller tube.

Pulses and circuits

Electrical pulses, especially square ones, are much used in electronic systems such as counters and computers. We will consider some simple ways of producing square pulses, both continuously and singly and then see what effect certain circuits have on them.

(a) *Producing a train of square pulses.* Examination of the output-input characteristic of the *transistor module*, shown earlier in Fig. 11.30*b* (p. 453) and repeated in Fig. 11.40*a*, indicates that an alternating p.d. applied to the input of the module will be 'squared'. Fig. 11.40*b* shows the effect on a sinusoidal input (2 V r.m.s.).

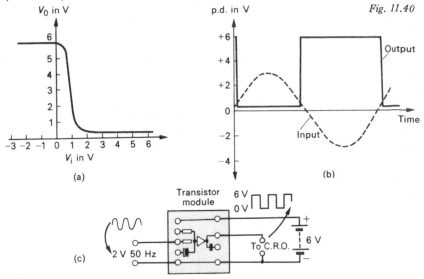

Fig. 11.40

(a)

(b)

(c)

The circuit of Fig. 11.40*c* can be used to demonstrate that a steady train of square pulses is obtained from a 2-V, 50-Hz a.c. supply. The output from an astable multivibrator, which gives roughly square pulses, may be squared further in the same way. The *transistor module* is thus a squarer, as well as a NOT gate, a NOR gate and a voltage amplifier.

(b) *Producing one square pulse.* The *transistor module* can again be used and single pulses of various lengths (up to several seconds) obtained. In the circuit of Fig. 11.41, connections have been made from one resistive input of the *transistor module* to +6 V and from the capacitive input to a *switch module*. If the switch is operated, it connects the capacitive input first to +6 V and then to 0 V and the lamp of the indicator flashes briefly. The pulse arises from changes in the charge on the input capacitor C which occur through the input resistor R; its duration depends on the time constant CR. How can the duration be (*i*) increased (*ii*) decreased? Replacing the indicator by an oscilloscope allows the pulse shape to be seen.

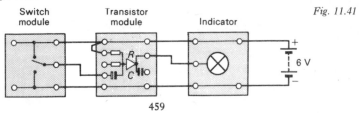

Fig. 11.41

The *transistor module* can be used as a pulse producer in a system that causes a lamp to flash say, six times in quick succession. The circuit also contains a slow *astable multivibrator* (made from two *transistor modules*), an *and* gate and an *indicator module*. Design and test the circuit.

(*c*) *Monostable multivibrator*. It has one stable state and is useful for producing a square pulse of fixed height (voltage) and length (time), whatever the shape and duration of the input pulse. The circuit is the same as that of an astable (Fig. 11.28, p. 451) but *one* of the capacitors is replaced by a resistor.

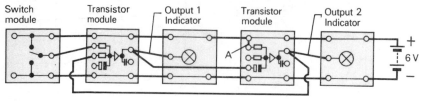

Monostable multivibrator   *Fig. 11.42*

In Figure 11.42 the input to the first transistor is ' low ' (0 V) and the output ' high ' (first lamp on). Output 1 is the input to the second transistor, output 2 is therefore 'low' (second lamp off) and, being fed back to the input of the first transistor, it holds the output of that transistor ' high '. This is the one stable state of the circuit.

Pressing the switch module makes the input to the first transistor high ($+6\,V$). Output 1 goes 'low', the first lamp goes off and the second transistor produces a pulse which brings on the second lamp while the pulse lasts. *On releasing* the switch the first lamp comes on again and the circuit returns to its stable state. The length of the output 2 pulse can be increased by having a resistor at A instead of a wire.

(*d*) *Effect of RC circuits*. Resistors and capacitors are often used to couple circuits and in doing so they may alter the signals they transmit.

Fig. 11.43   Differentiating circuit

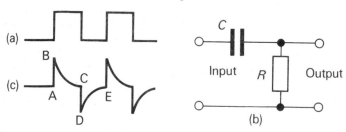

If a train of square pulses, Fig. 11.43*a*, is applied to the *RC* circuit (of small time constant *RC*) of Fig. 11.43*b*, their shape is changed to a series of spiky pulses, Fig, 11.43*c*. The circuit is said to *differentiate* the input.

The *RC* circuit of Fig. 11.44 is, if *RC* is large, an *integrating* circuit and a square wave input gives an output like that shown.

Fig. 11.44

Integrating circuit

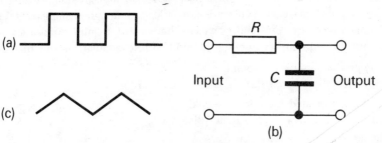

(a)

(c)

Input C Output

R

(b)

Differentiating and integrating circuits may be demonstr⸍ ⸜ with the arrangement of Fig. 11.45 in which the *transistor module* squ⸍ ⸜ the 50–Hz sinusoidal input before it is applied to the *RC* circuit.

Fig. 11.45

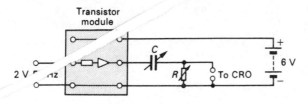

Transistor module

2 V ⸍ ⸝z

C

R

To CRO

6 V

+

–

Differentiating : $C = 0.022 \, \mu F$, $R = 10 \, k\Omega$
Integrating (interchange C and R) : $C = 0.22 \, \mu F$, $R = 100 \, k\Omega$

Analogue computers contain differentiating and integrating circuits which enable them to solve mathematical equations that represent models (or analogues) of physical situations. (See the booklet *Analogue computing—an introduction*, from Unilab Ltd, Clarendon Road, Blackburn.)

Electronic systems

(*a*) *The systems approach.* An electronic system such as a radio receiver, an oscilloscope or a computer can be regarded as consisting of a number of building bricks or *modules* each of which performs a certain task. The task may be to amplify, switch, count or shape etc. a signal. The main thing is then to know what a module does and how it can be assembled with other modules to achieve a particular end. An understanding of the *system* rather than of the individual working parts is required.

This is the so-called *systems approach* to electronics. It is becoming increasingly necessary to adopt it in practice because of the rapidity with which electronic devices have changed in recent years. On the other hand the jobs to be performed have not altered to the same extent. The design of devices and their associated circuitry has become much more specialized than before.

As an example of this approach we saw earlier how modules might be put together to give simple low-light level (Fig.11.32) and low-temperature (Fig. 11.36) warning systems. You were also asked to design a high-temperature warning system (p. 456) and a system that caused a lamp to flash six times (p. 460). In addition, try to design, build and test modular systems that

(*i*) produce six pips of sound,
(*ii*) make three lamps flash in traffic signals order.

If help is needed consult the booklet *Electronics kit and extension modules*, from Unilab Ltd, Clarendon Road, Blackburn.

In the systems approach the *block diagram* rather than the circuit diagram is generally adequate for most purposes. Its use will be illustrated for a radio transmitter and receiver.

(*b*) *Outline of radio.* The radiation from an aerial is appreciable only when the length of the aerial is comparable with the wavelength of the electromagnetic wave produced by the a.c. flowing in it. A 50-Hz alternating current corresponds to a wavelength of 6×10^6 m and so in practice r.f. currents (i.e. frequency $>$ 20 kHz) must be supplied. However, since speech and music generate a.f. currents (i.e. frequencies of 20 Hz to 20 kHz) some means of combining the two is necessary if intelligence is to be conveyed over a distance.

In a transmitter an *oscillator* generates an r.f. current which if applied to the aerial would produce an electromagnetic wave, called the *carrier wave*, of constant amplitude and having the same frequency as the r.f. current. If a normal receiver picked up such a signal nothing would be heard. The r.f. signal can be modified or modulated in various ways so that it carries the a.f. intelligence.

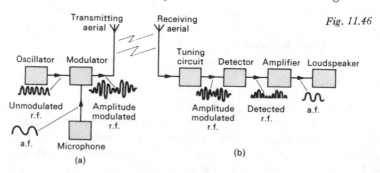

Fig. 11.46

In *amplitude modulation* (a.m.) the amplitude of the r.f. is varied so that it depends on the a.f. current from the microphone, the process occurring in a *modulator*. This type of transmission is used for medium and long-wave broadcasting in Great Britain. Fig. 11.46*a* shows the block diagram for an a.m. transmitter. In *frequency modulation* (f.m.) which is used in v.h.f. broadcasts, the frequency of the carrier is altered at a rate equal to the frequency of the a.f.

signal but the amplitude remains constant. Frequency modulated signals are relatively free from various kinds of electrical interference.

A block schematic diagram of a simple receiver of a.m. signals is given in Fig. 11.46*b*. The *tuning circuit* selects the wanted signal from the *aerial* and the *detector* (or demodulator) separates the a.f. intelligence (speech or music) from the r.f. carrier. The *amplifier* then boosts the a.f. so that an audible sound is produced in the *loudspeaker*.

Fig. 11.47

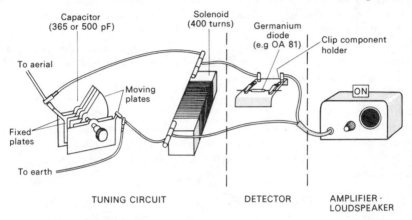

TUNING CIRCUIT | DETECTOR | AMPLIFIER - LOUDSPEAKER

A simple radio receiver that can be set up in a laboratory is shown in Fig. 11.47. A good aerial is necessary and an earth connection (e.g. to a metal water pipe) improves reception. The output waveform of each stage can be examined on an oscilloscope (time base about 1 ms cm^{-1}).

Other semiconductor devices

(*a*) *Photoelectric devices.* The photoemissive effect, which is given by semi-conductors as well as metals, was described earlier; here, photoconductive devices will be considered.

A *photoconductive cell*, Fig. 11.48 (and Fig. 11.9*l* p. 440), is a light-sensitive

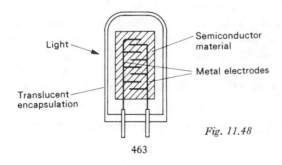

Fig. 11.48

463

ELECTRONICS

resistor whose resistance decreases as the intensity of the incident radiation increases (e.g. from 10 MΩ in the dark to 1 kΩ in the daylight). Cadmium sulphide is mainly suitable for light and lead sulphide for infrared; the sensitivity of the latter is such that an infrared source can be detected many miles away.

A *photodiode* works in a similar way to the photoconductive cell but the depletion layer at its *p-n* junction acts as the high resistance. It is usually reverse-biased so that the current (i.e. the leakage current) is due to minority carriers. When the junction is illuminated by sufficiently energetic photons, there is a marked decrease of resistance due to the production of more electron-hole pairs.

A *phototransistor* has a light-sensitive base into which incident photons (rather than an input signal) cause current flow. The collector-base junction is reverse-biased as usual and the device can be thought of as a photodiode giving current amplification due to transistor action.

A *photovoltaic* or *solar cell* generates an e.m.f. dependent on the intensity of the incident radiation. It is a silicon *p-n* diode, the *p*-layer being thin enough to allow the junction to be reached by light which supplies the energy for the potential barrier across the junction to become a continuous current source. Solar cells are used to power the electronic equipment in artificial satellites and space probes.

(*b*) *Thermistors.* These are devices whose resistance varies markedly with temperature change, brought about by direct heating or by the current flowing through them. Depending on their composition they can have *n*egative *t*emperature coefficients (n.t.c.) or *p*ositive *t*emperature coefficients (p.t.c) of resistance. Most thermistors are n.t.c. types, made from mixtures of oxides of nickel, copper, cobalt and other materials; these are used to control and measure temperature. The p.t.c. type is based on barium titanate and can show a resistance increase of 50 to 200 times for a temperature rise of a few degrees.

(*c*) *Solid state detector.* This detects alpha, beta, gamma and X-rays, protons and neutrons. Basically it is a reverse-biased silicon or germanium diode in which electrons and holes, created in the depletion layer by ionizing radiation (just as photons produce them in a photodiode), are rapidly swept apart to opposite electrodes. An electrical pulse of very short duration is produced and after amplification is counted by a scaler or ratemeter.

The amplitude of the pulse depends on the intensity of the ionization and a solid state detector can distinguish between different types of particle. (A G-M tube—p. 480—cannot do this.) The depletion layer must be close to the surface of the crystal especially if alpha particles are to reach it. The chance of detecting more weakly ionizing radiation, such as beta and gamma, is improved by increasing the reverse bias so that the thickness of the sensitive depletion layer becomes greater.

(*d*) *Integrated circuits.* An integrated circuit consists of a network of tiny resistors, capacitors, diodes, transistors and metal connections on a tiny chip of silicon. The manufacturing process is similar to that used for planar transistors and involves photographic masking, etching and diffusion into the surface of the silicon of n- and p-type impurities. On a wafer of silicon, 2–3 cm in diameter and 0.3 mm thick, 200 or more circuits, e.g. logic gates, binary counters, amplifiers, can be made. Fig. 11.49 shows an integrated circuit decade counter (before encapsulation) passing through the eye of a no. 5 needle; it contains over 120 components and the ' rope ' is an ordinary cotton thread.

Fig. 11.49

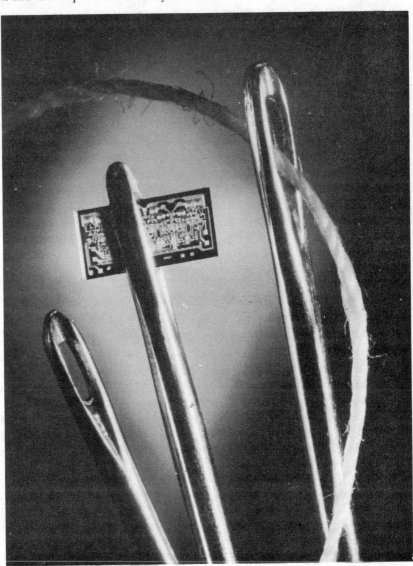

Microelectronics, as the use of integrated circuits is called, is a fast-growing technology and is well suited to the systems approach. Initially, applications were mainly in the computer and space industries where miniaturization and reliability are very important but many other uses are being developed, e.g. in telecommunications, radar systems, aircraft flight-data recorders, radio and television sets, car accessories, electronic calculators and medical electronics.

Thermionic diode valve

(*a*) *Construction and action.* The thermionic diode has a hot cathode and is surrounded by a metal anode, often in the form of a nickel cylinder, the whole being enclosed in a highly evacuated glass bulb. The construction and symbols of an indirectly heated diode are shown in Fig. 11.50.

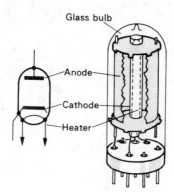

Fig. 11.50

When the anode is positive with respect to the cathode, the electrons emitted by the cathode are attracted to the anode and form a current from cathode to anode, called the *anode current*. This can be regarded as a flow of conventional current in the opposite direction. The circuit is completed through the power supply to the diode. If the anode is made negative with respect to the cathode, electrons are repelled by the anode and no current flows. The diode behaves like a mechanical valve allowing current in one direction only and before the advent of semiconductors was much used in rectifying circuits (p. 185).

(*b*) *Characteristics.* Characteristic curves show how the anode current depends on the anode potential and may be determined experimentally for a directly heated diode using the circuit of Fig. 11.51*a*. The low tension or l.t. supply maintains heating current through the filament and the high tension or h.t. supply maintains the anode at a positive potential. The l.t. can be a.c. or d.c. and is usually 6.3 V. The h.t. must be d.c. and variable up to 100 or 200 V

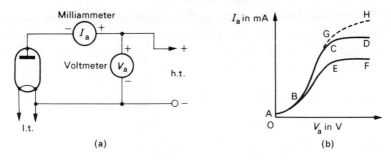

Fig. 11.51

depending on the valve. The anode current I_a is measured by a milliammeter and the p.d. V_a between anode and cathode by a high resistance voltmeter. The filament is joined to h.t. negative to provide a complete circuit for the anode current.

A typical curve ABCD is shown in Fig. 11.51b. When V_a is zero the electrons emitted by the filament tend to cluster round it since their emission speeds are mostly small. The negative charge of the electron cloud is called a *space charge* and it exerts a repulsive force on other electrons being emitted. Soon a condition of dynamic equilibrium is attained when the number of electrons returning to the filament per second equals the number emitted by it per second, so that the electron population of the space charge remains constant. A *very small* anode current does pass at this stage (even when the anode is made slightly negative, say a few tenths of a volt); this is due to the few electrons emitted with appreciable speed reaching the anode.

Along ABC when V_a is positive some of the outer electrons are attracted to the anode and the current rises. The repulsive effect of the space charge is now less and the number of electrons returning to the filament decreases. When V_a is sufficiently great, the space charge is dispersed, all electrons emitted reach the anode and I_a has its *saturation value* for that filament temperature. Along CD further increase of V_a has little effect on I_a for tungsten filament diodes. Along ABC I_a is said to be *space charge limited*.

At a lower filament current the emission temperature is smaller and the saturation current is less, as shown by ABEF in Fig. 11.51b. Oxide-coated valves saturate less abruptly (due to the emission depending on the electric field at the cathode surface as well as on the cathode temperature) and follow a curve similar to ABGH.

Triode valve

(*a*) *Construction and action*. The triode has three electrodes. The anode and cathode are as in the diode valve but the third electrode, called the *grid*, is a thin wire wound in a spiral close to but not touching the cathode so that electrons can pass through the spaces between its windings, Fig. 11.52.

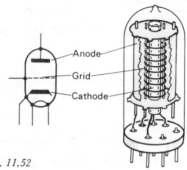

Fig. 11.52

In the diode, the anode current can be changed by changing the anode potential; in the triode, alteration of the potential of the grid with respect to the cathode provides an additional and more effective way of controlling the anode current. Thus the more negative the potential or *bias* of the grid the smaller the anode current. If the grid is positive, it attracts electrons causing grid current and certain undesirable effects (p. 471). Care is taken to ensure that the bias on the grid is always negative.

(*b*) *As a voltage amplifier.* The proximity of the grid to the cathode results in a small change of grid potential causing an appreciable change of anode current; the triode can therefore be used to amplify. To obtain voltage amplification a *load*, such as a resistor, is connected in the anode circuit. Anode current changes then produce voltage changes across the load which may be many times greater than those on the grid. The triode and load *together* act as a voltage amplifier.

Fig. 11.53

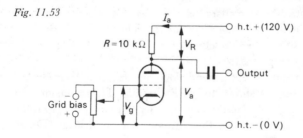

Consider the circuit of Fig. 11.53 with a resistor R of 10 kΩ as the load and an h.t. supply of 120 V. The anode current I_a flows (conventionally) from h.t.+, through R and the valve to h.t.− and causes a p.d. V_R across R. The anode potential is then less than 120 V by V_R. Thus if $I_a = 2$ mA when the grid bias $V_g = -3$ V, then $V_R = I_a \times R = 2 \times 10^{-3} \times 10 \times 10^3 = 20$ V. The anode potential $V_a = 120 - V_R = 100$ V. If V_g changes to -1 V and I_a increases to 4 mA, $V_R = 40$ V and V_a is now only 80 V. Thus if the grid potential changes by 2 V, the anode potential changes by 20 V. The voltage amplification or *gain* is defined by

$$gain = \frac{change\ in\ anode\ potential}{change\ in\ grid\ potential}$$

$$= 20/2 = 10.$$

In practice the grid potential is changed by applying the small alternating p.d. to be amplified, i.e. the input, between grid and cathode and to ensure that it never drives the grid positive a steady negative bias, represented by a battery in Fig. 11.54a, is applied in series with the input. If the input has a peak value of 1 V and the steady bias is -2 V, the grid potential varies from -3 V to -1 V, Fig. 11.54b. In this case the anode potential then fluctuates between 100 and 80 V.

Fig. 11.54

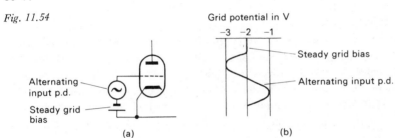

The anode potential variations form the output and may be tapped off by connecting across the valve as in Fig. 11.53. The output is a varying direct voltage which can be considered to consist of a steady direct p.d. with an alternating one superimposed on it. Here the steady component has a value of 90 V and the alternating one has a peak value of 10 V, Fig. 11.55. Only the alternating p.d. is required and it may be separated from the direct component by inserting a capacitor (to block the direct component) in the output connection from the anode. It can be seen from Figs. 11.54b and 11.55 that the input and output p.d.s are in antiphase, i.e. when the grid potential has its maximum negative value, the anode potential has its maximum positive value.

Fig. 11.55

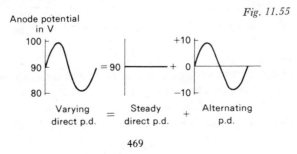

(*c*) *Characteristic curves.* The performance of a triode in a circuit can be predicted when its characteristics are known. The anode current I_a depends on both the anode potential V_a and the grid potential V_g and either V_a or V_g must be

Fig. 11.56

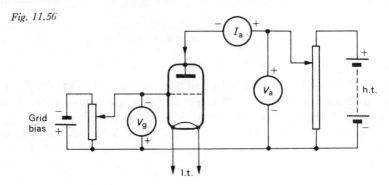

kept fixed while the other is varied when studying their effect on I_a. The circuit of Fig. 11.56 is suitable for investigating the relationships. Two sets of curves, called *static* characteristics, can be obtained.

The *anode characteristic* curves show how I_a varies with V_a when V_g is constant. Curves for the triode just considered are shown in Fig. 11.57*a* for four different values of V_g. Each one is found by keeping V_g fixed while V_a is increased up to the maximum value permitted for the valve and the corresponding values of I_a are measured.

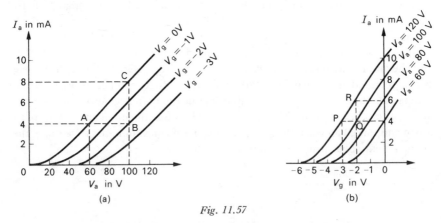

Fig. 11.57

It will be noted that (*i*) the curves are approximately straight and parallel over a good part of their length and (*ii*) a greater V_a is needed to obtain a given I_a the more negative is V_g, e.g. for $I_a = 4$ mA, $V_a = 60$ V when $V_g = 0$, but $V_a = 100$ A when $V_g = -2$ V.

The *mutual characteristics* curves relate I_a and V_g when V_a is constant. Four curves, also for the previous triode, are shown in Fig. 11.57b; their general form is similar to that of the I_a–V_a characteristics. Note that at higher values of V_a a more negative value of V_g is required to ' cut-off ' the valve, i.e. make $I_a = 0$.

The *dynamic* characteristic of a valve shows how I_a varies with V_g when there is a particular load in the anode circuit and the h.t. has a specified value. It can be obtained experimentally or derived from the mutual characteristics and is similar in form to the latter but of smaller slope. When used as an amplifier, the output p.d. is a true copy of the input p.d. (i.e. the amplification is distortion-less) only if the valve always works on the straight part of its dynamic charac-teristic. This means the steady negative grid bias has to be correctly selected and also that the input must never be so large as to cause it to go positive.

(*d*) *As an oscillator.* The oscillations set up by the discharge of a capacitor through an inductor (p. 182) can be maintained indefinitely by using a triode to feed back energy into the oscillatory circuit. The energy is provided by the h.t. supply to the valve; it compensates for that dissipated in the resistance of the circuit.

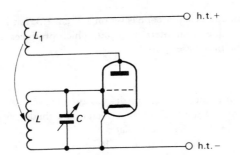

Fig. 11.58

A tuned grid oscillator is shown in Fig. 11.58 with the oscillatory circuit L–C connected between grid and cathode of the triode. The anode circuit contains a coil L_1, the feedback coil, arranged close to L so that it is inductively coupled to it. When the h.t. supply is switched on, the anode current rises to its steady value and a magnetic field builds up around L_1, inducing an e.m.f. in L which charges C. Once the anode current is steady, C discharges by oscillation. Left alone the oscillation would decay, but the alternating p.d. across C, being connected between grid and cathode, acts as the input to the triode and produces an a.c. component of anode current at the same frequency. If the coils are connected so that the e.m.f. induced in L is in the same direction as the oscillatory current, positive feedback occurs and undamped oscillations are obtained.

The circuit acts as an amplifier supplying its own input. The triode itself is not primarily responsible for the oscillations; the L–C circuit is, but by using the valve feedback is possible.

ELECTRONICS

1. Draw a sketch to show the essential parts of a cathode-ray oscilloscope having electrostatic deflection.

With the help of your sketch explain how in a cathode-ray oscilloscope: (a) the electrons are produced, (b) the electrons are focused, (c) the spot is made visible, and (d) the brightness of the spot is controlled.

What is meant by stating that a cathode-ray oscilloscope is fitted with a linear time base of variable frequency.
(*J.M.B.*)

2. (a) Draw a block diagram for an oscilloscope.

(b) Sketch and explain the forms of the traces seen on an oscilloscope screen when a p.d. alternating at 50 Hz is connected across the Y-plates if the time base is linear and has a frequency of (i) 10 Hz, and (ii) 100 Hz.

(c) What is the frequency of an alternating p.d. which is applied to the Y-plates of an oscilloscope and produces five complete waves on a 10-cm length of the screen when the time base setting is 10 ms cm^{-1}?

3. The gain control of an oscilloscope is set on 1 V cm^{-1}. What is (i) the peak value, and (ii) the r.m.s. value of an alternating p.d. which produces a vertical line trace 2 cm long when the time base is off?

4. With the help of a simple diagram discuss qualitatively the factors which determine the deflection of the spot on the screen of a cathode-ray oscilloscope when a given p.d. is applied between the Y plates.

What is meant by the statement that a CRO has a sensitivity of 40 V cm^{-1} for Y deflection?

A generator is believed to supply a sinusoidal voltage of 80 V r.m.s. at 200 Hz. Describe how you would use a CRO having the above sensitivity to test this. You may assume that a sinusoidal 50-Hz a.c. mains supply at any desired voltage is available.
(*J.M.B.*)

5. One sinusoidal voltage alternating at 50 Hz is connected across the X plates of a cathode-ray oscilloscope and another 50–Hz sinusoidal alternating voltage of approximately the same amplitude is connected across the Y plates. Sketch what you would expect to observe on the screen if the phase difference between the voltage is (a) zero, (b) $\pi/2$, and (c) $\pi/4$.

If the voltage on the X plates is replaced by a 100–Hz sinusoidal alternating voltage of similar amplitude, sketch what you would observe on the screen.

Explain briefly why figures of this type are useful in the study of alternating voltages.
(*J.M.B.*)

ELECTRONICS

6. (a) What is meant by (i) intrinsic and (ii) extrinsic conductivity?

(b) Explain the terms p-type and n-type semiconductors.

(c) Describe a p-n junction diode and draw a graph to show how the current through it varies with the p.d. across it.

7. Write a brief essay on the transistor and explain how transistors may be used to produce a simple amplifying circuit. (W.)

8. The resistance of a slice of semiconductor material falls when exposed to light of sufficiently short wavelength; explain briefly why this occurs.

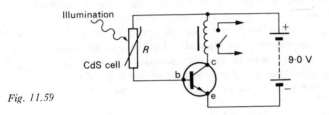

Fig. 11.59

The resistance, R, of the CdS photoconductive cell in Fig. 11.59 falls when the cell is illuminated. Find the approximate value of R at which the relay will close when the illumination is increased.

Relay closes at 20 mA. Current gain of the transistor = 80.

Assume that p.d. between base and emitter is negligible. (J.M.B. Eng. Sc.)

9. To find whether there was a minimum time for which a sound must persist to be heard a psychologist required an electronic device to produce single bursts of high-pitched sound, each lasting for a short time.

Here is a block diagram of the arrangement used, Fig. 11.60.

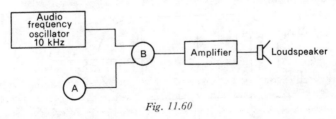

Fig. 11.60

(a) What is the function of A?

(b) What is the function of B?

(c) Suggest an addition or additions to the arrangement to enable the time for which each burst lasts to be timed to an accuracy of at least 1 millisecond. You may add to the sketch as well as explaining your idea below. (O. and C. Nuffield)

473

10. The diagram, Fig. 11.61, is of a very simple radio receiver which can be used for broadcasts from one station only.

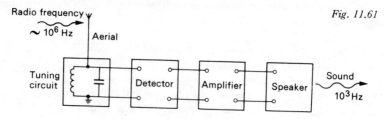

Fig. 11.61

(*a*) The tuning circuit selects one station (one frequency). Is the output of energy of the selected signal coming from the tuning circuit greater than the energy of the signal collected by the aerial? Explain your answer briefly.

(*b*) For which of the other three boxes is the energy of the output signal of the box larger than the energy of the input signal to the box? Give below the name of the box or boxes for which there is energy increase.

(*c*) How could the tuning circuit be altered so that it could select other stations?

(*d*) Which box or boxes is/are designed to transform energy from one form to another?

(*e*) The frequency of the signal received at the aerial is about 10^6 Hz, and the frequency of the speaker output is about 10^3 Hz. Would it matter if the amplifier could only amplify signals of frequency less than 10^5 Hz and so failed to amplify signals of 10^6 Hz? Explain your answer. (*O. and C. Nuffield*)

11. What is meant by *thermionic emission*?

Describe the structure of a diode valve and explain the action of a rectifying circuit using such a valve. Sketch curves showing the waveforms of (*a*) the input voltage and (*b*) the current in the circuit. (*L.*)

12. Sketch a typical graph relating the anode current I_a to the applied anode voltage V_a for a diode valve. Explain why (*a*) increasing V_a increases the flow of electrons across the valve; (*b*) the graph flattens off to a constant value for I_a no matter how much V_a is increased; (*c*) this maximum value of I_a depends on the p.d. applied to the filament.

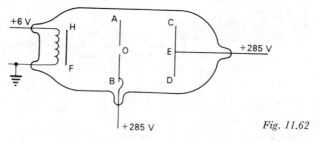

Fig. 11.62

Fig. 11.62 shows a special type of diode; the cathode HF is at earth potential. Show that for electrons to reach the anode AB at a velocity of 1.00×10^7 m s^{-1} the anode must be at a potential of about 285 V.

CD is a flat metal plate, similar to AB, and is also at a potential of 285 V. There is a small hole O at the centre of AB, through which a few electrons pass. Explain carefully why these electrons follow a straight path to E. With what velocity will these electrons arrive at E? Explain how you reach this result.

What arrangements could you make—having access to the inside of the valve and facilities to pump it out again, if necessary—to show that the stream of electrons was arriving at the point E? Give two possible ways of doing this.

(Electron mass $= 9.11 \times 10^{-31}$ kg, electron charge $= 1.60 \times 10^{-19}$ C.)　　　(*S.*)

13. A NOR gate ' opens ' and gives an output only if *both* its inputs are ' low ', whilst an OR gate ' closes '. An AND gate ' opens ' only if both inputs are ' high ', but a NAND gate ' closes '.

Copy and complete the truth table for each gate. ' High ' is represented by ' 1 ' and ' low ' by ' 0 '.

Input 1	*Input 2*	*NOR output*	*OR output*	*AND output*	*NAND output*
0	0				
1	1				
1	0				
0	1				

14. The system in Fig. 11.63*a* makes three lamps flash in the order of British traffic signals. The amber lamp is connected to the output of the slow astable, the red one to one of the outputs of the bistable and the green one to the output of the NOR gate.

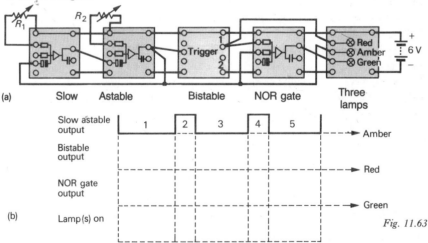

Fig. 11.63

In Fig. 11.63*b* the output pulses from the astable are shown. Copy it and add the corresponding outputs from the bistable and the NOR gate. Remember that (*i*) a bistable switches only when its input goes from ' high ' to ' low ' and (*ii*) a NOR gate ' opens ' only when both its inputs are ' low '.

Also show which lamp(s) is (are) on during intervals 1, 2, 3, 4 and 5. Write out the truth table for the system.

12 Nuclear Physics

Radioactivity

In 1896 the French scientist Becquerel found that uranium compounds emitted radiation which affected a photographic plate wrapped in black paper and, like X-rays, ionized a gas. The search for other *radioactive* substances was taken up by Marie Curie, who extracted from the ore pitchblende two new radioactive elements which she named polonium and radium.

(*a*) *Alpha, beta and gamma rays.* One or more of three types of radiation may be emitted which can be identified by their different penetrating power, ionizing ability and behaviour in a magnetic field.

Alpha (α) rays have a range of a few centimetres in air at s.t.p. and are stopped by a thick sheet of paper. They produce intense ionization in a gas and are deflected by a *strong* magnetic field in a direction which suggests they are relatively heavy, positively charged *particles*.

Beta (β) rays are usually more penetrating, having ranges which can be as high as several metres of air at s.t.p. or a few millimetres of aluminium. They

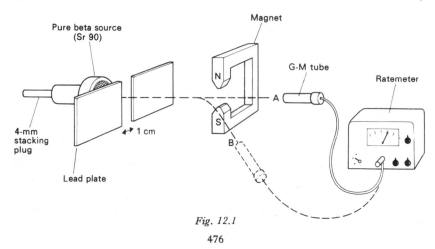

Fig. 12.1

476

cause much less intense ionization than alpha particles but are more easily deviated by a magnetic field and in a direction which indicates they are negatively charged *particles* of small mass. The magnetic deflection of beta particles may be demonstrated using the apparatus of Fig. 12.1. With the Geiger-Müller (G-M) tube in position A and without the magnet, the count-rate produced by the beam of beta particles is observed on the ratemeter. When the magnet is inserted, the count-rate decreases but rises again when the G-M tube is moved to some position such as B.

Gamma (γ) rays of high energy can penetrate several centimetres of lead. They ionize a gas weakly and are not deflected in a magnetic field. Their behaviour is not that of charged particles.

Alpha, beta and gamma rays are termed *nuclear radiation*, since, as we shall see later, they originate in atomic nuclei.

(*b*) *Nature of rays.* The specific charges of alpha and beta particles can be deduced from measurements of their deflections in electric and magnetic fields and give information about their nature.

Alpha particles have a specific charge which suggests they might be helium atoms which have lost two electrons, i.e. helium ions with a double positive charge—a helium nucleus $^4_2$He. In 1909 Rutherford and Royds confirmed this by compressing some of the radioactive gas radon in a tube A, Fig. 12.2, whose walls were thin enough to allow the alpha particles emitted by radon to escape into the evacuated space enclosed by the thicker-walled tube B. After a week the mercury level was raised so that any gas which had collected in B was forced into C. On passing a current through C the line spectrum of helium was observed. Each alpha particle penetrating A had collected two electrons and changed into a helium atom.

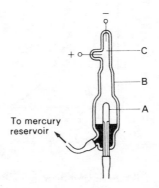

Fig. 12.2

Specific charge measurements for beta particles show they are high-speed electrons. (Allowance has to be made for the increase of mass with speed as predicted by relativity since in some cases beta particles are emitted with speeds (v) very close to that of light (c) and if the relativistic mass m is not used, the specific charge decreases with increasing speed: $m = m_0(1 - v^2/c^2)^{-\frac{1}{2}}$ where m_0 is the rest mass of an electron.)

Gamma rays were the subject of controversy until they were shown to be diffracted by a crystal, thus establishing their wave-like nature. Their wavelengths are those of very short X-rays, and like X-rays they are a form of electromagnetic radiation travelling with the speed of light (since among other things they give the photoelectric effect). Whilst diffraction gives the most accurate method of finding gamma ray wavelengths, it is difficult and rarely done. Nowadays the rays are usually absorbed by a solid state detector calibrated by rays of known energy.

(*c*) *Energy and speed of emission.* The energies of alpha and beta particles are found from measurements of their paths in magnetic fields and in the case of gamma rays as explained above.

(*i*) Alpha particles. In many cases the alpha particles emitted by a particular nuclide all have the same energy and are said to be monoenergetic. Energies vary from 4 to 10 MeV, corresponding to emission speeds of 5 to 7 per cent of the speed of light.

(*ii*) Beta particles. They exhibit quite a different behaviour. Their energy spectrum is a continuous one in which all energies are present from quite small values up to a certain maximum as shown by Fig. 12.3. The maxima are characteristic of the nuclide and vary from 0.025 to 3.2 MeV for natural radioactive sources. The highest represent beta particle emission speeds of 99 per cent of the speed of light.

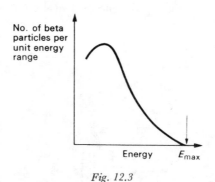

No. of beta
particles per
unit energy
range

Energy E_{max}

Fig. 12.3

(*iii*) Gamma rays. These fall into several distinct monoenergetic groups, giving a 'line' spectrum. The gamma rays from cobalt 60 (see below) have two different energies of 1.2 and 1.3 MeV.

(*d*) *Sources.* Those used for instructional purposes are usually supplied mounted in a holder with a 4-mm plug. The active material is sealed in metal foil which is

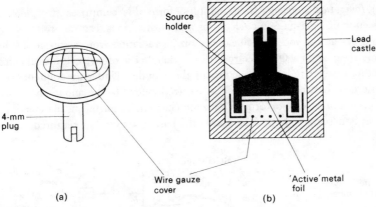

Source
holder

Lead
castle

4-mm
plug

Wire gauze
cover

'Active' metal
foil

(a)

(b)

Fig. 12.4

protected by a wire gauze cover, Fig. 12.4a. When not in use they are stored in a small lead castle, Fig. 12.4b, in a wooden box.

The health hazard is negligible when using the weak closed sources listed shortly, provided they are (*i*) *always lifted with forceps*, (*ii*) *held so that the open window is directed away from the body and* (*iii*) *never brought close to the eyes for inspection.*

Approved *closed* sources have low activities, about 5 microcuries (5 μCi: see p. 487) and include

(*i*) Radium 226 for α, β and γ rays.

(*ii*) Americium 241 or plutonium 239 for α particles only (but the former also emits some low energy γ rays).

(*iii*) Strontium 90 for β particles only.

(*iv*) Cobalt 60 for γ rays. (An aluminium cover disc absorbs the beta particles also emitted.)

(*v*) Some other very weak sources (0.1 μCi) are not completely enclosed but the radioactive material is secured to a support. Such sources are used in school cloud chambers.

Sources (*ii*), (*iii*) and (*iv*) are prepared in nuclear reactors.

Nuclear radiation detectors

In a nuclear radiation detector *energy* is transferred from the radiation to atoms of the detector and may cause

(*i*) ionization of a gas as in an ionization chamber, a Geiger-Müller tube, a cloud (or similar) chamber,

(*ii*) exposure of a photographic emulsion,

(*iii*) fluorescence of a phosphor as in a scintillation counter, or

(*iv*) mobile charge-carriers in a semiconducting solid state detector.

The radiation is thus detected by the effects it produces.

(a) *Ionization chamber.* In its simplest form this comprises two electrodes between which ion-pairs, i.e. electrons and positive ions, can be produced from neutral gas atoms and molecules by ionizing radiation from a source inside or outside the chamber. One electrode of the chamber is often a cylindrical can and the other a metal rod along the axis of the cylinder. Under the influence of an electric field between the electrodes, electrons move to the anode and positive gas ions to the cathode to form an ionization current. Some means of detecting the current is necessary. Fig. 12.5a shows the basic arrangement required.

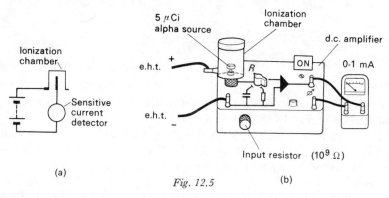

Fig. 12.5

The current depends on the nature of the radiation and the volume of the chamber. A 5–μCi alpha source creates a current of the order of 10^{-9} A in a small chamber; beta particles and particularly gamma rays cause very much smaller currents. A school-type d.c. amplifier (p. 57) can detect the currents due to alpha particles but is insensitive to beta and gamma radiation.

To measure the ionization current the d.c. amplifier is first calibrated (p. 58) then arranged as in Fig. 12.5b. The e.h.t. is increased until the milliammeter reading reaches a maximum value which is recorded. All ions produced are then being collected by the electrodes of the chamber and the ionization current has its saturation value I. This is a measure of the intensity of the radiation from the 5–μCi alpha source in the chamber (p. 406). It is calculated from $I = V/R$ where V is the p.d. across the input resistor R (probably 10^9 Ω) at saturation; V is obtained from the calibration measurement. A very rough estimate of the energy of an alpha particle may also be made (see Qn. 1, p. 515).

Other experiments with an ionization chamber and d.c. amplifier are given on pp. 490 and 493.

(b) *Geiger–Müller tube.* The G–M tube is a very sensitive type of ionization chamber which can detect single ionizing events. It consists of a cylindrical metal cathode (the wall of the tube) and a coaxial wire anode, containing argon at low pressure, Fig. 12.6a. The very thin mica end-window allows beta and gamma radiation to enter, as well as more energetic alpha particles; gamma rays can also penetrate the wall.

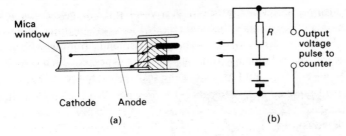

Mica window

Cathode   Anode

(a)

R   Output voltage pulse to counter

(b)

Fig. 12.6

A p.d. of about 450 V is maintained between anode and cathode and, since the anode is very thin, an intense electric field is created near it. If an ionizing ' particle ' passing through the tube produces an ion-pair from an argon atom, the resulting electron is rapidly accelerated towards the anode and when close to it has sufficient energy itself to produce ion-pairs in encounters. The electrons thus freed produce additional ionization and an avalanche of electrons spreads along the whole length of the wire, which absorbs them to produce a large pulse of anode current. In this way, one electron freed in one ionizing event can lead to the release, in a few tenths of a microsecond, of as many as 10^8 electrons.

During the electron avalanche the comparatively heavy positive ion members of the ion-pairs have been almost stationary round the anode. After the avalanche has occurred they move towards the cathode under the action of the electric field, taking about 100 microseconds to reach it. They now have appreciable energy and would cause the emission of electrons from the cathode by bombardment. A second avalanche would follow, maintaining the discharge and creating confusion with the effect of a later ionizing particle entering the tube. The presence in the tube of a small amount of a *quenching agent* such as bromine tends to prevent this, since the positive ion energy is used to decompose the molecules of the quencher. In a halogen quenched tube these subsequently recombine and are available for further quenching.

A Geiger-Müller tube has a *dead time* of about 200 microseconds due to the time taken by the positive ions to travel towards the anode. Ionizing particles arriving within this period will not give separate pulses, i.e. are not resolved. If radioactive substances emitted particles at regular intervals a maximum of 5000 pulses per second could be detected, but this is not so and in practice the counting rate is less. Almost every beta particle entering a Geiger-Müller tube is detected. By contrast, the detection efficiency for gamma rays is less than 1 per cent. Gamma photons produce ion-pairs indirectly in the gas of the tube as a result of the secondary electrons they create when absorbed by the tube wall (cathode). The number of such electrons is small since gamma rays interact weakly with matter and this accounts for the low detection efficiency.

If a resistor R is connected in series with a G-M tube as in Fig. 12.6b, a voltage pulse (of about 1 V) is created which can be applied to an electronic counter such as a scaler or a ratemeter. A *scaler*, Fig. 12.7*a*, counts the pulses and indicates on a series of dials the total received in a certain time. The *ratemeter*, Fig. 12.7*b*, has a meter marked in ' counts per second ' from which the average pulse rate can be read directly. It usually has a loudspeaker which gives a ' click ' for every pulse. For quantitative work the scaler is more accurate.

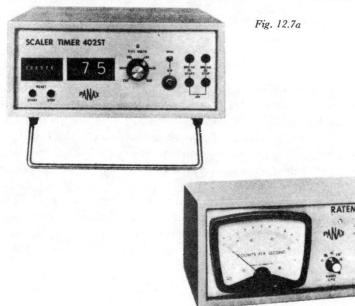

Fig. 12.7a

Fig. 12.7b

The characteristic curve of a G-M tube shows how its response depends on the applied p.d. and allows the selection of an appropriate working p.d. The curve is obtained by placing a beta source at a short distance from the tube and noting the count-rates on a scaler (or ratemeter) as the p.d. is increased. A typical curve is given in Fig. 12.8. When the p.d. reaches the *threshold* value, the count-rate remains almost constant over a range called the *plateau* (of 100 V or more). Here a full avalanche is obtained along the entire length of the anode and all particles whatever their energy (and nature) produce the same output pulse. Beyond the plateau the count-rate for the same intensity of radiation increases rapidly with p.d. and a continuous discharge occurs which damages the tube.

Normally a tube is operated about the middle of the plateau where the count-rate is unaffected by p.d. fluctuations. The plateau usually has a slight upward slope, calculated from

$$\text{percentage slope} = \frac{C_2 - C_1}{V_2 - V_1} \times \frac{100}{C_M} \text{ per cent per volt}$$

where $C_M = (C_1 + C_2)/2$. As the condition of a tube deteriorates, the length of the plateau decreases and the slope increases. The latter should be less than 0.15 per cent per volt.

Other experiments with a G-M tube are outlined on pp. 491 and 493.

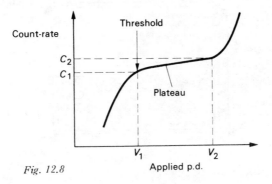

Fig. 12.8

(*c*) *Cloud chambers.* There are two types, and in both saturated vapour (alcohol) is made to condense on air ions created by the radiation and the resulting white line of tiny liquid drops shows up as a track in the chamber when suitably illuminated. Note that what we see is the track, not the ionizing radiation.

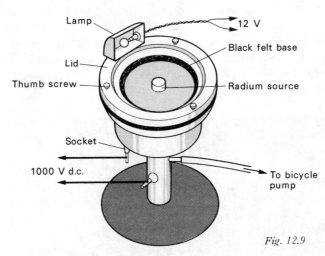

Fig. 12.9

In the *expansion cloud chamber* the vapour is cooled by a rapid expansion and condensation occurs. This is achieved in the small chamber of Fig. 12.9 by withdrawing sharply the piston of a bicycle pump having the cup washer reversed so that it removes air. Ions are being produced all the time (from a very weak source inside the chamber) and would cause blurred tracks at the moment of expansion. A high p.d. (e.g. 1 kV) between the top and bottom of the chamber provides a clearing field which removes them. Consequently only the tracks of radiation which has just left the source are observed.

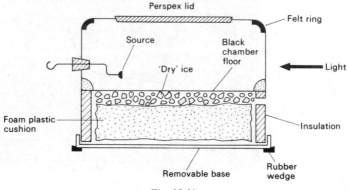

Perspex lid

Felt ring

Source

Black chamber floor

'Dry' ice

Light

Foam plastic cushion

Insulation

Removable base

Rubber wedge

Fig. 12.10

In the *diffusion cloud chamber* tracks are produced continuously. The upper compartment in the simple chamber of Fig. 12.10 contains air which is at room temperature at the top and at about $-78\ °C$ at the bottom due to the layer of ' dry ice ' (solid carbon dioxide) in the lower compartment. The felt ring at the top of the chamber is soaked with alcohol, which vaporizes in the warm upper region, diffuses downwards and is cooled. About 1 cm from the floor of the chamber, the air contains a layer of saturated alcohol vapour where conditions are suitable for the condensation of droplets on air ions produced along the path of the radiation from the source. Tracks are seen in this sensitive layer which are well-defined if an electric field is created by frequently rubbing the Perspex lid of the chamber with a cloth.

Fig. 12.11a

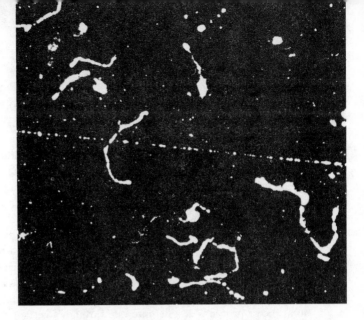

Fig. 12.11b

Alpha particles give bright, straight tracks like those in Fig. 12.11a. Very fast beta particles produce thin, straight tracks whilst those travelling more slowly give short, thicker, tortuous ones, Fig. 12.11b. Gamma rays (like X-rays) cause electrons to be ejected from air molecules and give tracks similar to X-rays; Fig. 12.11c is due to an intense beam of X-rays.

The *bubble chamber*, in which ionizing radiation leaves a trail of bubbles in liquid hydrogen, has replaced the cloud chamber and is now used extensively in high-energy nuclear physics. The photographs which they give of atomic collisions yield information about the particles involved. Estimates of mass and speeds can be obtained assuming momentum and energy are conserved.

Fig. 12.11c

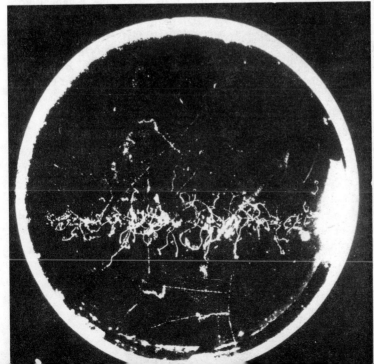

NUCLEAR PHYSICS

Radioactive decay

(a) *Disintegration theory.* The changes accompanying the emission of radiation from radioactive substances are unlike ordinary chemical changes in certain fundamental respects. They are spontaneous, they cannot be controlled and they are unaffected by chemical combination and physical conditions such as temperature and pressure. Energy considerations too suggest that they are different. Thus the energy released by a radioactive substance emitting alpha particles is several million electron-volts per atom compared with a few electron-volts per atom in any chemical change.

We believe that radioactivity involves the nucleus of an atom—not its extra-nuclear electrons as do chemical changes—and is an attempt by an *unstable* nucleus to become more stable. The emission of an alpha or beta particle from the nucleus of a radioactive atom produces the nucleus of a different atom— called the *daughter* or *decay atom*—which may itself be unstable. The disintegration process proceeds at a definite rate through a certain number of stages until a stable end-product is formed. When first advanced by Rutherford and others the theory conflicted with existing ideas about the permanency of atoms but it is a view which accounts satisfactorily for the known facts.

An alpha particle is a helium nucleus consisting of 2 protons and 2 neutrons and when an atom decays by alpha emission, its mass number decreases by 4 and its atomic number by 2. It becomes the atom of an element two places lower in the periodic table. For example, when radium of mass number 226 and atomic number 88 emits an alpha particle, it decays to radon of mass number 222 and atomic number 86. The change may be written

$$^{226}_{88}\text{Ra} \rightarrow {}^{222}_{86}\text{Rn} + {}^{4}_{2}\text{He}.$$

Radon in turn disintegrates and after seven more disintegrations, a stable isotope of lead is formed.

When beta decay occurs a neutron changes into a proton and an electron. The proton remains in the nucleus and the electron is emitted as a beta particle. The new nucleus has the same mass number, but its atomic number increases by one since it has one more proton. It becomes the atom of an element one place higher in the periodic table. Radioactive carbon, called carbon 14, decays by beta emission to nitrogen.

$$^{14}_{6}\text{C} \rightarrow {}^{14}_{7}\text{N} + {}^{0}_{-1}\text{e}.$$

Beta emitters have a higher proportion of neutrons; few occur naturally in appreciable quantities but many are obtained artificially by irradiating matter with neutrons.

The emission of gamma rays is explained by considering that nuclei (as well as atoms) have energy levels and if an alpha or beta particle is emitted, the nucleus is left in an excited state. A gamma ray photon is emitted when the nucleus returns to the ground state. The existence of different nuclear energy levels would account for the ' line-type ' energy spectrum of gamma rays (p. 478).

Naturally-occurring radioactive nuclides of high atomic number fall into three decay series, known as the *thorium* series, the *uranium* series and the *actinium* series. Radium is a member of the uranium series and at any time all the daughter nuclides will be present—which explains why a radium source emits alpha particles and gamma rays of several energies, as well as beta particles.

(*b*) *Decay law.* Radioactive decay is a completely haphazard or random process in which nuclei disintegrate quite independently. Since there is always a very large number of active nuclei in a given amount of radioactive material we can apply the methods of statistics to the process and obtain an expression for the certain fraction of the nuclei originally present that will have decayed on average in a given time interval.

We assume that the *rate of disintegration of a given nuclide at any time is directly proportional to the number of nuclei N of the nuclide present at that time;* that is, in calculus notation,

$$- \frac{dN}{dt} \propto N.$$

The negative sign indicates that N decreases as t increases. The *radioactive decay constant* λ is defined as the constant of proportionality in this expression, giving

$$- \frac{dN}{dt} = \lambda N. \tag{1}$$

If there are N_0 undecayed nuclei at some time $t = 0$ and a smaller number N at a later time t, then integrating (1),

$$\int_{N_0}^{N} \frac{dN}{N} = -\lambda \int_{0}^{t} dt$$

$$\therefore \quad (\ln N)_{N_0}^{N} = -\lambda t \qquad (\ln = \log_e)$$

$$\ln (N - N_0) = \ln (N/N_0) = -\lambda t \tag{2}$$

$$\therefore \quad N = N_0 \, e^{-\lambda t}. \tag{3}$$

This is the decay law and states that *a radioactive substance decays exponentially with time*—a fact confirmed by experiment. The law is a statistical one, it does not tell us when a particular nucleus will decay but only that after a certain time a certain fraction will have decayed. On the microscopic scale the process is purely random as is evident from the variations which occur when particles from a source are counted. On the macroscopic scale, however, where large numbers of particles are concerned, a definite law is followed.

The rate of decay of a source, i.e. the number of disintegrations per second, is called its *activity* and is often expressed in curies (Ci). One curie is the activity of a source which on average undergoes 3.70×10^{10} disintegrations per second. Sources used in school laboratories have activities of about 5 μCi which is very small compared with some medical sources.

From equation (1) we see that $\lambda = -dN/(N \cdot dt)$ and so the decay constant is the fraction of the total number of nuclei present which decays in unit time, provided the unit of time is small.

(c) *Half-life*. An alternative but more convenient term to the decay constant λ is the *half-life* $t_{\frac{1}{2}}$. This is the *time for the number of active nuclei present in a source at a given time to fall to half its value*. Whereas it is often difficult to know when a substance has lost practically all its activity, it is less difficult to find out how long it takes for the activity to fall to half the value it has at some instant.

Half-lives vary from millionths of a second to thousands of millions of years. Radium 226 has a half-life of 1622 years, therefore starting with 1 g of pure radium, $\frac{1}{2}$ g remains as radium after 1622 years, $\frac{1}{4}$ g after 3244 years and so on. An exponential decay curve, like that given on p. 53 for the discharge of a capacitor through a high resistor, is shown in Fig. 12.12 and illustrates the idea of half-lives again.

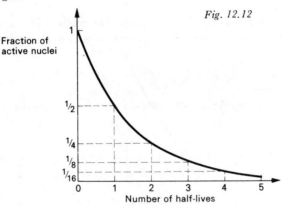

Fig. 12.12

The relationship between λ and $t_{\frac{1}{2}}$ can be derived from equation (3) for the decay law. Thus

$$N = N_0 e^{-\lambda t}$$

$$\therefore \quad \frac{N_0}{N} = e^{\lambda t}.$$

When $N = N_0/2$, $t = t_{\frac{1}{2}}$

$$\therefore \quad 2 = e^{\lambda t_{\frac{1}{2}}}.$$

Taking logs to base e, $\qquad \ln 2 = \lambda t_{\frac{1}{2}}$

$$\therefore \quad t_{\frac{1}{2}} = \frac{0.693}{\lambda}.$$

$t_{\frac{1}{2}}$ (and λ) is characteristic for each radionuclide and is an important means of identification.

Note. An analogue experiment using dice to illustrate the random decay law is outlined in Appendix 12, p. 542.

(*d*) *Nuclear stability.* Whilst the chemical properties of an atom are governed entirely by the number of protons in the nucleus (i.e. the atomic number Z), the stability of an atom appears to depend on both the number of protons and the number of neutrons. In Fig. 12.13 the number of neutrons ($A-Z$, where A is the mass number) has been plotted against the number of protons for all known nuclides, stable and unstable, natural and man-made. A continuous line has been drawn approximately through the stable nuclides (only a few are labelled) and the shading on either side of this line shows the region of unstable nuclides.

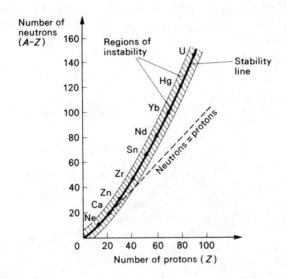

Fig. 12.13

For *stable* nuclides the following points emerge:

(*i*) The lightest nuclides have almost equal numbers of protons and neutrons.

(*ii*) The heavier nuclides require more neutrons than protons, the heaviest having about 50 per cent more.

(*iii*) Most nuclides have both an even number of protons and an even number of neutrons. The implication is that two protons and two neutrons, i.e. an alpha particle, form a particularly stable combination and in this connection, it is worth noting that oxygen ($^{16}_{8}$O), silicon ($^{28}_{14}$Si) and iron ($^{56}_{28}$Fe) together account for over three-quarters of the earth's crust.

For *unstable* nuclides the following points can be made:

(*i*) Disintegrations tend to produce new nuclides nearer the 'stability' line and continue until a stable nuclide is formed.

(*ii*) A nuclide above the line decays so as to give an increase of atomic number, i.e. by beta emission (in which a neutron changes to a proton and an electron). Its neutron-to-proton ratio is thereby increased.

(*iii*) A nuclide below the line disintegrates in such a way that its atomic number decreases and its neutron to proton ratio increases. In heavy nuclides this can occur by alpha emission.

Measuring half-lives

The next two experiments can be performed in a school laboratory.

(*a*) *Half-life of radon 220* ($^{220}_{86}$Rn). This is an alpha-emitting radioactive gas, usually called *thoron*, which collects, above its solid, parent nuclides in an enclosed space and is readily separated from them. If a polythene bottle containing thorium hydroxide is connected to an ionization chamber mounted on a d.c. amplifier, Fig. 12.14, a small quantity of thoron enters the chamber when the bottle is squeezed.

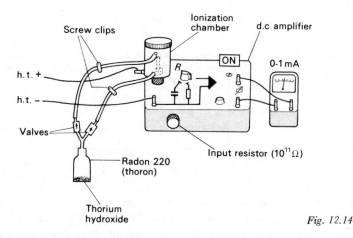

Fig. 12.14

As the thoron decays the ionization current decreases and is always a measure of the number of alpha particles present (i.e. the quantity of thoron remaining) so long as the p.d. across the chamber produces the saturation value of the current. If the time is determined for the reading of the output meter on the d.c. amplifier to fall to a half (or better still, a quarter) of some previous value, the half-life can be found. Alternatively the meter may be read every ten seconds and the half-life found from a graph of meter reading against time.

The chief constituent of the thoron source is thorium 232 and part of its decay series is shown below.

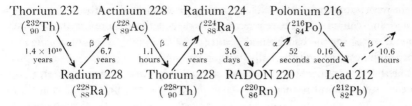

Thorium 232   Actinium 228   Radium 224   Polonium 216
$(^{232}_{90}Th)$ $(^{228}_{89}Ac)$ $(^{224}_{88}Ra)$ $(^{216}_{84}Po)$,

α β β α α α β
1.4 × 10¹⁰   6.7   1.1   1.9   3.6   52   0.16   10.6
years   years   hours   years   days   seconds   second   hours

Radium 228   Thorium 228   RADON 220   Lead 212
$(^{228}_{88}Ra)$ $(^{228}_{90}Th)$ $(^{220}_{86}Rn)$ $(^{212}_{82}Pb)$

The decay products following radon 220 do not affect the result appreciably since their half-lives are so very different from that of radon 220.

If some thoron is puffed into a closed diffusion cloud chamber, the two tracks can be observed which are due to the two alpha particles emitted almost simultaneously (Why?) when radon 220 decays to polonium 216 and then to lead 212. V-shaped tracks are obtained.

(*b*) *Half-life of protactinium 234* $(^{234}_{91}Pa)$. This is a beta-emitting daughter product in the decay of uranium 238; part of the series is given below.

low-energy   high-energy

Uranium α Thorium β PROTACTINIUM β Uranium
238    $\xrightarrow[\text{years}]{4.5 \times 10^9}$ 234 $\xrightarrow[\text{days}]{24}$ 234 $\xrightarrow[\text{seconds}]{72}$ 234
$(^{238}_{92}U)$ $(^{234}_{90}Th)$ $(^{234}_{91}Pa)$ $(^{234}_{92}U)$

The protactinium 234 is extracted almost completely by an organic solvent such as amyl acetate, from an acidified aqueous solution of uranyl nitrate (proportions—1 g uranyl nitrate, 3 cm³ water, 7 cm³ concentrated hydrochloric acid). Only the high-energy beta particles from the decay of protactinium 234 are detected by a G-M tube.

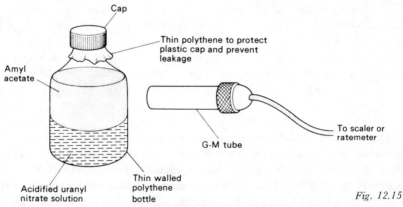

Cap

Thin polythene to protect plastic cap and prevent leakage

Amyl acetate

To scaler or ratemeter

G-M tube

Acidified uranyl nitrate solution

Thin walled polythene bottle

Fig. 12.15

Equal volumes of the organic and aqueous solutions are contained in a thin-walled, stoppered, polythene bottle so that each layer is as deep as the width of the G-M tube window. The bottle is well shaken and then arranged as in Fig. 12.15 with the G-M tube opposite the top half of the bottle. As soon as the

491

layers have separated, the scaler or ratemeter connected to the G-M tube is started and counts taken at ten-second intervals without stopping the counter. After allowing for the background count (p. 499), a graph of count-rate against time can be plotted and the half-life found.

The thorium from which the protactinium 234 has been extracted is left in the aqueous solution and immediately starts to generate the protactinium again as it decays. A second experiment may be carried out with the G-M tube opposite the lower half of the bottle to observe the growth of $^{234}_{91}$Pa.

Absorption of alpha, beta and gamma rays

(a) *Absorption processes.* In their passage through matter alpha and beta particles are ' absorbed ' by losing kinetic energy in ionizing encounters with atoms of the absorbing medium, i.e. an electron is ' knocked out ' of an atom to form an ion-pair. After travelling a certain distance, called the *range*, they have insufficient energy to produce any more ion-pairs and are then considered to have been absorbed. An alpha particle can also have an encounter with a nucleus (p. 414), and beta particles, because of their small mass, are readily ' back-scattered ' by atoms of the medium to emerge from the incident surface.

A measure of the intensity of the ionization produced when a charged particle passes through a gas is given by the *specific ionization*. This is defined as the number of ion-pairs formed per centimetre of path. It increases with the size of the charge on the particle and decreases as the speed of the particle increases. A fast-moving particle spends less time near an atom of the gas through which it is travelling and so there is less chance of an ion-pair being formed. An alpha particle may produce 10^5 ion-pairs per centimetre in air at atmospheric pressure while a beta particle of similar energy produces about 10^3 on account of its higher speed (on average about ten times greater) and smaller charge. It should, however, be noted that for alpha and beta particles of the same energy the *total* number of ion-pairs formed would be of the same order since a beta particle travels about one hundred times farther in air than an alpha particle. For each ion-pair produced in the track of any type of charged particle in air the *average* energy loss is 34 eV.

Gamma rays are usually most strongly absorbed by elements of high atomic number such as lead, and the absorption process is complex and differs from that occurring with charged particles. Whereas the alpha or beta particle gradually loses kinetic energy by a series of ionizing encounters with electrons belonging to atoms of the absorber, a gamma ray photon may interact with either an electron or, if it has enough energy, with a nucleus in several ways. The energy given up by the photon produces one or more high-speed ' secondary ' electrons and it is these which are responsible for the ionization created in a gas by gamma rays. They enable gamma rays to be detected by a G-M tube. The specific ionization of gamma rays therefore depends on the energy of the secondary electrons.

(b) *Range of alpha particles in air.* An ionization chamber with a d.c. amplifier can be used with the source (e.g. 5 μCi americium 241) supported centrally above the chamber which is fitted with a wire gauze lid, Fig. 12.16a. The ionization current (saturation) in the chamber is large when the source is close to the lid and decreases as d increases by, say, 5-mm steps. The corresponding readings of the d.c. amplifier output meter are noted and from a graph like that in Fig. 12.16b the range in air can be estimated.

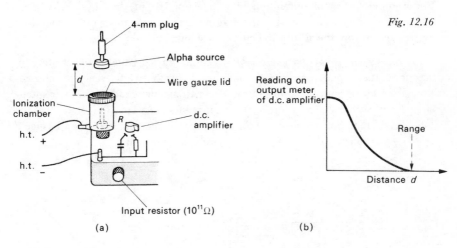

Fig. 12.16

(a)　(b)

(c) *Range of beta particles in aluminium.* Beta particles have a continuous energy spectrum and the number emerging from an absorber falls off gradually as the thickness of the absorber increases; this behaviour contrasts with the fairly sharp cut-off given by alpha particles. For practical purposes the range of beta particles is defined as the thickness of aluminium beyond which very few particles can be detected.

The range may be determined by inserting an increasing number of aluminium sheets between a pure beta source (e.g. 5 μCi strontium 90) and a G-M tube connected to a scaler or ratemeter, Fig. 12.17a. Count-rates are measured and an absorption curve is drawn.

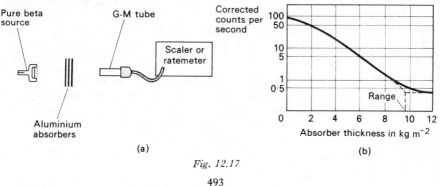

(a)　(b)

Fig. 12.17

493

A typical curve is given in Fig. 12.17*b* in which the wide range of count-rates is accommodated by plotting the logarithm of the count-rate. This is most conveniently done by plotting the count-rate (after subtracting the background count-rate) directly on the log scale of semi-log graph paper. The absorber thickness can be expressed either in millimetres of aluminium or, as is more common, in terms of the *surface density* of the absorber. This is the mass per unit area and equals the product of the density and the thickness of the absorber. For example, a sheet of aluminium 2.0 mm (2.0×10^{-3} m) thick of density 2.7×10^3 kg m^{-3} has a surface density of 5.4 kg m^{-2}. The range in kg m^{-2} is found by extrapolation from the absorption curve, as shown.

(*d*) *Inverse square law for gamma rays.* Gamma rays are highly penetrating on account of their small interaction with matter. In air they suffer very little absorption or scattering and, like other forms of electromagnetic radiation, their intensity falls off with distance according to the inverse square law. This states that the intensity of radiation I is inversely proportional to the square of the distance d from a *point* source. That is, $I = k/d^2$ where k is a constant.

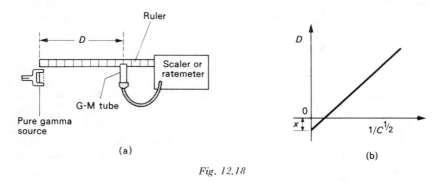

Fig. 12.18

The law may be investigated by placing a G-M tube at various distances from a pure gamma source (e.g. 5 μCi cobalt 60) and measuring the corresponding count-rates, Fig. 12.18*a*. If after correction for background, C is the count-rate, then C is directly proportional to I and we can write $C = k_1/d^2$, where k_1 is another constant. A graph of C against $1/d^2$ should be a straight line through the origin but small errors occur in the measurement of d and become important as d decreases. Non-linearity of the graph results. The errors can be eliminated by an alternative procedure.

Let $d = D + x$, where $D = $ distance measured from the source to any point on the G-M tube and $x = $ an unknown correction term which gives the true distance d. The law may then be written

$$C = k_1/(D + x)^2$$

$$\therefore D + x = (k_1/C)^{\frac{1}{2}}.$$

494

Hence
$$D = (k_1/C)^{\frac{1}{2}} - x.$$

A graph of D against $1/C^{\frac{1}{2}}$ should be a straight line of slope $k_1^{\frac{1}{2}}$ and intercept $-x$ on the D-axis, as shown in Fig. 12.18b.

In practice, departure from the law arises because of (i) the finite size of the source and (ii) counting losses due to the G-M tube having a ' dead-time ' when the source is close to the tube and count-rates high.

(e) *Half-thickness of lead for gamma rays.* The ' half-thickness ', denoted by $x_{\frac{1}{2}}$, is a convenient term for dealing with the absorption of gamma rays. It is defined as the thickness of absorber, usually lead, which reduces the intensity of the gamma radiation to half its incident value.

The half-thickness can be determined by inserting sheets of lead between a gamma source (e.g. 5 μCi cobalt 60) and a G-M tube and counter in a similar way to that described for beta absorption.

The count-rate C, corrected for background, is directly proportional to the intensity of the radiation, and it is again convenient to plot C directly on semi-log graph paper. The thickness of lead is usually expressed in the surface density unit of kg m^{-2}. An absorption curve for a parallel beam of monoenergetic gamma rays is shown in Fig. 12.19; by considering two (or more) half-thicknesses, i.e. the thickness required to halve the count-rate twice, a more accurate result is obtained for $x_{\frac{1}{2}}$.

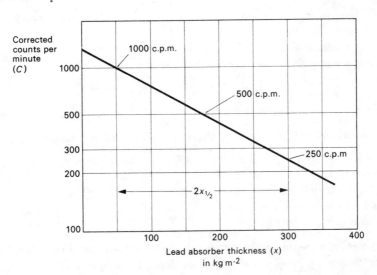

Fig. 12.19

The graph of log C against absorber thickness is seen to be a *straight line*. This implies that the intensity of the gamma radiation decreases *exponentially* with increasing absorber thickness. Thus if I_0 is the intensity of the gamma rays from the source in the absence of absorbing material and I is the intensity after passing through a thickness x of absorber, then

$$I = I_0\, e^{-\mu x}$$

where μ is a constant, called the *linear absorption coefficient*. Taking count-rate as a measure of intensity

$$C = C_0\, e^{-\mu x}$$

$$\therefore\ \ln C = \ln C_0 - \mu x$$

and
$$\log C = \log C_0 - \frac{\mu}{2.303}\cdot x.$$

The graph of $\log C$ against x should thus be a straight line if the absorption is exponential. This is only true for a parallel beam of monochromatic (i.e. of one wavelength and energy) gamma rays. μ, which is an alternative term to $x_{\frac{1}{2}}$, can be found from the slope of the line and is smaller for high energy (more penetrating) rays than for low energy rays.

Radioisotopes and their uses

(*a*) *Man-made radioactivity.* The first artificial radioactive substance was produced in 1934 by Irène Joliot-Curie (daughter of the discoverer of radium) and her husband. They bombarded aluminium with alpha particles and, as a result of a nuclear reaction (p. 499), obtained an unstable isotope of phosphorus:

$$^{27}_{13}\text{Al} + {}^{4}_{2}\text{He} \rightarrow {}^{30}_{15}\text{P} + {}^{1}_{0}\text{n} \qquad [{}^{27}_{13}\text{Al}\,(\alpha,\,\text{n})\,{}^{30}_{15}\text{P}].$$

The expression in brackets is the symbolic way of writing a nuclear reaction, i.e.

initial nuclide (incoming particles, outgoing particles) final nuclide.

Since then artificial radioisotopes (i.e. radioactive isotopes) of every element have been produced and today about 1500 are known. They are made by bombarding a stable element with neutrons in a nuclear reactor (p. 506) or with charged particles in a particle accelerator (p. 510). Their use in industry, research and medicine grows annually.

(*b*) *Some uses.* Some applications use the fact that the extent to which radiation is absorbed when passing through matter depends on its thickness and density. For example, in the manufacture of paper the thickness can be checked by having a beta source below the paper and a G-M tube and counter above it. A thickness gauge of this type may be adapted for automatic control of the manufacturing process. Level indicators also depend on absorption and are used to check the filling of toothpaste tubes and packets of detergents.

Leaks can be detected in underground pipe-lines carrying water, oil etc. by adding a little radioactive solution to the liquid being pumped. Temporary activity gathers in the soil around the leak which can be detected from the ground above, Fig. 12.20a. This technique is now used widely when studying river pollution, sand-wave movement in river estuaries and in the accurate

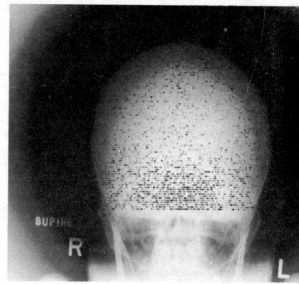

Fig. 12.20a and b

measurement of fluid flow. In research into wear in machinery a small amount of radioactive iron is introduced into the bearings and the rate of wear found from the resulting radioactivity of the lubricating oil.

The use of radioactive ' tracers ' in medicine, agriculture and biological research depends on the fact that the radioisotope of an element can take part in the same processes as its non-active isotope since it is chemically identical (and only slightly different physically). A scan of a normal brain is shown in Fig. 12.20b. It can be used to detect the concentration of a radioisotope given earlier to the patient and brain tumours can be located in this way. The use of radioactive phosphorus as a tracer in agriculture has provided information about the best type of phosphate fertilizer to supply to a particular crop and soil.

Gamma rays from high-activity cobalt 60 sources have many applications. In radiotherapy they are replacing X-rays from expensive X-ray machines in the treatment of cancer. The rapidly growing cells of the diseased tissue which cause cancer are even more affected by radiation than are healthy cells. Medical instruments and bandages are sterilized after packing by brief exposure to gamma rays. Food may be similarly treated and meat made to stay fresh for fifteen days instead of three. This is perfectly safe since no radioactivity is produced in the material irradiated by gamma rays.

(c) *Carbon 14 dating.* A natural radioisotope interesting from an archaeological point of view is carbon 14, formed in a nuclear reaction when neutrons ejected from nuclei in the atmosphere by cosmic rays collide with atmospheric nitrogen.

$$^{14}_{7}N + {}^{1}_{0}n = {}^{14}_{6}C + {}^{1}_{1}H \qquad [^{14}_{7}N\,(n, p)\,{}^{14}_{6}C].$$

Subsequently carbon 14 forms radioactive carbon dioxide and may be taken in by plants and trees for the manufacture of carbohydrates by photosynthesis. The normal activity of living carbonaceous material is 15.3 counts per minute per gram of carbon, but when the organism dies (or when part of it ceases to have any interaction with the atmosphere, as in the heartwood of trees) then no fresh carbon is taken in and carbon 14 starts to decay by beta emission with a half-life of 5.70×10^3 years. By measuring the residual activity, the age of any ancient carbon-containing material such as wood, linen or charcoal may be estimated within the range 1000 to 50 000 years. This has been done with the Dead Sea Scrolls (about 2000 years old) and charcoal from Stonehenge (about 4000 years old).

The ages of radioactive rocks have been similarly estimated.

(d) *Transuranic elements.* Uranium has the highest atomic number (92) of any naturally-occurring element, but traces of elements with atomic numbers from 93 to 104 have been made artificially and, like other artificial isotopes made from stable atoms in particle accelerators or nuclear reactors, they are radioactive; they are called the *transuranic* elements.

Radiation hazards

The radiation hazards to human beings arise from (i) exposure of the body to external radiation and (ii) ingestion or inhalation of radioactive matter.

The effect of radiation depends on the nature of the radiation, the part of the body irradiated and the dose received. The hazard from alpha particles is slight (unless the source enters the body) since they cannot penetrate the outer layers of skin. Beta particles are more penetrating in general: most of their energy is absorbed by surface tissues and adequate protection is afforded by a sheet of Perspex or aluminium a few millimetres thick. Gamma rays present the main external radiation hazard since they penetrate deeply into the body and may require substantial lead or concrete shielding.

Radiation can cause immediate damage to tissue and, according to the dose, is accompanied by radiation burns (i.e. redness of the skin followed by blistering and sores which are slow to heal), radiation sickness and, in extremely severe cases, by death. Delayed effects such as cancer, leukaemia and eye cataracts may appear many years later. Hereditary defects may also occur in succeeding generations due to genetic damage. The most susceptible parts are the reproductive organs, blood-forming organs such as the liver, and to a smaller extent the eyes.

NUCLEAR PHYSICS

The hands, forearms, feet and ankles are less vulnerable. Damage to human cells is due to the creation of ions which upset or destroy them.

The *radiation dose* is the energy absorbed (in $J\ kg^{-1}$) by unit mass of the irradiated material. Equal doses of different ionizing radiations provide the same amount of energy in a given absorber but they do not have the same biological effect on the human body. Thus

$$\text{effective dose} = \text{total dose} \times \text{r.b.e.}$$

where r.b.e. is the *r*elative *b*iological *e*ffectiveness. For beta particles, X- and gamma rays, the r.b.e. is about 1; for alpha particles, protons and fast neutrons it is about 10.

We all unavoidably absorb background radiation due to cosmic rays, radioactive minerals, radon in the atmosphere, potassium 40 in the body, X-rays from television screens. The dose rate from one of the weak sources used for experimental work in school physics is very small. In industry and research strong sources have to be handled by a ' master-slave ' manipulator and protection obtained for the operator from concrete and lead walls. The effective dose for radiation workers must not exceed a certain maximum in a certain period.

Nuclear reactions

Much information concerning the nucleus has been obtained by studying the effects of bombarding atomic nuclei with fast-moving particles. Three historic nuclear reactions will be considered.

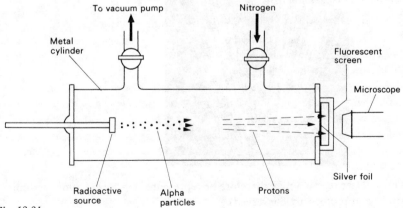

Fig. 12.21

(*a*) *Bombardment of nitrogen by alpha particles.* In 1919 Rutherford bombarded gases with alpha particles. His apparatus is shown in Fig. 12.21. One end of the metal cylinder was closed by thin silver foil capable of stopping most of the alpha particles from a radioactive source near the other end. Any which penetrated fell on a fluorescent screen and their scintillations were observed through a microscope. Various gases were admitted: with nitrogen, scintillations were obtained. Evidently more penetrating particles were produced and further experiments involving the deflection of these longer range particles in a magnetic field showed they were protons of high energy.

Two explanations seem plausible. Either an alpha particle ' chips ' a proton off a nitrogen nucleus, in which case the alpha particle survives the encounter; or the alpha particle actually enters the nitrogen nucleus and the latter immediately ejects a proton. In either case the proton is energetic enough to cause a scintillation on the screen. Since only one alpha particle in about a million caused a disintegration the chance of identifying the residual nucleus seemed small.

In 1925, however, about 20 000 photographs of the tracks of alpha particles passing through nitrogen in a cloud chamber were taken. In eight of these, a forked track like that of Fig. 12.22 was obtained. The alpha source is at the bottom and the collision occurs near the top. The short thick track at one o'clock is due to the residual nucleus and the long thin one at eight o'clock is that of a proton. There is nothing to indicate the existence of the alpha particle after the disintegration.

Fig. 12.22

This nuclear reaction may therefore be represented by the equation (assuming neutrons and protons are conserved),

$$\underset{\substack{\text{nitrogen}\\\text{nucleus}}}{{}^{14}_{7}\text{N}} \quad + \quad \underset{\substack{\text{alpha}\\\text{particle}}}{{}^{4}_{2}\text{He}} \quad \rightarrow \quad \underset{\substack{\text{residual}\\\text{nucleus}}}{{}^{17}_{8}\text{O}} \quad + \quad \underset{\text{proton}}{{}^{1}_{1}\text{H}} \qquad [{}^{14}_{7}\text{N}\ (\alpha,\ \text{p})\ {}^{17}_{8}\text{O}].$$

The residual nucleus is an isotope of oxygen. This was the first occasion on which one element (nitrogen) was changed into another (oxygen) although the number of atoms transformed was small. The event is sometimes referred to as the ' splitting of the atom '. One other point requires comment.

If the inverse square law holds right up to the nucleus the repulsive electro-static force would become infinitely large as the distance approaches zero. A charged particle would never enter a nucleus, however great its energy. Nuclear reactions do occur, however, as Rutherford demonstrated and so the repulsive force must be replaced by an extremely powerful attractive force acting over a range of about 10^{-15} m. It is in fact this force which keeps nucleons together in a nucleus and is responsible for the existence of matter; it is called the *nuclear force*.

The tracks produced by protons in a diffusion cloud chamber (p. 484) may be observed by making a small pin-hole in a thin sheet of polythene and placing it over a 5-μCi americium source, Fig. 12.23a, in the chamber. Alpha particles emerging from the pinhole produce short tracks all about four centimetres long, but occasionally a much longer track due to the ejection of a proton from the hydrogen-rich polythene appears as a result of alpha particle bombardment, Fig. 12.23b.

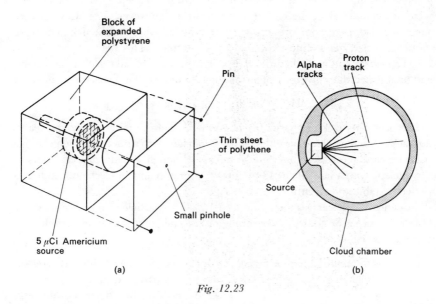

Block of expanded polystyrene

Pin

Alpha tracks

Proton track

Thin sheet of polythene

Source

Small pinhole

5 μCi Americium source

Cloud chamber

(a)

(b)

Fig. 12.23

(b) *Discovery of the neutron.* After the transmutation of nitrogen many other light elements were found to emit protons when bombarded by alpha particles. Beryllium, however, behaved unexpectedly and was found to emit radiation capable of penetrating many centimetres of lead. It could also cause protons of high energy to be shot out from materials such as paraffin wax which contain hydrogen, Fig. 12.24.

At first it was suggested that ' beryllium radiation ' might be very energetic

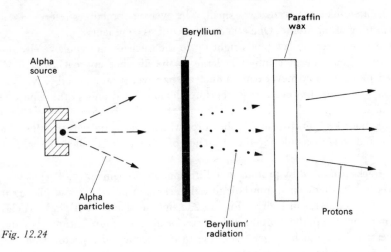

Fig. 12.24

gamma rays until it was realized that if this were so then energy and momentum were not conserved in the collision producing it. However, in 1932 Chadwick suggested that there would be conservation if the radiation consisted of uncharged particles with a mass almost the same as a proton's. This neutral particle was called a *neutron* ($_0^1$n) and Chadwick was able to account for all the observed effects and established that the action of alpha particles on beryllium was represented by

$$_4^9\text{Be} + {}_2^4\text{He} \rightarrow {}_6^{12}\text{C} + {}_0^1\text{n} \qquad [{}_4^9\text{Be}\,(\alpha,\,\text{n})\,{}_6^{12}\text{C}].$$

Because of its lack of charge the neutron penetrates matter easily and leaves no tracks in a cloud chamber unless it has a collision with a nucleus, when charged particles can be released that do give tracks. Although a neutron decays outside the nucleus with a half-life of 13 minutes to a proton and an electron, inside it is perfectly stable and an entity in its own right.

In general, nuclear reactions induced by alpha particles cause the emission of either protons or neutrons depending on the energy of the alpha particles and the nucleus under attack.

(*c*) *Cockcroft and Walton's experiment.* Alpha particles from radioactive sources have limitations as atomic projectiles. This is due partly to the fact that, because their energy is limited, they cannot overcome the powerful repulsion of the large positive nuclear charge of a heavy atom. In addition only a very small proportion of particles are generally able to cause disintegration, and so radioactive sources of high activity which emit a large number of particles per second would be necessary. These are both difficult to obtain and to manipulate. Consequently, in the early 1930s attention was given to the problem of building machines, called *particle accelerators*, to accelerate charged particles such as protons to high speeds by means of p.d.s of hundreds of thousands of volts.

NUCLEAR PHYSICS

The first nuclear reaction to be induced by artificially-accelerated particles was achieved in 1932 by Cockcroft and Walton. They bombarded lithium with protons and obtained alpha particles. The reaction is represented by the equation

$$^7_3Li + {}^1_1H \to {}^4_2He + {}^4_2He \qquad [^7_3Li\ (p,\ \alpha)\ {}^4_2He].$$

Mass and energy

(a) *Einstein's relation.* In 1905, while developing his special theory of relativity, Einstein made the startling suggestion that energy and mass are equivalent. He predicted that if the energy of a body changes by an amount E, its mass changes by an amount m given by the equation.

$$E = mc^2$$

where c is the speed of light.

Everyday examples of energy gain are much too small to produce detectable changes of mass. For example, when 1.0 kg of water absorbs 4.2×10^3 J to produce a temperature rise of 1 K, according to $E = mc^2$ the increase in mass of the water is

$$m = \frac{E}{c^2} = \frac{4.2 \times 10^3}{(3.0 \times 10^8)^2}$$

$$= 4.7 \times 10^{-14}\,kg.$$

The changes of mass accompanying energy changes in chemical reactions are not much greater and cannot be used to prove Einstein's equation.

However, radioactive decay, which is a spontaneous nuclear reaction, is more helpful. Thus for a radium atom, the combined mass of the alpha particle it emits and the radon atom to which it decays is, by atomic standards, appreciably less than the mass of the original radium atom. Atomic masses can now be measured to a very high degree of accuracy by mass spectrographs and the mass decrease m for the decay of one radium atom is 8.8×10^{-30} kg. The energy equivalent E is given by

$$E = mc^2 = 8.8 \times 10^{-30} \times (3.0 \times 10^8)^2\,J$$

$$= 7.9 \times 10^{-13}\,J.$$

But $$1\,eV = 1.6 \times 10^{-19}\,J$$

$$\therefore\ E = \frac{7.9 \times 10^{-13}}{1.6 \times 10^{-19}}\,eV$$

$$= 4.9 \times 10^6\,eV$$

$$= 4.9\,MeV.$$

The alpha particle carries off 4.8 MeV of this energy; some appears as k.e of the recoiling nucleus, and the rest is emitted soon afterwards as a γ-ray photon. Mass therefore appears as energy and the two can be regarded as equivalent. In nuclear physics mass is measured in *unified atomic mass units* (u), 1 u being one-twelfth of the mass of the carbon 12 atom and equals 1.66×10^{-27} kg. It can readily be shown, using $E = mc^2$, that

$$931 \text{ MeV has mass } 1 \text{ u.}$$

A unit of energy may therefore be considered to be a unit of mass, and in tables of physical constants the masses of various atomic particles are often given in MeV as well as in kg and u. For example, the electron has a rest mass of about 0.5 MeV.

If the principle of conservation of energy is to hold for nuclear reactions it is clear that mass and energy must be regarded as equivalent.

The implication of $E = mc^2$ is that any reaction producing an appreciable mass decrease is a possible source of energy. Shortly we will consider two types of nuclear reaction in this category.

(b) *Binding energy.* The mass of a nucleus is found to be less than the sum of the masses of the constituent protons and neutrons. This is explained as being due to the binding of the nucleons together into a nucleus and the mass defect represents the energy which would be released in forming the nucleus from its component particles. The energy equivalent is called the *binding energy* of the nucleus; it would also be the energy needed to split the nucleus into its individual nucleons if this were possible.

Consider an example. The helium atom, $_2^4$He, has an atomic mass of 4.0026 u;

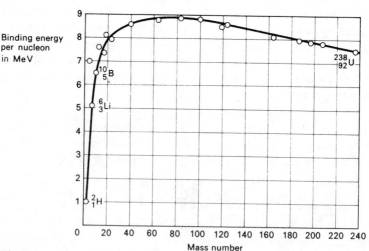

Fig. 12.25

this includes the mass of its two electrons. Its constituents comprise two neutrons each of mass 1.0087 u and two hydrogen atoms (i.e. two protons and two electrons) each of mass 1.0078 u; the total mass of the particles is thus $(2 \times 1.0087 + 2 \times 1.0078) = 4.0330$ u. The *mass defect* is $(4.0330 - 4.0026) = 0.0304$ u and since 1 u is equivalent to 931 MeV, it follows that the binding energy of helium is 28.3 MeV.

The binding energy is derived in a similar manner for other nuclides and is found to increase as the mass number increases. For neon, $^{20}_{10}\text{Ne}$, it is 160 MeV. If the binding energy of a nucleus is divided by its mass number (i.e. the number of nucleons), the *binding energy per nucleon* is obtained. The graph of Fig. 12.25 shows how this quantity varies with mass number; in most cases it is about 8 MeV. Nuclides in the middle of the graph have the highest binding energy per nucleon and are thus the most stable since they need most energy to disintegrate. The smaller values for higher and lower mass numbers imply that potential sources of nuclear energy are reactions involving the disintegration of a heavy nucleus or the fusing of particles to form a nucleus of higher mass number. In both cases nuclei are produced having a greater binding energy per nucleon and there is consequently a mass transfer during their formation.

Nuclear energy

(a) *Fission.* The discovery of the neutron provided nuclear physicists with an important new missile. Being uncharged, it is not repelled on approaching the large positive charge on the nucleus of a heavy atom as are protons and alpha particles. In 1939, following some work by Fermi in Italy and Hahn and Strassmann in Germany, it was found that the bombardment of uranium by neutrons may split the uranium nucleus into two large nuclei—for example, those of barium and krypton.

This nuclear reaction, called *nuclear fission*, differs from earlier nuclear reactions in three respects.

(*i*) The nucleus is deeply divided into two large *fission fragments* of roughly equal mass.

(*ii*) The *mass decrease* is appreciable.

(*iii*) Other neutrons, called *fission neutrons*, are emitted in the process.

Fission occurs in several heavy nuclides as well as in the two main isotopes of uranium, $^{238}_{92}\text{U}$ and $^{235}_{92}\text{U}$. The latter is the more useful and one reaction which occurs is

$$^{235}_{92}\text{U} + ^{1}_{0}\text{n} \rightarrow ^{144}_{56}\text{Ba} + ^{90}_{36}\text{Kr} + 2\,^{1}_{0}\text{n} \qquad [^{235}_{92}\text{U (n, 2n)}\,^{144}_{56}\text{Ba},\,^{90}_{36}\text{Kr}].$$

The mass decrease in this reaction is about 200 MeV per fission, which is 45 million times greater per atom of fuel (uranium 235) than in a chemical reaction where the energy change is no more than a few electron-volts per atom. The uranium 235 nucleus is evidently a vast storehouse of energy. The fission energy appears mostly as kinetic energy of the fission fragments (e.g. barium and krypton nuclei) which fly apart at great speed. The kinetic energy of the fission neutrons also makes a slight contribution. In addition, one or both of the large fragments are highly radioactive and a small amount of energy takes the form of beta and gamma radiation.

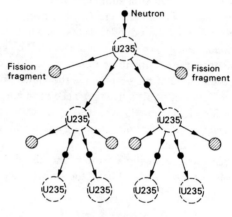

Fig. 12.26

The production of fission neutrons (two in the above case) raises the possibility of having a *chain reaction* in which fission neutrons cause further fission of uranium 235 and the reaction keeps going, Fig. 12.26. In practice only a proportion of the fission neutrons is available for new fissions since some are lost by escaping from the surface of the uranium before colliding with another nucleus. The ratio of neutrons escaping to those causing fission decreases as the size of the piece of uranium 235 increases and there is a *critical size* (about the size of a cricket ball) which must be attained before a chain reaction can start.

In the *atomic bomb* an increasing uncontrolled chain reaction occurs in a very short time when two pieces of uranium 235 (or plutonium 239) are rapidly brought together to form a mass greater than the critical size. The reaction is started either by having a small neutron source in the bomb or by stray neutrons from the occasional spontaneous fission of a uranium 235 nucleus.

(b) *Nuclear reactors.* In a nuclear reactor the chain reaction is steady and controlled so that on average only one neutron from each fission produces another fission. The reaction rate is adjusted by inserting neutron–absorbing rods of boron steel into the uranium 235. The basic arrangement is shown in Fig. 12.27. The graphite core is called the *moderator* and is needed to maintain a chain reaction when the fuel is not fairly pure uranium 235.

Natural uranium contains over 99 per cent of uranium 238 and less than 1 per cent of uranium 235 and unfortunately the former captures the medium-speed

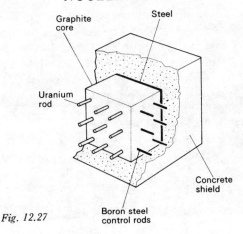

Graphite core

Steel

Uranium rod

Concrete shield

Fig. 12.27

Boron steel control rods

fission neutrons without fissioning. The reaction which occurs is

$$^{238}_{92}U + ^{1}_{0}n \rightarrow ^{239}_{92}U \rightarrow ^{239}_{93}Np + ^{0}_{-1}e \qquad [^{238}_{92}U\,(n,\,e)\,^{239}_{93}Np].$$

Neptunium 239 is the first transuranic element and decays, also by beta emission, to plutonium, the second transuranic element.

$$^{239}_{93}Np \rightarrow ^{239}_{94}Pu + ^{0}_{-1}e.$$

Uranium 238 only fissions with very fast neutrons. On the other hand, uranium 235 (and plutonium 239) fissions with *slow neutrons* and the job of the moderator is to slow down the fission neutrons very quickly so that most escape capture by uranium 238 and then cause the fission of uranium 235. A bombarding particle gives up most energy when it has an elastic collision with a particle of similar mass. For neutrons, hydrogen atoms would be most effective but unfortunately absorption occurs. However, deuterium (in heavy water) and carbon (as graphite) are both suitable.

In a nuclear power station a nuclear reactor provides the heat required to produce steam instead of a coal- or oil-burning furnace. In the reactor core the large fission fragments share their kinetic energy with surrounding atoms as a result of collisions and raise the internal energy of the system. In a gas-cooled reactor a current of high pressure carbon dioxide gas is pumped through the core of the reactor and transfers heat to the heat exchanger where water is converted to steam which is then used as in a conventional power station to drive a turbo-generator. A thick concrete shield gives protection from neutrons and gamma rays.

Nuclear reactors like that just described are called *thermal reactors* because fission is caused by slow neutrons with thermal energies, i.e. energies equal to the average kinetic energy of the surrounding atoms. The *fast breeder reactor* has a core made of highly enriched fuel (i.e. natural uranium from which uranium 238 has been extracted), fission occurs by *fast* neutrons and no modera-

Fig. 12.28a and b

tor is required. The core is surrounded by a blanket of natural uranium which is converted to plutonium 239 by fission neutrons that escape from the core and would otherwise be lost. New fissionable material is thus ' bred ' for the reactor. Fig. 12.28*a* shows an external view of the Prototype Fast Reactor (PFR) at Dounreay in the north of Scotland; it is designed to operate a 250–MW electricity generator. Fig. 12.28*b* shows the reactor top and the charge machine console.

Important by-products of nuclear reactors are artificial radioisotopes; they are made when stable nuclides, inserted in the core of the reactor, are bombarded by neutrons.

(c) *Fusion.* The union of light nuclei into heavier nuclei can also lead to a transfer of mass and a consequent liberation of energy. Such a reaction has been achieved in the ' hydrogen bomb ' and it is believed to be the principal source of the sun's energy.

A reaction with heavy hydrogen or deuterium which yields 3.3 MeV per fusion is

$$^{2}_{1}\text{H} + {}^{2}_{1}\text{H} \rightarrow {}^{3}_{2}\text{He} + {}^{1}_{0}\text{n} \qquad [{}^{2}_{1}\text{H}\,({}^{2}_{1}\text{H, n})\,{}^{3}_{2}\text{He}].$$

By comparison with the 200 MeV per fission of uranium 235 this seems small, but per unit mass of material it is not. Fusion of the two deuterium nuclei, i.e. deuterons, will only occur if they overcome their mutual electrostatic repulsion. This may happen if they collide at very high speed when, for example, they are raised to a very high temperature. If fusion occurs, enough energy is released to keep the reaction going and since heat is required, the process is called *thermonuclear fusion.*

Temperatures between 10^{8} and 10^{9} K are required and in the uncontrolled thermonuclear reaction occurring in the hydrogen bomb, the high initial temperature is obtained by using an atomic (fission) bomb to trigger off fusion. If a controlled fusion reaction can be achieved an almost unlimited supply of energy will become available from deuterium in the water of the oceans. The problem is, first, how to achieve the high temperature necessary and, second, how to keep the hot reacting gases (called the plasma) from touching the vessel holding it. Research continues on a world-wide scale into this difficult but challenging task. Fig. 12.29 shows some of the equipment being used for research into fusion at the United Kingdom Atomic Energy Authority's laboratories at Culham.

Fig. 12.29

Particle physics

Insight into the structure of the nucleus has been obtained by bombarding matter with high-speed particles, thereby causing nuclear reactions. As we have seen, alpha particles from radioactive sources were the earliest projectiles but these have limitations (p. 502), among them a maximum energy of 10 MeV. The various types of particle accelerator are designed to provide a copious but controlled supply of a variety of particles such as electrons, protons, deuterons etc. of known, high energy. Before describing some of these in outline, another natural source of high-speed particles will be considered.

(a) *Cosmic rays*. Ionizing radiations from outer space, called *cosmic rays*, were first postulated to explain the gradual discharge of a well-insulated electroscope. The rays have been of great interest to physicists, not only because of the mystery of their origin but also on account of their energies which are far in excess of any that present-day accelerators can produce. Their use as projectiles for bombarding nuclei has led to the discovery of other particles such as the positron (a positive electron) and various mesons (short-lived and most with masses between that of the electron and the proton).

Those rays arriving at the outer limit of the earth's atmosphere are called *primary cosmic rays* and consist mostly of protons. Their mean energy is 10 GeV (10^{10} eV) but some particles have energies greater than 10^{10} GeV. Information about the primary rays has been obtained by sending stacks of special photographic plates to heights of 30 km using balloons, and more recently rockets and artificial satellites have been used. The high-energy primary rays react with

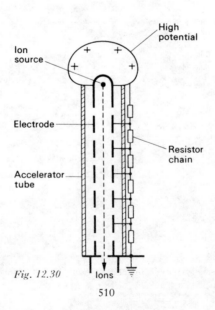

Fig. 12.30

510

atomic nuclei in the earth's upper atmosphere, causing nuclear reactions in which the nucleons themselves change and are not just rearranged within the nucleus as in lower energy reactions. As a result *secondary cosmic rays* are produced comprising electrons, mesons, gamma rays, positrons, neutrons and protons and these create most of our ' background ' radiation.

(*b*) *Electrostatic accelerator.* A high p.d. is used to accelerate charged particles (protons, deuterons, alpha particles etc.) between two electrodes in an evacuated tube. Cockcroft and Walton (p. 502) produced a p.d. of 2 MV from an a.c. input using a voltage multiplying circuit. A van de Graaff generator may also be used and a maximum p.d. of 14 MV obtained.

Ions from an appropriate source enter the accelerator tube, Fig. 12.30, which contains a series of cylindrical electrodes at decreasing potentials. The ions are accelerated while travelling between the electrodes (since the electric field well inside a charged conductor is zero) and, because the field lines are curved there, the ions are focused into a narrow beam. They emerge through a fine slit at the bottom on to the target.

(*c*) *Linear accelerator.* It is possible to produce high-energy particles without especially large p.d.s by a method called *synchronous acceleration* and which is used in the linear accelerator. The electrodes are again a series of coaxial cylinders but they increase in length towards the target and alternate ones are connected to the same output terminal of a high-frequency alternating p.d. as in Fig. 12.31.

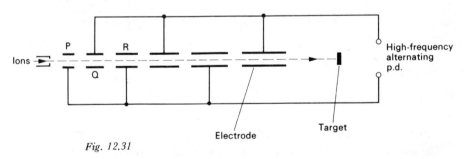

Fig. 12.31

Acceleration occurs in the small gaps between the electrodes, the electric field inside an electrode being zero. Positive ions from the source which reach the gap between P and Q when Q is at a high negative potential relative to P will be greatly accelerated. If the length of Q is such that the time taken by the ion (travelling with constant velocity) to travel through Q is half the period of oscillation of the supply p.d. then the field between Q and R will produce further acceleration, i.e. R will be at a negative potential. If the peak value of the supply p.d. is 200 kV, the ions gain 200 keV of energy at each gap and because their speed increases as they travel along the tube, the electrodes must progressively increase in length. What is the advantage of using a very high frequency supply? Electron linear accelerators working on the same principle are also in use.

(*d*) *Cyclotron*. The cyclotron also uses synchronous acceleration but a magnetic field makes the charged particles traverse a spiral of increasing radius rather than a long straight path. The machine consists of two semi-circular boxes, D_1 and D_2 in Fig. 12.32a, called 'dees' because of their shape, enclosed in a chamber C containing gas at low pressure. C is arranged between the poles of an electromagnet so that a nearly uniform magnetic field acts at right angles to the plane of the dees. A hot filament F emits electrons which ionize the gas present, producing protons from hydrogen, deuterons from deuterium etc. An alternating electric field is created in the gap between the dees by using them as electrodes to apply a high-frequency alternating p.d. Inside the dees there is no electric field, only the magnetic field.

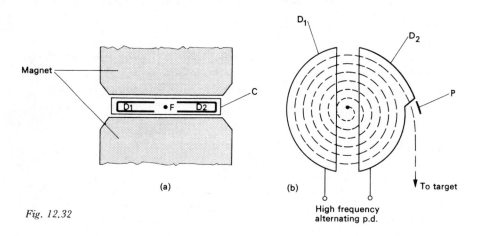

Fig. 12.32

(a)

(b)

High frequency
alternating p.d.

To target

Suppose that at a certain instant D_1 is positive and D_2 negative. A positively charged particle starting from F will be accelerated towards D_2 and when inside this dee it describes a semi-circular path at constant speed since it is under the influence of the magnetic field alone. The radius r of this path is given by

$$BQv = \frac{mv^2}{r}$$

where B is the magnetic flux density, Q the charge on the particle, v its speed inside D_2 and m its mass. Hence

$$r = \frac{mv}{BQ}. \tag{1}$$

If the frequency of the alternating p.d. is such that the particle reaches the gap again when D_1 is negative and D_2 positive, it accelerates across the gap and describes another semi-circle inside D_1 but of greater radius since its speed has increased. The particle thus gains kinetic energy and moves in a spiral of increasing radius, Fig. 12.32b, provided that the time of one complete half-oscillation of the p.d. equals the time for the particle to make one half-revolution. We shall now show that this condition generally holds.

Let T be the time for the particle to describe a semi-circle of radius r with speed v; then

$$T = \frac{\pi r}{v}.$$

Substituting for r from (1), we get

$$T = \frac{\pi m}{BQ}.$$

T is therefore independent of v and r and constant if B, m and Q do not alter. Hence, for paths of larger radius the increased distance to be covered is exactly compensated by the increased speed of the particle. After about 100 revolutions, a plate P, at a high negative potential, draws the particles out of the dees before it bombards the target under study.

(e) *Synchrotron*. It is not possible to obtain protons with energies greater than about 20 MeV using a cyclotron. The limit arises when the speed of the particle is sufficiently great for its relativistic increase of mass to increase the time of revolution and upset the synchronization. The synchrotron, on the other hand, uses this effect.

The speed of the particle is first increased to be close to that of light, either by a subsidiary accelerator or in some other way. The particle then traverses the same circular path under the action of a magnetic field and its energy increased

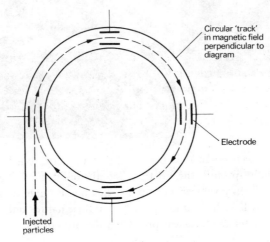

Fig. 12.33

by applying synchronous electrical pulses from a high-frequency alternating p.d. connected to several cylindrical electrodes spaced round the track, Fig. 12.33. As a result, the relativistic mass of the particle increases but its speed remains more or less the same.

Synchrotrons have been built to accelerate both electrons and protons but the initial acceleration is more easily achieved for electrons. Fig. 12.34a shows part of the 30 GeV proton synchrotron at CERN (The European Organization for Nuclear Research) near Geneva, whose energy is now being increased to 300 GeV. The whole laboratory covers a large area, Fig. 12.34b and is devoted to pure research into the fundamental structure of matter by groups of visiting scientists from the twelve European member countries.

Fig. 12.34a and b

(*f*) *Elementary particles.* In 1932 it seemed that all matter was built from three basic particles—the proton, the neutron and the electron. Today about 200 'particles' have been produced in nuclear reactions brought about by particle accelerators. Many have lives of less than 10^{-10}s, their nature and function is one of the exciting mysteries of nuclear physics at the present time.

NUCLEAR PHYSICS

QUESTIONS

1. Estimate the energy in MeV of an alpha particle from a source of activity 1.0 μCi which creates a saturation current of 1.0×10^{-9} A in an ionization chamber. Assume $e = 1.6 \times 10^{-19}$ C, 1 Ci $= 3.7 \times 10^{10}$ disintegrations s^{-1} and 30 eV is needed to produce one ion pair.

2. Part of the uranium decay series is shown below.

$$\overset{(1)}{\underset{92}{^{238}}\text{U} \rightarrow} \overset{(2)}{\underset{90}{^{234}}\text{Th} \rightarrow} \overset{(3)}{\underset{91}{^{234}}\text{Pa} \rightarrow} \overset{(4)}{\underset{92}{^{234}}\text{U} \rightarrow} \overset{(5)}{\underset{90}{^{230}}\text{Th} \rightarrow} \underset{88}{^{226}}\text{Ra}$$

(a) What particle is emitted at each decay?
(b) How many pairs of isotopes are there?
(c) If the stable end-product of the complete uranium series is lead 206, how many alpha particles are emitted between radium 226 and the end of the series?

3. A radioisotope of silver has a half-life of 20 minutes. (a) How many half-lives does it have in one hour? (b) What fraction of the original mass would *remain* after one hour? (c) What fraction would have *decayed* after two hours?

4. Taking the half-life of radium 226 to be 1600 years
(a) what fraction of a given sample remains after 4800 years
(b) what fraction has decayed after 6400 years, and
(c) how many half-lives does it have in 9600 years?

5. State the law governing the rate of decay of a radioactive substance, and explain the terms decay constant (λ) and half-life (T). Show that these two quantities are related by the equation

$$\lambda T = \ln 2.$$

Describe briefly how the decay law may be verified experimentally for a source of half-life of about one hour.

Two radioactive sources A and B initially contain equal numbers of radioactive atoms. Source A has a half-life of one hour, and source B a half-life of two hours. What is the ratio of the rate of disintegration of source A to that of source B (a) initially, (b) after two hours, and (c) after ten hours?   (O. and C.)

6. Discuss the assumption on which the law of radioactive decay is based.
What is meant by the *half-life* of a radioactive substance?
A small volume of a solution which contained a radioactive isotope of sodium had an activity of 12 000 disintegrations per minute when it was injected into the bloodstream of a patient. After 30 hours the activity of 1.0 cm^3 of the blood was found to be 0.50 disintegrations per minute. If the half-life of the sodium isotope is taken as 15 hours, estimate the volume of blood in the patient.   (J.M.B.)

7. Describe, with the aid of a diagram, the structure of a Geiger-Müller tube. Why does the tube contain a small quantity of a halogen or an organic gas?
Explain in what important respect a Geiger-Müller tube suitable for detecting beta particles differs from a tube used for detecting gamma radiation. State, giving your reasons, whether either tube would be suitable for detecting alpha particles.

A Geiger-Müller tube in conjunction with a scaler was used to investigate the rate of decay of a radioactive isotope of protactinium. Counts were made over periods of ten seconds, each count starting 30 seconds after the previous count was completed. The results, corrected for background radiation, were recorded as follows:

Time interval from start:	0–10	40–50	80–90 seconds
Count:	3410	2310	1620
Time interval from start:	120–130	160–170	220–210 seconds
Count:	1110	770	505

Determine, graphically or otherwise, the half-life of the isotope. (L.)

8. What is gamma-radiation? Explain *one* way in which it originates.

An experiment was conducted to investigate the absorption by aluminium of the radiation from a radioactive source by inserting aluminium plates of different thicknesses between the source and a Geiger tube connected to a ratemeter (or scaler). The observations are summarized in the following table:

Thickness of aluminium (cm)	Corrected mean count rate (min^{-1})
2.3	1326
6.9	802
11.4	496
16.0	300

Use these data to plot a graph and hence determine for this radiation in aluminium the *linear absorption coefficient*, μ (defined by $\mu = -\dfrac{dI}{I} \cdot \dfrac{1}{dx}$ where I is the intensity of the incident radiation and dI is the part of the incident radiation absorbed in thickness dx).

Draw a diagram to illustrate the arrangement of the apparatus used in the experiment and describe its preliminary adjustment.

What significance do you attach to the words *corrected* and *mean* underlined in the table? (*J.M.B.*)

9. Write short notes on *atomic number, mass number, neutron*.
Discuss briefly the reaction represented by

$$^{9}_{4}\text{Be} + {}^{4}_{2}\text{He} = {}^{12}_{6}\text{C} + {}^{1}_{0}\text{n}.$$

What is the nature and possible origin of the particle $^{4}_{2}\text{He}$? (L.)

10. A particle of mass m and speed v has a head-on collision with a stationary particle of mass M. Assuming the collision is elastic, derive an expression for the velocity of each particle after impact.

Hence determine what happens if the moving particle is an alpha particle and the stationary particle is (*a*) an electron, (*b*) a helium atom, and (*c*) a gold atom.

Under what conditions is there maximum energy transfer?

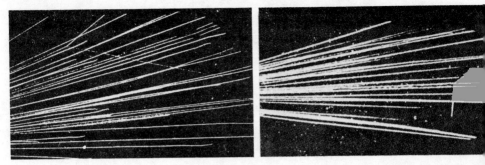

Fig. 12.35a and b

11. Two collisions between atomic particles are shown in the cloud chamber photographs of Figs. 12.35a and b. In both photographs most of the tracks are due to alpha particles travelling from left to right. In one case there is hydrogen in the chamber and in the other nitrogen and in each an elastic collision has occurred giving a ' split ' track.

(a) Which photograph shows the nitrogen collision? Why?

(b) Copy the hydrogen collision tracks (three parts) and label each part to show the tracks of the alpha particle before and after the collision and of the hydrogen atom set into motion.

(c) Draw the collision tracks for an alpha particle having an oblique elastic collision with a helium atom. Mark the size of any angle which you think is important.

12. Compare nuclear fission and nuclear fusion. How many fissions must occur per second to generate a power of 1.00 MW if each fission of uranium 235 liberates 200 MeV. (1 eV = 1.60×10^{-19} J.)

Objective-type revision questions

The first figure of a question number gives the relevant chapter, e.g. **3.2** is the second question for chapter 3.

Multiple choice

Select the response which you think is correct.

1.1. A sphere carrying a charge of Q coulombs and having a weight of W newtons falls under gravity between a pair of vertical plates distance d metres apart. When a potential difference of V volts is applied between the plates the path of the sphere changes as shown in Fig. 1, becoming linear along CD. The value of Q is

A $\dfrac{W}{V}$ **B** $\dfrac{W}{2V}$ **C** $\dfrac{Wd}{V}$ **D** $\dfrac{2Wd}{V}$ **E** $\dfrac{Wd}{2V}$ (*J.M.B.*)

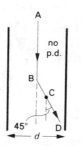

Fig. 1

2.1. A parallel plate capacitor is to be made by inserting one of the sheets of dielectric material indicated below, between and in contact with two plates of copper.

	Relative permittivity	Thickness (mm)
Teflon	2	0.4
Quartz	3	0.8
Glass	4	1.0
Mica	5	1.2
Porcelain	6	1.3

The maximum capacitance will be obtained by using the sheet of

A Teflon   **B** Quartz   **C** Glass   **D** Mica   **E** Porcelain   (*J.M.B.*)

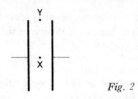

Fig. 2

2.2. Fig. 2 represents a parallel plate capacitor whose plate separation of 10 mm is very small compared with the size of the plates. If a potential difference of 5.0 kV is maintained across the plates, the intensity of the electric field in V m^{-1} at (*a*) X and (*b*) Y, is

A 50   **B** 2.5×10^2 **C** 5.0×10^2 **D** 2.5×10^5 **E** 5.0×10^5

3.1. Two long solenoids, P and Q, have n_1 and n_2 turns per unit length respectively and carry currents I_1 and I_2 respectively. The magnetic flux density on the axis of P at a point near the middle is four times that at a corresponding point in Q. The value of I_1/I_2 is

A $\dfrac{2n_1}{n_2}$ **B** $\dfrac{2n_2}{n_1}$ **C** $\dfrac{4n_1}{n_2}$ **D** $\dfrac{4n_2}{n_1}$ **E** $\dfrac{16n_1}{n_2}$ (*J.M.B.*)

3.2. The wire X on Fig. 3 is at right angles to the plane of the paper and carries a current into the paper. The magnetic flux density due to this current will be in the same direction as that of the horizontal component of the earth's field at a point on the diagram

A labelled 1   **B** labelled 2   **C** labelled 3   **D** labelled 4
E whose position is not deducible from the data without a knowledge of the value of the current.   (*J.M.B.*)

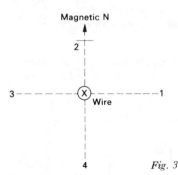

Fig. 3

4.1. The e.m.f. induced in a coil of wire which is rotating in a magnetic field does not depend on

A the angular speed of rotation   **B** the area of the coil
C the number of turns on the coil   **D** the resistance of the coil
E the magnetic flux density.

OBJECTIVE-TYPE REVISION QUESTIONS

4.2. When the speed of an electric motor is increased due to a decreasing load, the current flowing through it decreases. Which of the following is the best explanation of this?

    A The resistance of the coil changes.
    B Frictional forces increase as the speed increases.
    C Frictional forces decrease as the speed increases.
    D The induced back e.m.f. increases.
    E At high speeds it is more difficult to feed current into the motor.

4.3. When an ammeter is well damped,

    A large currents can be measured.
    B small currents can be measured.
    C readings can be taken quickly.
    D it is very accurate.
    E it is very robust.

4.4. Which feature of a moving coil ammeter is of most importance in making it well damped?

    A The strength of the hairsprings.
    B The number of turns on the coil.
    C The former on which the coil is wound.
    D The size of the iron cylinder.
    E The material used for the coil.

4.5. Alternating current is preferable to direct current for the transmission of power because

    A it can be rectified.
    B it is easier to generate.
    C thinner conductors can be used.
    D no question of polarity arises with equipment.
    E it is safer.

5.1. The direct current which would give the same heating effect in an equal constant resistance as the current shown in Fig. 4, i.e. the r.m.s. current, is

    A Zero   **B** $\sqrt{2}$ A   **C** 2 A   **D** $2\sqrt{2}$ A **E** 4 A   (*J.M.B.*)

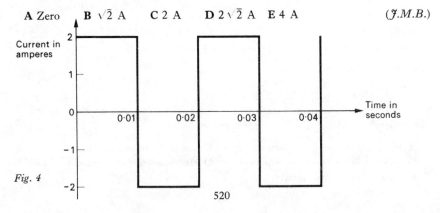

Fig. 4

520

5.2. What value of L, in henries, will make the circuit in Fig. 5 resonate at 100 Hz?

$\mathbf{A} \; \dfrac{1}{4\pi^2}$ $\quad \mathbf{B} \; \dfrac{1}{2\pi}$ $\quad \mathbf{C} \; 1$ $\quad \mathbf{D} \; 4\pi^2$

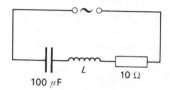

100 μF $\qquad L \qquad$ 10 Ω $\qquad$ *Fig. 5*

(J.M.B. Eng. Sc.)

5.3. The diagram in Fig. 6 shows a half wave rectifier with reservoir capacitor and load resistance.

The time constant of C and R should be

A small compared with the time of one cycle.
B independent of the time of one cycle.
C large compared with the time of one cycle.
D the same as the time of one cycle.

(J.M.B. Eng. Sc.)

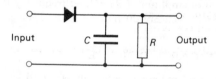

Input $\qquad$ C $\qquad$ R $\qquad$ Output

Fig. 6

6.1. Which one of the following statements is *not* correct? For interference to occur between two sets of waves

A each set must have a constant wavelength.
B the two sets must have the same wavelength.
C the two sets must not be polarized or must have corresponding polarizations.
D the waves must be transverse.
E the waves must have similar amplitudes.

6.2. Which of the graphs in Fig. 7 best represents the variation of the frequency of waves with wavelength if their speed remains constant?

Fig. 7

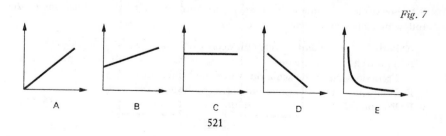

A $\qquad$ B $\qquad$ C $\qquad$ D $\qquad$ E

521

6.3. Fig. 8 shows two coherent sources of waves vibrating in phase and separated by a distance of 8.0 cm. P is the nearest point to the axis at which constructive interference occurs in a plane 24 cm from the line joining the sources. If S_1P is 26 cm, the wavelength of the waves in cm is

A 4.0   **B** 8.0   **C** 10   **D** 16   **E** 18

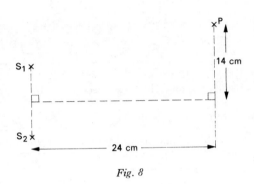

Fig. 8

6.4. Two waves each of amplitude 1.5 mm and frequency 10 Hz are travelling in opposite directions with velocity 20 mm s⁻¹. The distance in mm between adjacent nodes is

A 1.0   **B** 1.5   **C** 2.0   **D** 5.0   **E** 10

7.1. A steel piano wire 0.5 m long has a total mass of 0.01 kg and is stretched with a tension of 800 N. The frequency when it vibrates in its fundamental mode is

A 2 Hz   **B** 4 Hz   **C** 100 Hz   **D** 200 Hz   **E** 20 000 Hz   (*J.M.B.*)

7.2. A closed pipe resonates at its fundamental frequency of 300 Hz. Which one of the following statements is *not* correct?

A If the pressure rises the fundamental frequency increases.
B If the temperature rises the fundamental frequency increases.
C The first overtone is of frequency 900 Hz.
D An open pipe with the same fundamental frequency has twice the length.
E If the pipe is filled with a gas of lower density the fundamental frequency increases.

8.1. For a monochromatic light wave passing from air to glass which one of the following statements is true?

A Both frequency and wavelength decrease.
B The frequency increases and wavelength decreases in the same proportion.
C Frequency stays the same but wavelength increases.
D Frequency stays the same but wavelength decreases.
E Frequency and wavelength are unchanged.   (*J.M.B.*)

8.2. In a Young's double-slit interference experiment using green light the fringe width was observed to be 0.20 mm. If red light replaces green light the fringe width becomes

A 0.31 mm   **B** 0.25 mm   **C** 0.20 mm   **D** 0.16 mm   **E** 0.13 mm

(Wavelength of green light $= 5.2 \times 10^{-7}$ m; of red light $= 6.5 \times 10^{-7}$ m.)

(J.M.B.)

8.3. Newton's rings are formed in the air gap between a plane glass surface and the convex surface of a lens which is in contact with it. If the first bright ring has a radius of 2.0×10^{-4} m, the radius of the second bright ring in m is

A 1.4×10^{-4} **B** 2.8×10^{-4} **C** 3.5×10^{-4} **D** 4.0×10^{-4}
E 6.0×10^{-4}

9.1. At pressure P and absolute temperature T a mass M of an ideal gas fills a closed container of volume V. An *additional* mass $2M$ of the same gas is introduced into the container and the volume is then reduced to $V/3$ and the temperature to $T/3$. The pressure of the gas will now be

A $\dfrac{P}{3}$ **B** P **C** $3P$ **D** $9P$ **E** $27P$ *(J.M.B.)*

9.2. If, at a pressure of 10^5 Pa (N m^{-2}), the density of oxygen is 1.4 kg m^{-3}, it follows that the root mean square velocity of oxygen molecules in m s^{-1} is

A 5   **B** 18   **C** 120   **D** 270   **E** 460   *(J.M.B.)*

9.3. An ideal gas at 300 K is adiabatically expanded to twice its original volume and then heated until the pressure is restored to its initial value. What is the final temperature?

A 300 K   **B** 400 K   **C** 450 K   **D** 600 K   *(J.M.B. Eng. Sc.)*

9.4. A Carnot engine operates between temperatures of 600 K and 300 K and accepts a heat input of 1000 joules. The work output is

A 300 J   **B** 400 J   **C** 500 J   **D** 600 J   *(J.M.B. Eng. Sc.)*

9.5. A heat pump is to be installed to heat a greenhouse; the heat is to be extracted from a neighbouring stream at a maximum rate of 20 kW. The maximum temperature of the greenhouse is 300 K.

When the maximum temperature difference between the greenhouse and the stream is 40 K assuming a Carnot cycle the input power required for the pump will be of the order of

A 1 kW   **B** 3 kW   **C** 10 kW   **D** 30 kW   *(J.M.B. Eng. Sc.)*

OBJECTIVE-TYPE REVISION QUESTIONS

10.1. An electrostatic field E and a magnetic flux density B act over the same region, and an electron enters the region. Which one of the combinations of E and B in Fig. 9 can be made to cause the electron to pass undeflected?

(*J.M.B. Eng. Sc.*)

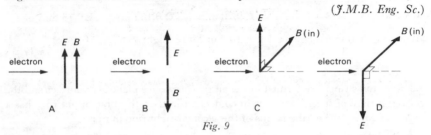

Fig. 9

10.2. Which one of the following phenomenon cannot be explained by the wave theory of light?

A Refraction   **B** Interference   **C** Diffraction   **D** Polarization
E Photoelectric effect

10.3. A photocell is illuminated with ultraviolet. The intensity of the illumination is reduced resulting in

A a reduction in the average kinetic energy of the electrons but no change in their rate of emission.
B no change in either the rate at which electrons are emitted or in their average kinetic energy.
C a reduction in the rate at which electrons are emitted but no change in their average kinetic energy.
D a reduction in both the rate at which they are emitted and their average kinetic energy. (*J.M.B. Eng. Sc.*)

10.4. Which one of the following statements is *not* correct? The value obtained by Thomson for the ratio of the charge to the mass of cathode rays

A did not depend on the gas in the tube.
B did not depend on the electrode material.
C did not depend on the tube material.
D did not depend on the accelerating voltage.
E was lower than the value obtained for protons.

11.1. When a linear time base of 25 Hz is applied to the X-plates of an oscilloscope and a sinusoidal voltage of 50 Hz is applied to the Y-plate, the trace could be, Fig. 10,

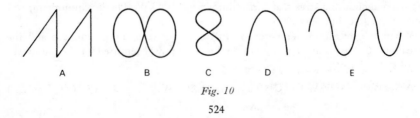

Fig. 10

524

OBJECTIVE-TYPE REVISION QUESTIONS

12.1. Two radioactive elements X and Y have half-value periods ('half-lives') of 50 minutes and 100 minutes respectively. Samples of A and B initially contain equal numbers of atoms. After 200 minutes the value of the fraction

$$\frac{\text{number of atoms of X unchanged}}{\text{number of atoms of Y unchanged}}$$

is   **A** 4   **B** 2   **C** 1   **D** $\frac{1}{2}$ **E** $\frac{1}{4}$ (*J.M.B.*)

12.2. When a lithium nucleus ($^7_3$Li) is bombarded with certain particles, two alpha particles only are produced. The bombarding particles are

A electrons   **B** protons   **C** deuterons   **D** neutrons
E photons

Multiple selection

In each question one or more of the responses may be correct. Choose one letter from the answer code given.

*Answer **A** if (i), (ii) and (iii) are correct*
*Answer **B** if only (i) and (ii) are correct*
*Answer **C** if only (ii) and (iii) are correct*
*Answer **D** if (i) only is correct*
*Answer **E** if (iii) only is correct*

1.2. When a positive charge is given to an isolated hollow conducting sphere

(*i*) a positively charged object anywhere inside the sphere experiences an electric force acting towards the centre of the sphere.

(*ii*) the potential to which it is raised is proportional to its radius.

(*iii*) the potential gradient outside the sphere is independent of its radius.

1.3. The intensity of an electric field

(*i*) is a vector quantity.

(*ii*) can be measured in N C^{-1}.

(*iii*) can be measured in V m^{-1}.

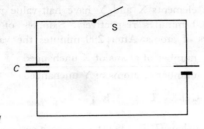

Fig. 11

2.3. A parallel plate capacitor *C* is charged by connection to a battery using the switch S as shown in Fig. 11. S is opened and the plate separation is then increased. Increases occur in the following

(*i*) the charge stored

(*ii*) the potential difference between the plates

(*iii*) the energy stored.

3.3. The magnetic flux density well inside a long uniformly wound solenoid depends on

(*i*) the number of turns per unit length

(*ii*) the area of cross-section

(*iii*) the distance from the axis of the solenoid.

4.6. The circuit of Fig. 12 shows a cell of negligible internal resistance in series with a pure inductor and a pure resistor. When the switch is closed

(*i*) the current rises initially at the rate of 6.0 A s⁻¹.

(*ii*) the final value of the current is 1.5 A.

(*iii*) the final energy stored in the space in and around the coil is 2.3 (2.25) J.

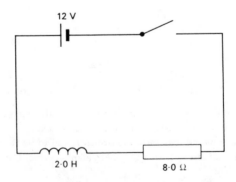

Fig. 12

10.5. If the accelerating voltage across an X-ray tube is doubled

(*i*) the wavelengths of the characteristic lines are halved.

(*ii*) the minimum wavelength of the X-rays is halved.

(*iii*) the X-rays are most probably more penetrating.

OBJECTIVE-TYPE REVISION QUESTIONS

Assertion-reason

Each question consists of an assertion (statement) followed by a reason. Only if you decide that both are true do you need to consider if the reason is a valid explanation of the assertion. Select your answer according to the code below.

Answer	Assertion	Reason	
A	True	True	Reason correct
B	True	True	Reason incorrect
C	True	False	—
D	False	True	—
E	False	False	—

5.4. When the frequency of an a.c. supply connected across a perfect capacitor is increased the average power dissipated in the capacitor increases *because* the reactance of a capacitor increases with frequency. (*J.M.B.*)

7.3. Sound waves can be diffracted *because* they are longitudinal.

7.4. The pitch of a train whistle is lower when the train is approaching the listener than when it is moving away *because* the velocity of sound in air is independent of the motion of the source. (*J.M.B.*)

9.6. The specific heat capacity of a gas at constant pressure is greater than at constant volume *because* at constant pressure the molecules have to travel greater distances in between collisions with one another.

11.2. When the filament current in a diode valve is increased the saturation current increases *because* more electrons flow through the filament in a certain time.

12.3. X-rays cannot be detected by a G-M tube *because* X-rays do not ionize gases.

Appendix 1

Construction and action of a flame probe

(*a*) *Construction*. Details are shown in Fig. A1.1. In use the flame should be as small as possible by *slowly* reducing the gas supply (a screw clip helps).

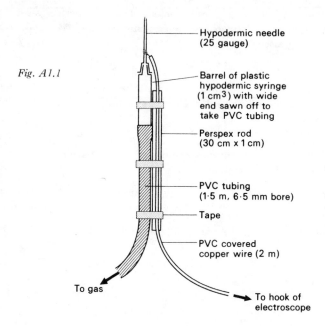

Fig. A1.1

Hypodermic needle
(25 gauge)

Barrel of plastic
hypodermic syringe
(1 cm³) with wide
end sawn off to
take PVC tubing

Perspex rod
(30 cm x 1 cm)

PVC tubing
(1·5 m, 6·5 mm bore)

Tape

PVC covered
copper wire (2 m)

To gas

To hook of
electroscope

(*b*) *Action*. The deflection of the electroscope is a measure of the potential at the point where the flame probe is situated. Roughly, the action is as follows. When brought near to a positively charged body the metal probe (i.e. the needle) has a negative charge induced in it whilst a positive charge appears on the electroscope movement. However, the flame is producing positive and negative ions; the former are attracted to the probe and neutralize its charge whilst the latter are repelled. The electroscope remains positively charged and is at the same potential as the uncharged probe since they are connected. The closer the probe is to the charged body the higher is its potential and the greater is the positive charge induced in the electroscope and so the greater the deflection.

Appendix 2

Capacitor experiments using a d.c. amplifier

A calibrated d.c. amplifier can replace the reed switch in certain capacitor experiments considered earlier.

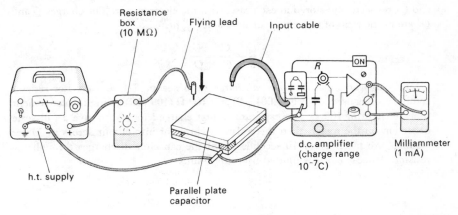

Fig. A2.1

(a) *Parallel plate capacitor investigation* (also see p. 36). The arrangement is shown in Fig. A2.1, the capacitor being two thick metal plates, each of side 25 cm, kept apart by four small polythene spacers (5 mm × 5 mm) about 1.5 mm thick, at the corners.

The capacitor is charged to a known p.d. using a flying lead from a protective resistance (10 MΩ) in series with the output from an h.t. power supply. The flying lead is removed and the charge on the capacitor measured by touching the upper plate with the tip of the screened input cable from the calibrated electrometer set on an appropriate charge range (usually 10^{-7} C). The input capacitor of the electrometer should be discharged (by switching to ' rest ').

By varying the charging p.d. V in 50-V steps up to 250 V and noting the corresponding charges Q on the capacitor, it will be seen that $Q \propto V$.

If V is kept constant (at, say, 200 V) and the separation d of the plates changed by using more spacers at the corners, $Q \propto 1/d$ can be tested.

Finally if V and d are fixed (e.g. at 200 V and one spacer thickness) and the area of overlap A of the plates is varied, a test of $Q \propto A$ is possible.

APPENDIX 2

(b) *Measurement of* ϵ_0 *and* ϵ_r (also see p. 41). To find ϵ_0, the permittivity of free space, we use the expression for the capacitance of a parallel plate capacitor, $C = A\epsilon_0/d$ (p. 39) and experimental results from (a). Since $Q = VC$ we have

$$\epsilon_0 = \frac{Qd}{AV}.$$

A mean value of Q/V can be obtained from a graph of Q against V; d and A are readily measured.

The relative permittivity ϵ_r of, say, polythene is found by charging the parallel plate capacitor to the same p.d. (e.g. 100 V) first with a sheet of polythene filling the space between the plates (giving capacitance C), then with the same thickness of vacuum (or air) between them (capacitance C_0) and measuring the charges (Q and Q_0 respectively) stored in each case, with the electrometer. The charges Q and Q_0 are in the ratio of the capacitances C and C_0, hence

$$\epsilon_r = \frac{C}{C_0} = \frac{Q}{Q_0}.$$

(c) *Decay curve for capacitor discharge* (also see p. 52). A capacitor-resistor combination is used, 10^{-9} F (0.001 μF) and 10^{11} Ω (100 GΩ) giving the convenient time constant of 100 s. The capacitor is first charged to 1 V to give a full-scale deflection on the meter, then the resistor is brought into circuit across it. Meter readings are taken every 10 seconds whilst the capacitor discharges through the resistor. A decay curve is plotted from the results.

Appendix 3

Calculation of B for an infinitely long straight wire using the Biot-Savart law

We require to find the flux density at P, perpendicular distance a in air from the long straight wire carrying current I.

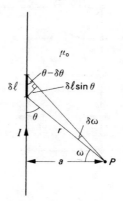

Fig. A3.1

The small element δl in Fig. A3.1 contributes flux density δB at P. By Biot and Savart,

$$\delta B = \frac{\mu_0 I \delta l \sin \theta}{4\pi r^2} = \frac{\mu_0 I}{4\pi} \cdot \frac{\delta \omega}{r} \text{ (since } \delta l \sin \theta = r \cdot \delta \omega\text{).}$$

The total flux density B at P is obtained by integrating this expression over the whole length of wire between the limits $-\pi/2$ and $+\pi/2$, where these are the angles subtended at P by the ends of the wire. We have $\cos \omega = a/r$, hence

$$B = \int_{-\pi/2}^{+\pi/2} dB = \frac{\mu_0 I}{4\pi a} \int_{-\pi/2}^{+\pi/2} \cos \omega \, d\omega$$

$$= \frac{\mu_0 I}{4\pi a} \left(\sin \omega \right)_{-\pi/2}^{+\pi/2}$$

$$= \frac{\mu_0 I}{2\pi a}.$$

531

Appendix 4

Demonstration of rectification and smoothing using a CRO

The circuit of Fig. A4.1a can be connected on a circuit board and the various waveforms studied on a CRO.

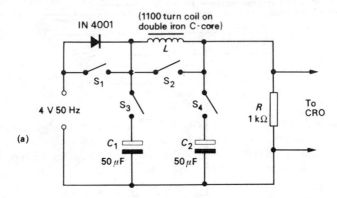

Fig. A4.1a

(a) With S_1 and S_2 closed, S_3 and S_4 open, the diode is short-circuited and the alternating input p.d. is developed across the load R.

(b) With S_2 closed, S_1, S_3 and S_4 open, half-wave rectification occurs but no smoothing.

(c) With S_2 and S_3 closed, S_1 and S_4 open, smoothing is produced by the reservoir capacitor C_1.

(d) With S_2, S_3 and S_4 closed, S_1 open, the capacitor-input filter LC_2 supplements the smoothing due to C_1.

(e) To show full-wave rectification and smoothing the 1N4001 diode is replaced by a bridge rectifier (e.g. BY 164), Fig. A4.1b.

Fig. A4.1b

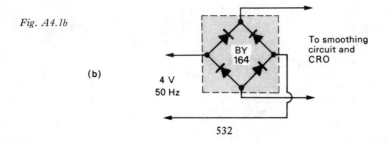

Appendix 5

Speed of sound in a solid

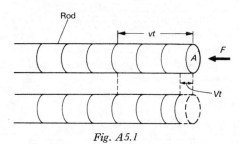

Fig. A5.1

Let an external force F be applied to the end of a solid rod of cross-sectional area A, setting it in motion with speed V and causing a compression pulse to travel along the rod with speed v. In time t the pulse covers a distance vt and this length of rod is compressed Vt, Fig. A5.1. Hence

$$\text{stress} = \frac{\text{force}}{\text{area}} = \frac{F}{A}$$

$$\text{strain} = \frac{\text{compression}}{\text{length compressed}} = \frac{Vt}{vt} = \frac{V}{v}.$$

If the material has Young's modulus E then

$$E = \frac{\text{stress}}{\text{strain}} = \frac{Fv}{AV}.$$

Therefore
$$F = \frac{EAV}{v}.$$

But, *force × time* equals *change of momentum* and momentum of mass of length vt of rod set into motion with speed V is $(vtA\rho)V$ where ρ is the density of the material. Thus

$$Ft = \left(\frac{EAV}{v}\right) t = vtA\rho V$$

$$\therefore v^2 = \frac{E}{\rho}$$

or,
$$v = \sqrt{\frac{E}{\rho}}.$$

Appendix 6

Ruling optical slits

Using a paint brush with fine bristles, one side of a microscope slide is coated with a smooth, thin paste of colloidal graphite (Aquadag) and water. The slide is allowed to dry, preferably overnight, and then inserted in the holder, Fig. A6.1. A blunt needle is held against the cross-piece and a slit ruled by drawing it across the slide so that it removes the graphite.

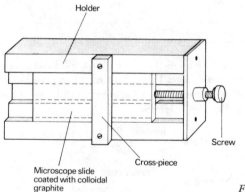

Holder

Screw

Cross-piece

Microscope slide
coated with colloidal
graphite

Fig. A6.1

If a second slit is required, the screw on the end of the holder is turned, thus moving the slide along slightly. A separation of 0.5 mm (usually one revolution of the screw) between the slits is suitable.

In the double-slit experiment the closer the slits are together the greater is the fringe spacing and the easier are the fringes seen. Also, wider slits give brighter fringes but fewer are obtained.

Appendix 7

Bromine diffusion experiment (pp. 342 and 344)

Bromine liquid and vapour blister the skin and the vapour will cause a sore throat. They also attack most materials but not glass.

(a) *Precautions*. A 500 cm³ beaker containing 0.88 concentrated ammonia solution diluted with its own volume of water should be to hand when preparing and performing the experiment. This solution must be poured at once on any bromine spilt on the bench or skin (harmless ammonium bromide is formed) but for the eyes use plenty of cold water.

(b) *Cleaning the apparatus*. Using rubber gloves the whole apparatus is placed in a plastic bucket half full of weak ammonia solution (200 cm³ of 0.88 ammonia to half a bucket of water). The rubber stopper is eased out from the side tube *under the solution*. As the solution enters, white clouds of ammonium bromide are formed. The apparatus is manipulated until all signs of bromine have gone. The diffusion tube is emptied, washed out with hot water and dried in a current of air.

The rubber tubing from the tap is removed *under the solution* and the tap dismantled before washing, drying and lightly lubricating with Vaseline.

(c) *Use of rubber stoppers and rubber tubing*. The rubber tubing should be discarded after use *once*. The stopper on the tap may be used two or three times within a day or two; thereafter it hardens and must be replaced.

Appendix 8

The 'random walk' game (p. 343)

Six arrows each making 60° with its neighbour are drawn in pencil in the centre of a piece of isometric grid paper (i.e. graph paper ruled with 60° triangles) and labelled 1 to 6, Fig. A8.1.

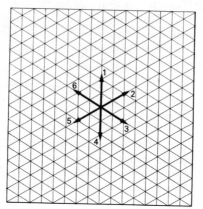

Fig. A8.1

A die is thrown and one step marked with a pen from the centre of the sheet in the direction given by the number on the die. The die is thrown again and a further step of one triangle-side taken *following on from the end of the first*. This is repeated for a total of 25 throws.

The distance in 'steps' from the starting point to the finishing point 'as the crow flies' is measured and compared with the theoretical value of √25.

If the game is played by the members of a class the scatter of results can be examined to see the large probable error in the result.

Appendix 9

Evidence of energy levels from the absorption spectrum of iodine vapour

Fig. A9.1a

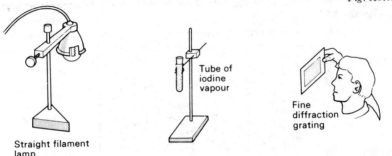

Tube of iodine vapour

Fine diffraction grating

Straight filament lamp

The experimental arrangement is shown in Fig. A9.1a, and what is seen in Fig. A9.1b. The test-tube (hard glass) containing one or two small iodine crystals, after being warmed along its length and lightly corked, is heated at the bottom until the iodine vaporizes and colours the tube strongly. As it cools it is observed in front of the straight filament lamp through a fine diffraction grating (about 300 lines per millimetre).

(a) I_2 Absorption

5000 5500 6000 A

Fig. A9.1b

The iodine molecules absorb from the light, those frequencies whose quanta have the correct amount of energy to enable them to jump from one energy level to another. Certain frequencies are therefore missing in the continuous spectrum of the light from the lamp and show up as dark lines. The presence of so many of these lines indicates that a whole 'ladder' of energy levels exists. In a gas or vapour the molecules are sufficiently far apart not to affect each other's energy levels to any extent.

Appendix 10

Photoelectric effect and rough estimate of Planck's constant

The apparatus is shown in Fig. A10.1. Before the electrometer/d.c. amplifier is connected to the photoelectric unit (*i*) its ' sensitivity' control is adjusted to give a full-scale deflection on its display meter (1 mA) when a 1.5-V cell is across its input, and (*ii*) any input resistor is removed so that it acts as a voltmeter whose resistance is as high as possible.

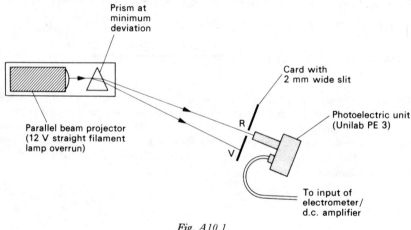

Prism at
minimum
deviation

Card with
2 mm wide slit

Photoelectric unit
(Unilab PE 3)

Parallel beam projector
(12 V straight filament
lamp overrun)

R

V

To input of
electrometer/
d.c. amplifier

Fig. A10.1

The photoelectric unit contains a photocell (see p. 396) with a photocathode of potassium (in a vacuum) which loses electrons when light falls on it. These travel to a collecting wire (of material that does not emit photoelectrons for light) which now becomes negatively charged and soon stops the arrival of further electrons. A steady p.d. is thus created between the cathode and the collecting wire and is measured by the electrometer connected across the photocell. If this is, say, 1.0 V then this must be the maximum k.e. of the electrons emitted by the cathode, otherwise they would continue to reach the collecting wire. The maximum k.e. of the photoelectrons is thus indicated by the electrometer reading (provided its input resistance is sufficiently high).

(*a*) *Effect of colour*. The spectrum is rotated *slowly* so that the colours from red to violet and beyond fall in turn on the slit in front of the photocell. The reading on the electrometer meter rises steadily indicating that the higher the frequency of the light, the greater the maximum k.e. of the photoelectrons.

The p.d. readings for red and violet light should be noted.

538

APPENDIX 10

(b) *Effect of intensity.* With violet light on the slit and the electrometer reading set to near full-scale (using ' sensitivity ' control), a stop is placed in front of the lamp to about halve the light intensity. The electrometer reading should remain *almost* constant showing that the maximum k.e. of the photoelectrons is more or less independent of the brightness of the light. (What is the effect of decreased intensity?)

(c) *Planck's constant.* An estimate of h may be obtained if we assume that for red light $f = 4.5 \times 10^{14}$ Hz and for violet light $f = 7.5 \times 10^{14}$ Hz. If the corresponding p.d.s are, say, 0.25 V and 1.45 V then the maximum k.e.s are 0.25 eV (i.e. $0.25 \times 1.6 \times 10^{-19}$ J) and 1.45 eV (i.e. $1.45 \times 1.6 \times 10^{-19}$ J) respectively. Hence from Einstein's photoelectric equation (p. 393) we have,

$$hf - W = \tfrac{1}{2}mv_{\text{max}}^2.$$

For violet light $\quad h \times 7.5 \times 10^{14} - W = 1.45 \times 1.6 \times 10^{-19}.$

For red light $\quad h \times 4.5 \times 10^{14} - W = 0.25 \times 1.6 \times 10^{-19}.$

Subtracting, $\qquad h = \dfrac{(1.45 - 0.25)\, 1.6 \times 10^{-19}}{(7.5 - 4.5)\, 10^{14}}$

$$= 6.4 \times 10^{-34} \text{ J s.}$$

Appendix 11

Electron collision experiments using a commercial xenon-filled thyratron valve (EN 91)

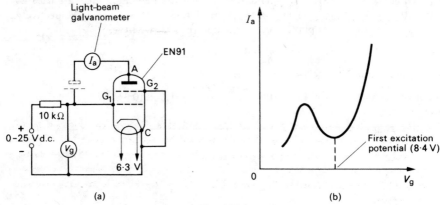

Fig. A11.1

(a) *Excitation potential.* The circuit is shown in Fig. A11.1a. Electrons emitted by the cathode C of the thyratron are accelerated by the positive potential V_g on the grid G_1, and initially are able to reach the anode A since, for small values of V_g, any collisions they have with atoms of the xenon gas are elastic. The electron flow I_a through the thyratron is recorded by the light beam galvanometer on its most sensitive range and maintains A at a small negative potential with respect to G_1.

As V_g is increased some electrons have just enough energy to cause excitation of xenon atoms and, having lost all their kinetic energy in the inelastic collision, are unable to overcome the retarding p.d. between A and G_1 causing I_a to fall. Further increase of V_g causes more electrons to lose their energy after excitation and I_a decreases further. When most of the electrons passing through G_1 produce excitation, I_a is a minimum and the corresponding value of V_g gives the *first excitation potential* of xenon. Increasing V_g beyond this value enables electrons, even after they have had inelastic collisions, to overcome the retarding p.d., reach A and I_a rises again. The form of the $I_a - V_g$ graph is shown in Fig. A11.1b.

The 10 kΩ resistor in the grid circuit prevents the current exceeding 10 mA (and destroying the thyratron) in the event of the gas ionizing during the experiment. With some thyratrons it may be necessary to *apply* a small retarding p.d. to prevent electrons reaching A. This is done by connecting a 1.5-V cell between A and G_1, in series with the galvanometer, as shown by the dotted symbol in Fig. A11.1a.

540

APPENDIX 11

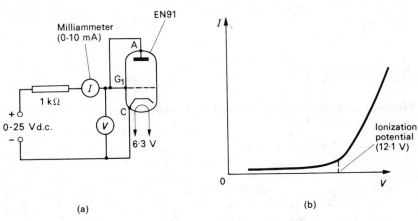

(a)   (b)

Fig. A11.2

(b) *Ionization potential.* The thyratron, used as a diode with A joined to G_1, is connected as in Fig. A11.2a. When V equals the ionization potential of xenon, the current recorded by the milliammeter increases due to electrons from C having inelastic collisions with xenon atoms and ionizing them. The positive xenon ions created act as a new source of current. Fig. A11.2b shows the form of the current-p.d. graph.

Appendix 12

Radioactive decay analogue experiment using dice

Radioactive atoms are represented by small wooden cubes having one face marked in some way. If 100 of these are placed in a large can or jar and then thrown on to the bench so that they are in a single layer, the cubes with the marked face uppermost are considered to have 'decayed' and are removed and counted. The remaining 'undecayed' cubes are returned to the can, thrown again and the 'decayed' cubes removed and counted as before. The process is repeated at least 15 times until only a few cubes have not 'decayed'.

The whole procedure is repeated *another four times*, always starting with 100 cubes and making the same number of throws so that the effect is the same as if 500 cubes had been thrown initially. The results can be recorded as shown, the total number N of surviving cubes being obtained for each throw t.

Number of throw	Number of 'decayed' cubes						Number of 'surviving' cubes					
t	(1)	(2)	(3)	(4)	(5)	Total	(1)	(2)	(3)	(4)	(5)	Total N
0 1 2	0	0	0	0	0	0	100	100	100	100	100	500

A graph of N against t is plotted and a smooth curve drawn through the points. An estimate of the half-life for cube decay, i.e. the number of throws necessary for half the original number of cubes to decay, can be made.

Assuming the decay law is exponential we have

$$N = N_0 e^{-\lambda t}. \qquad \text{(p. 487)}$$

Hence

$$\ln N = \ln N_0 - \lambda t. \qquad (\ln = \log_e)$$

But $\ln N = 2.303 \log_{10} N$,

$$\therefore \quad \log_{10} N = \log_{10} N_0 - \frac{\lambda}{2.303} \cdot t.$$

What should be the form of a graph of $\log_{10} N$ against t?

Appendix 13

Converting a single beam into a double beam CRO using an electronic beam splitter

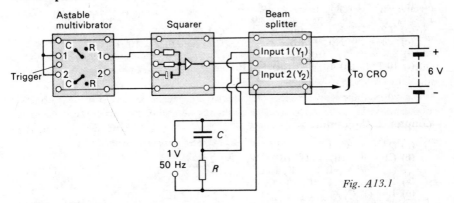

Fig. A13.1

The circuit is shown in Fig. A13.1. It consists of an astable multivibrator producing pulses, at about 2.5 kHz, that are fed into a transistor unit which 'squares' them before they are fed into the beam splitter unit. This switches the CRO beam to and fro rapidly between the two inputs to the unit. If the switching frequency is appreciably higher than the signal frequency, both traces seem almost continuous when a suitable time base speed is chosen.

Answers

Chapter 1. Electric fields

1. **(a)** 8.9×10^4 V m^{-1} **(b)** 8.9×10^3 V
2. **(a)** 29 V **(b)** 4.6×10^{-18} J **(c)** 4.6×10^{-18} J
3. $_BE_A = 1.0 \times 10^3$ V m^{-1}; $_CE_B = 2.0 \times 10^3$ V m^{-1}; $_CE_D = 3.0 \times 10^3$ V m^{-1}
4. **(a)** 2.9×10^{11} N C^{-1} **(b)** 4.3×10^6 N C^{-1}; ratio 0.68×10^5:1
   (c) 1.4×10^{11} N C^{-1} **(d)** 5.8×10^7 N C^{-1}; ratio 0.25×10^4:1
5. **(i)** 8.0×10^{-16} J **(ii)** 1.6×10^{-15} J
6. **(i)** 10^{-12} J **(ii)** 10^{-12} J **(iii)** 5×10^{-11} N
   (iv) 5×10^4 N C^{-1} **(v)** 5×10^4 V m^{-1}

Chapter 2. Capacitors

1. **(a)** 6 V; 18 μC; 3 μF
   (b) **(i)** 4 μC **(ii)** 4 μC **(iii)** 4 V **(iv)** 2 V
   (v) 2 μF **(vi)** 2/3 μF; 4 μC
2. **(a)** 10^{-12} F (1 pF) **(b)** 10^{-9} C
   (c) $V = Q_1/C_1 = Q_2/C_2 \therefore Q_1/Q_2 = C_1/C_2 = 10^{-9}/10^{-12} = 10^3/1$
   (d) **(i)** 10^{-9} C **(ii)** zero **(e)** equally
3. **(a)** 28.6 V; 71.4 V **(b)** 143 μC **(c)** 2.05×10^{-3} J; 5.11×10^{-3} J
4. **(a)** **(i)** 3.7×10^{-8} C **(ii)** 9.3×10^{-8} C **(b)** 2.3 m
5. 1.07×10^{-4} m; 2.1×10^3 V
6. 50 J; 33 J
7. 4 banks of 4 capacitors in series
8. 1.1×10^{-9} F; 8.8×10^{-12} F m^{-1}; 1.8×10^9 Ω
9. **(a)** 1.2×10^{-5} C **(b)** 8.0×10^5 V **(c)** 20 cm
10. 11 s

Chapter 3. Magnetic fields

1. **(b)** downwards **(c)** upwards
2. **(i)** 1×10^{-2} N **(ii)** 0.5×10^{-2} N
3. 2.0×10^{-5} T; 2.8×10^{-5} T
4. 1.6 A $(5/\pi)$
5. 0.23 A
6. 3.6 A
7. 1.1 kV
8. $Be/(2\pi m)$; 2.8×10^7 revs s^{-1}
9. **(a)** No
   (c) $Ee = Bev \therefore E = Bv$. But $E = V_H/d \therefore V_H = Bdv$
   (d) 1 mm s^{-1}
10. 1.0×10^{-4} T; 1.0×10^{-3} N
11. 4.0×10^{-5} T; 4.0×10^{-3} N
12. 1.0×10^{-3} N m
13. 1.1×10^{-3} A
14. **(a)** 10 **(b)** $\frac{1}{4}$

ANSWERS

Chapter 4. Electromagnetic induction

2. 1.5 mA
3. 0.38 V
4. 16 mV
5. (a) 50 Hz (b) 23.6 V
6. 0.31 mV
9. 237 V, 948 W; 195 V, 11.7 × 10³ W; 329 r.p.m.
10. 4 A
11. 0.20 V
12. 4.0 A; 2.0 A s⁻¹
13. 0.50 T
14. 1.43 T

Chapter 5. Alternating current

1. (a) 12 V (b) 17 V (c) 2.8 cm
2. (a) 99 kV, 0 (b) 19.8 MW, 0; 141 kV
3. (a) 7.0×10^{-4} C (c) (i) Current a maximum when rate of change of charge is a maximum, i.e. when $Q = 0$ (ii) Current a minimum when rate of change of charge is zero, i.e. when $Q = \pm 7.0 \times 10^{-4}$ C
   (d) See Fig. 5.10b (e) 1.6×10^{-1} A
4. 6.37 μF
5. (a) 0 (b) 10^{-4} A (c) 2 V s⁻¹
6. (a) $V_R = 3.0$ V; $V_L = 4.0$ V (c) 37°
7. 120 Ω; 0.66 H; 15 μF
8. 225 Hz (a) $10^4/\sqrt{2}$ V (b) $10^4/\sqrt{2}$ V (c) 0
9. (a) 40 Ω (b) 10 H (c) 1.6×10^3 V
10. (a) 0.10 A, 40 V (b) 3.0 J s⁻¹
11. (i) 4.7 A, 89 W; $V_R = 19$ V; $V_L = 15$ V
12. 2 A; 40 W
13. Damped oscillations occur of frequency about 5 Hz

Chapter 6. Wave motion

1. 19°
2. 47°; 42° with the normal to the oil surface
3. (a) $\lambda/2 = S_2Q - S_1Q$; $\lambda = S_2R - S_1R$
   (b) The spacings PQ and QR (i) decrease (ii) increase
4. 4.7 m s⁻¹
5. 20.0 m to 17.0 mm
6. (a) 200 kHz (b) 60.0 cm; 59.4 cm (c) 5×10^{14} Hz
7. (a) In phase and polarized in the same direction
   (b) same signal
   (c) more than 100 m
8. $(7.16 \pm 0.21) \times 10^{-2}$ N m⁻²
9. (a) 3.3×10^2 m s⁻¹ (b) 6.6×10^{-4} m s⁻¹

Chapter 7. Sound

1. At the point C of zero intensity the path difference for sound from A and B is $(13.2 - 11.0)/0.40 = 5\frac{1}{2}$ wavelengths.
   If I_A and I_B are the intensities due to A and B respectively at C then $I_A = I_B$ but $I_A \propto P_A/AC^2$ and $I_B \propto P_B/BC^2$ (inverse square law holds) where P_A and P_B are the rates of emission of sound energy from A and B respectively.

$$\therefore \quad \frac{P_A}{AC^2} = \frac{P_B}{BC^2} \text{ or } \frac{P_A}{P_B} = \frac{AC^2}{BC^2} = \frac{13.2^2}{11.0^2} = 1.44.$$

The sound intensity is therefore zero at C because the path difference is an odd number of half wavelengths and the distances are such that the intensities (and therefore the amplitudes) are equal.

2. Stationary wave system set up with wavelength of 3.0 m; nodes occur every 1.5 m. A beat note due to Doppler effect of 1.3 Hz.
3. 344 Hz
4. 215 Hz; 645 Hz; 1075 Hz
5. 306 m s^{-1}
6. 0.78 m; 2.8
8. 190 Hz; 202 N
9. Resonant vibration at 115 Hz
10. $\frac{10}{9}f; \frac{11}{10}f$
11. 320 Hz; 282 Hz; 1.5 s
12. 136 cm s^{-1}
13. 120 Hz
14. (a) 400 m s^{-1} (b) 408 m s^{-1}
15. Stationary wave pattern formed of wavelength 2 × (5.66 − 2.99)/4 mm = 1.34 mm; 1.34 × 10^3 m s^{-1}
16. 0.266 m
17. 280, 350, 420 or 490 Hz

Chapter 8. Physical optics

1. 3.0 × 10^8 m s^{-1}
2. (a) 2.0 × 10^8 m s^{-1} (b) 5.0 × 10^{14} Hz (c) 4.0 × 10^{-7} m
4. 5.0 × 10^{-7} m
6. (i) 0.200 mm (ii) 0.300 mm (iii) 0.600 mm
7. 1.5; to side of covered slit
8. 100
9. 0.20 mm
10. 5.87 × 10^{-7} m
11. 1.51 m; 1.33
12. 9°
13. 2.5 × 10^5 m^{-1}
14. 4.8 × 10^{-6} m
15. 20.1°
16. 0.950°; 3.32 mm
20. 9.0 × 10^2 °C
21. 0.96 °C s^{-1}
22. 16 (15.6) J s^{-1}
23. 1.8 × 10^2 K

Chapter 9. Kinetic theory: thermodynamics

1. 8.20 J mol^{-1} K^{-1}
2. (b) 2.66 × 10^{25}
3. 289 mm (*Hint.* Apply $p_1/T_1 = p_2/T_2$ to the *air* in the mixture of air and saturated water vapour.)
4. 6.8 × 10^2 mm
5. 6.63 × 10^2 m s^{-1}
6. 450 K; 2.40 × 10^3 m s^{-1}
8. (a) 1.2 × 10^{16} (b) 1.6 × 10^{-7} m (c) 1.9 × 10^3 m s^{-1}

9. **(iv)** 200 cmHg **(v)** -205 °C (68 K); 273 °C (546 K)
   (vi) 136 J at constant pressure
10. $\frac{3}{2}nRT$ (n = no. of moles) $= \frac{3}{2} \times \frac{1}{4} \times 8.3\,T = 3.1\,T$;
    $c_v = 3.1 \times 10^3$ J kg^{-1} K^{-1}; $c_p = 5.2 \times 10^3$ J kg^{-1} K^{-1}
11. 0.167×10^6 J
12. 5.3×10^2 J kg^{-1} K^{-1}; 1/2
13. 1.49 atmospheres
14. 0.53 atmosphere; -130 °C (143 K); 6.3 litres

Chapter 10. Atomic physics

1. 1.60×10^{-19} J; 3.20×10^{-14} J; 6.19×10^6 m s^{-1}
2. **(a)** 0.39×10^{-15} N **(b)** 4.2×10^{14} m s^{-2} **(c)** 5.6×10^6 m s^{-1}
3. 2.7×10^7 m s^{-1}; 3.2×10^{-4} W
4. **(a)** 2.65×10^7 m s^{-1} **(b)** 4.55×10^{-4} T
5. 2.01×10^3 V
6. 3; 19.6 m s^{-2}
7. **(a)** 4.1×10^{-7} m **(b)** 8.8×10^5 m s^{-1}
8. **(a)** 1.5 eV **(b)** 3.5×10^{14} Hz
9. 6.6×10^{-19} J
10. 5.1×10^{14} Hz
11. 8.5×10^7 m s^{-1}
12. **(a)** 6.3×10^{15} electrons s^{-1} **(b)** 2.4×10^{-16} J
13. 1.2×10^{-11} m
16. A: 0.65 μm; B: 0.48 μm; C: 0.43 μm; D: 0.41 μm
17. **(a)** 4.9 eV; 6.7 eV; 8.8 eV; 10.4 eV
    (b) 10.4 eV
    (c) **(i)** 5 eV, $(5 - 4.9) = 0.1$ eV
    (ii) 10 eV, $(10 - 4.9) = 5.1$ eV, $(10 - 6.7) = 3.3$ eV
    (d) **(i)** absorbed and disappears
    (ii) scattered

Chapter 11. Electronics

2. **(b)** **(i)** 5 waves **(ii)** $\frac{1}{2}$ wave
   (c) 50 Hz
3. 1 V; 0.7 V
8. 36 kΩ
12. 1.00×10^7 m s^{-1}

13.

Input 1	Input 2	NOR output	OR output	AND output	NAND output
0	0	1	0	0	1
1	1	0	1	1	0
1	0	0	1	0	1
0	1	0	1	0	1

14.

R	A	G
1	0	0
1	1	0
0	0	1
0	1	0

Chapter 12. Nuclear physics

1. 5.1 MeV
2. (a) alpha, beta, beta, alpha, alpha (b) 2 (c) 5
3. (a) 3 (b) 1/8 (c) 63/64
4. (a) 1/8 (b) 15/16 (c) 6
5. (a) 2:1 (b) 1:1 (c) 1:16
6. 6.0×10^3 cm^3
7. 75 s
8. 0.11 cm^{-1}
10. $v_m = (m - M)v/(m + M)$; $v_M = 2mv/(m + M)$
    (a) alpha v; electron $2v$
    (b) alpha zero; helium atom v
    (c) alpha $- v$; gold atom zero
    When there is a head-on collision between a moving and a stationary particle of equal mass
11. (a) Fig. 12.35b (c) 90°
12. 3.13×10^{16}

Objective-type questions

1.1. C 1.2. E 1.3. A
2.1. A 2.2. (a) E (b) D 2.3. C
3.1. D 3.2. C 3.3. D
4.1. D 4.2. D 4.3. C 4.4. C 4.5. C 4.6. A
5.1. C 5.2. A 5.3. C 5.4. E
6.1. D 6.2. E 6.3. A 6.4. A
7.1. D 7.2. A 7.3. B 7.4. D
8.1. D 8.2. B 8.3. C
9.1. C 9.2. E 9.3. D 9.4. C 9.5. B 9.6. B
10.1. C 10.2. E 10.3. C 10.4. E 10.5. C
11.1. E 11.2. B
12.1. E 12.2. B 12.3. E

A

Absolute
  thermodynamic scale, 333
  zero, 331
absolute measurement
  current, 93
  resistance, 146
absorption
  of radiation, 311–12
  selective, 305
  spectra, 300
a.c. circuits, 165–77, 180
accelerators, particle, 511–14
acceptor atom, 442
action at points, 63–5
activity of a source, 487
adiabatic process, 361–3
aerial, 221
alpha particles
  absorption, 492
  charge, 477
  deflection of, 476
  energy, 478
  ionizing effect, 476, 492
  nature, 477
  radiation hazard, 498
  range in air, 476, 492–3
  scattering of, 414
  speeds, 478
alternating current
  effects, 160

generator, 122–3
  measurement, 163–5
  motors, 127–9
alternator, 122–3
ammeter
  moving coil, 97–101
  moving iron, 163
  rectifier-type, 164
  thermocouple, 164
ampere, the, 93
amplifier
  d.c., 57–63, 490, 493, 529
  transistor, 449–50
  triode, 468–9
amplitude modulation, 462
AND gate, 455–6
Andrews' curves, 350
angle
  of dip, 73
  polarizing, 305–6
antinode, 216
armature, 124, 126
artificial radioactivity, 496
atom
  and wave mechanics, 417, 427
  Bohr model, 416–17
  energy levels in, 417–19, 438, 537
  excited states of, 418
  Rutherford model, 414
atomic
  bomb, 506
  mass, 411